Motorcycle Dynamics

Second English Edition

Vittore Cossalter

Importante notice

This book should not be seen as a guide for modifying, designing or manufacturing a motorcycle. Anyone who uses it as such does so at his own risk and peril. Street testing motorcycles can be dangerous. The author and publisher are not responsible for any damage caused by the use of any information contained in this book.

2nd English edition, 2006

ISBN 978-1-4303-0861-4

Design and illustrations by the author.

For Annalia, Fabrizio, Flavio

Acknowledgment

I am deeply indebted particularly to Roberto Lot, Mauro Da Lio, Alberto Doria, who helped to make this book possible.

This book was written thanks to the enthusiastic participation of PhD students of the Motorcycles Engineering Course: Alessandro Bellati , Roberto Berritta, Francesco Biral, Daniele Bortoluzzi, Mario Dalla Fontana, Giovanni Dalla Torre, Davide Fabris, Pasquale De Luca, Stefano Garbin, Giuseppe Lisciani, Fabiano Maggio, Massimo Maso, Matteo Massaro, Luigi Mitolo, Martino Peretto, Nicolay Ruffo, Jim Sadauckas, Mauro Salvador, Riccardo Toazza ....

Foreword

In today's globalized and hyper-technological world all you have to do to buy a motorcycle is go on-line and give your credit card number and you've got a motorcycle. However, this type of transaction takes place without the emotions and special relationship which have always bound me to motorcycles.

I remember watching my neighbor get his Parilla ready. Dressed in black leather, he would slowly put his gloves on, push down on the pedal and finally drive off. As the motorcycle disappeared into the distance I could hear the symphony created by its engine slowly fade away among the clouds: mine was true passion.
It was the Sixties. There were the elegant and refined Mods with their shining scooters, and the Rockers, both feared and respected, with their motorcycles. England was the homeland of motorcycles. When I got off the ship in Dover, I remember seeing a group of motorcycles next to some scooters. They were the wonderful English motorcycles of the Sixties and Seventies that left the sign of their passing with drops of motor oil on the road wherever they went.

I immediately knew that I was a motorcyclist. As soon as I got home, I managed to buy an old Guzzi Falcone 500. I worked all winter, every evening, to perfectly restore it. My desire to hear the engine roar, to smell the air and to feel the wind blow across my cheeks drove me in my mission until one day in early spring everything was ready. My Falcone never betrayed me, it always gave me incomparable emotions. As I rode it, curve after curve, the engine pushed on almost as if it were a hammer strong enough to forge any type of steel with violent blows of metal on metal.

This book is the result of this past and present passion of mine for motorcycles. I have tried to offer a new approach to technical-scientific writing by combining the exact and often aseptic nature of scientific discourse with my passion for this perfect vehicle. I realize that this is no small challenge, but it is this very passion, of a man who feels more at ease on a motorcycle than behind a desk, which has motivated my research in the field of motorcycles. Together with its thorough technical discussion, this book also takes into account the fascinating history of the motorcycle and motorcyclists. No business will ever be able to take away the adventuresome, and somewhat crazy, nature of the motorcycle.

Vittore Cossalter

Padova, spring 2002

Contents

Motorcycle Bianchi "Freccia Celeste" 350 cc of 1924

1 Kinematics of Motorcycles

The kinematic study of motorcycles is important, especially in relation to its effects on the dynamic behavior of motorcycles. Therefore, in this chapter, in addition to the kinematic study, some simple examples of the dynamic behavior of motorcycles are reported in order to show how kinematic peculiarities influence the directional stability and maneuverability of motorcycles.

1.1 Definition of motorcycles

Although motorcycles are composed of a great variety of mechanical parts, including some complex ones, from a strictly kinematic point of view, by considering the suspensions to be rigid, a motorcycle can be defined as simply a spatial mechanism composed of four rigid bodies:

- the rear assembly (frame, saddle, tank and motor-transmission drivetrain group),
- the front assembly (the fork, the steering head and the handlebars),
- the front wheel,
- the rear wheel.

These rigid bodies are connected by three revolute joints (the steering axis and the two wheel axles) and are in contact with the ground at two wheel/ground contact points as shown in Fig. 1-1.

Each revolute joint inhibits five degrees of freedom in the spatial mechanism, while each wheel-ground contact point leaves three degrees of freedom free. If we consider the hypothesis of the pure rolling of tires on the road to be valid, it is easy to ascertain that each wheel, with respect to the fixed road, can only rotate around:

- the contact point on the wheel plane (forward motion),
- the intersection axis of the motorcycle and road planes (roll motion),
- the axis passing through the contact point and the center of the wheel (spin).

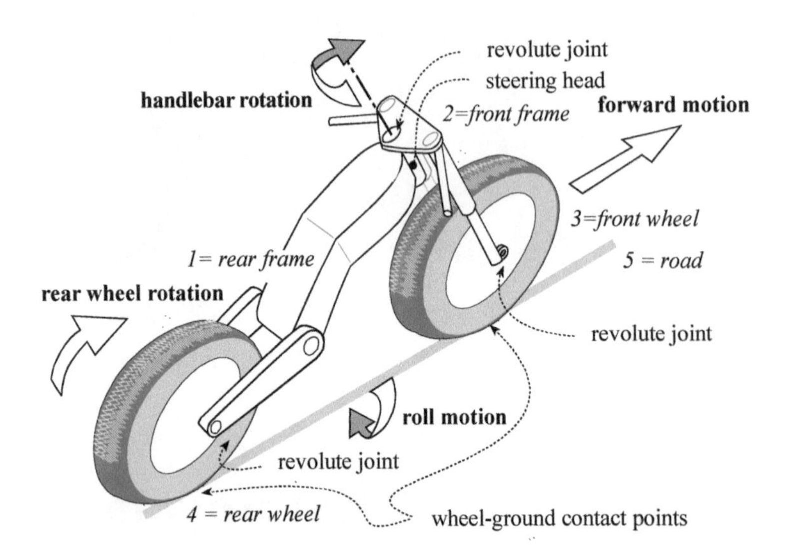

Fig. 1-1 Kinematic structure of a motorcycle.

In conclusion, a motorcycle's number of degrees of freedom is equal to 3, given that the 15 degrees of freedom inhibited by the 3 revolute joints and the 6 degrees of freedom eliminated by the 2 wheel-ground contact points must be subtracted from the 4 rigid bodies' 24 degrees of freedom, as summarized in Fig. 1-2.

A motorcycle's three degrees of freedom may be associated with three principal motions:

- forward motion of the motorcycle (represented by the rear wheel rotation);
- roll motion around the straight line which joins the tire contact points on the road plane;
- steering rotation.

While he drives, the rider manages all three major movements, according to his personal style and skill: the resulting movement of the motorcycle and the corresponding trajectory (e.g. a curve) depend on a combination, in the time domain, of the three motions related to the three degrees of freedom. This generates one maneuver, among the thousands possible, which represents the personal style of the driver.

These considerations have been formulated assuming that the tires move without slippage. However, in reality, the tire movement is not just a rolling process.

The generation of longitudinal forces (driving and braking forces) and lateral forces requires some degree of slippage in both directions, longitudinally and laterally, depending on the road conditions. The number of degrees of freedom is therefore seven:

- forward motion of the motorcycle,
- rolling motion,
- handlebar rotation,
- longitudinal slippage of the front wheel (braking),
- longitudinal slippage of the rear wheel (thrust or braking),
- lateral slippage of the front wheel,
- lateral slippage of the rear wheel.

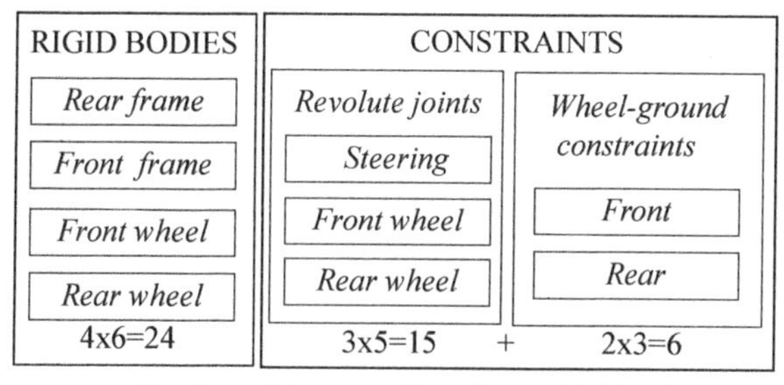

Fig. 1-2 Degrees of freedom of a motorcycle.

1.2 The geometry of motorcycles

This kinematic study refers to a rigid motorcycle, i.e. one without suspensions with the wheels fitted to nondeformable tires, and schematized as two toroidal solid bodies with circular sections (Fig. 1-3).

Motorcycles can be described using the following geometric parameters:

- p wheelbase;
- d fork offset: perpendicular distance between the axis of the steering head and the center of the front wheel;
- ε caster angle;
- R_r radius of the rear wheel;
- R_f radius of the front wheel;
- t_r radius of the rear tire cross section;
- t_f radius of the front tire cross section.

Some important geometric parameters can be expressed in terms of these variables:

- $\rho_r = (R_r - t_r)$ radius of the front torus center circle;
- $\rho_f = (R_f - t_f)$ radius of the rear torus center circle;
- $a_n = R_f \sin\varepsilon - d$ normal trail;
- $a = a_n / \cos\varepsilon = R_f \tan\varepsilon - d / \cos\varepsilon$ mechanical trail.

The geometric parameters usually used to describe motorcycles are the following:

- the wheelbase p;
- the caster angle ε ;
- the trail a .

These parameters are measured with the motorcycle in a vertical position and the steering angle of the handlebars set to zero.

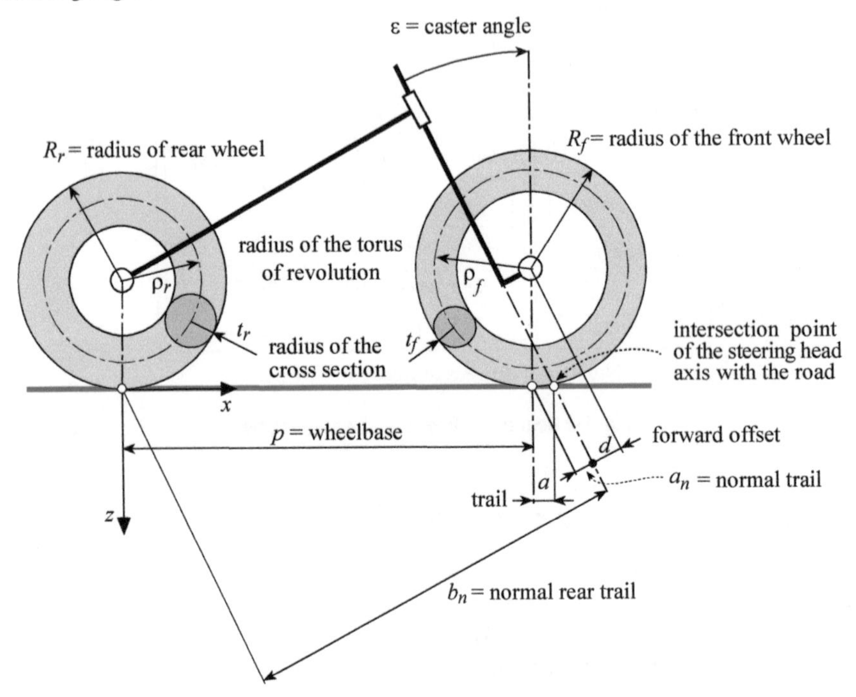

Fig. 1.3 Geometry of a motorcycle.

The wheelbase p is the distance between the contact points of the tires on the road. The caster angle ε is the angle between the vertical axis and the rotation axis of the front section (the axis of the steering head). The trail a is the distance between the contact point of the front wheel and the intersection point of the steering head axis with the road measured in the ground plane.

Together these parameters are important in defining the maneuverability of the motorcycle as perceived by the rider. It is not practical, however, to examine the effects produced by only one geometric parameter, independently of the others, because of the strong interaction between them. Here we will present some considerations regarding the way in which these parameters influence the kinematic and dynamic behavior of motorcycles.

The value of the wheelbase varies according to the type of motorcycle. It ranges from 1200 mm in the case of small scooters to 1300 mm for light motorcycles (125 cc displacement) to 1350 mm for medium displacement motorcycles (250 cc) up to

1600 mm, and beyond, for touring motorcycles with greater displacement.

In general, an increase in the wheelbase, assuming that the other parameters remain constant, leads to:

- an unfavorable increase in the flexional and torsional deformability of the frame. These parameters are very important for maneuverability (frames that are more deformable make the motorcycle less maneuverable),
- an unfavorable increase in the minimum curvature radius, since it makes it more difficult to turn on a path that has a small curvature radius,
- in order to turn, there must be an unfavorable increase in the torque applied to the handlebars,
- a favorable decrease in transferring the load between the two wheels during the acceleration and braking phases, with a resulting decrease in the pitching motion; this makes forward and rearward flip-over more difficult,
- a favorable reduction in the pitching movement generated by road unevenness,
- a favorable increase in the directional stability of the motorcycle.

The trail and caster angle are especially important inasmuch as they define the geometric characteristics of the steering head. The definition of the properties of maneuverability and directional stability of motorcycles depend on them, among others.

The caster angle varies according to the type of motorcycle: from 19° (*speedway*) to 21-24° for competition or sport motorcycles, up to 27-34° for touring motorcycles. From a structural point of view, a very small angle causes notable stress on the fork during braking. Since the front fork is rather deformable, both flexionally and torsionally, small values of the angle will lead to greater stress and therefore greater deformations, which can cause dangerous vibrations in the front assembly (oscillation of the front assembly around the axis of the steering head, called wobble).

The value of the caster angle is closely related to the value of the trail. In general, in order to have a good feeling for the motorcycle's maneuverability, an increase in the caster angle must be coupled with a corresponding increase in the trail.

The value of the trail depends on the type of motorcycle and its wheelbase. It ranges from values of 75 to 90 mm in competition motorcycles to values of 90 to 100 mm in touring and sport motorcycles, up to values of 120 mm and beyond in purely touring motorcycles.

1.3 The importance of trail

One of the peculiarities of motorcycles is the steering system, whose function is essentially to produce a variation in the lateral force needed, for example, to change the motorcycle's direction or assure equilibrium.

According to this point of view , the steering system could hypothetically be made up of two little rockets placed perpendicular to the front wheel which, when appropriately activated, could, although not without significant if not insurmountable difficulties for the rider, generate lateral thrusts, that is, perform the same function as the steering system.

From a geometrical point of view, the classic steering mechanism is described by three parameters:

- the caster angle ε ;
- the fork offset d ;
- the radius of the wheel R_f .

These parameters make it possible to calculate the value of the normal trail a_n , which is the perpendicular distance between the contact point and the axis of the motorcycle's steering head. This is considered positive when the front wheel's contact point with the road plane is behind the point of the axis intersection of the steering head with the road itself, as presented in Fig. 1-4. As we have previously seen, the trail measured on the road is related to the normal trail by the equation:

$$a = a_n / \cos\varepsilon$$

The value of the trail is most important for the stability of the motorcycle, especially in rectilinear motion.

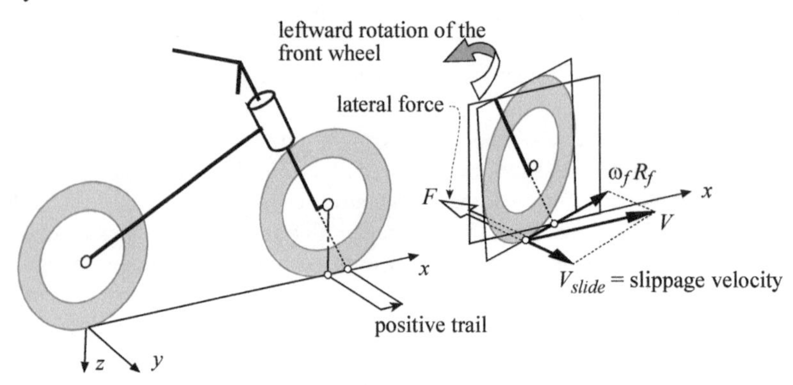

Fig. 1-4 Stabilizing effect of the positive trail during forward movement.

To develop this concept, let us consider a motorcycle driving straight ahead, at constant velocity V , and let us suppose that an external disturbance (for example, an irregularity in the road surface or a lateral gust of wind) causes a slight rotation of the front wheel to the left. For the time being, let us ignore the fact that the motorcycle starts to turn to the left and that because of centrifugal forces, begins at the same time to lean to the right, concentrating our attention instead on the lateral friction force F generated by the contact of the tire with the ground.

In other words, let us suppose that the motorcycle is driving at constant velocity V and that the front wheel contact point also has velocity V in the same direction. The vector V may be divided into two orthogonal components:

- the component $\omega_f R_f$, which represents the velocity due to rolling: it is placed in the plane of the wheel and rotated to the left at an angle which depends on the steering angle;
- the component V_{slide}, which represents the sliding velocity of the contact point with respect to the road plane.

A frictional force, F, therefore acts on the front tire. F is parallel to the velocity of slippage but has the opposite sense, as illustrated in Fig. 1-4. Since the trail is positive, friction force F generates a moment that tends to align the front wheel. The straightening moment is proportional to the value of the normal trail.

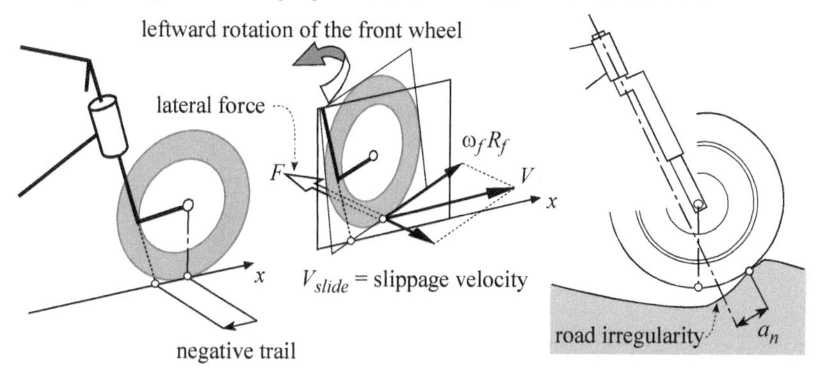

Fig.1-5 Destabilizing effect of the negative trail during forward movement.

If the value of the trail were negative (the contact point in front of the intersection point of the steering head axis with the road plane) and considering that friction force F is always in the opposite direction of the velocity of slippage, a moment around the steering head axis that would tend to increase the rotation to the left would be generated. In Fig. 1-5 one can observe how friction force F would amplify the disturbing effect, seriously compromising the motorcycle's equilibrium. Figure 1-5 demonstrates that the road profile can make the trail negative, for example, when the wheel goes over a step or bump.

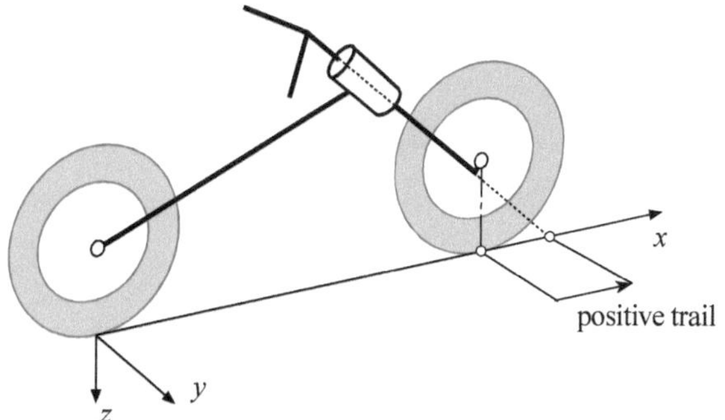

Fig. 1-6 Motorcycle with a high value of trail.

Small trail values generate small aligning moments of the lateral friction force. Even if the rider has the impression that the steering movement is easy, the steering mechanism is very sensitive to irregularities in the road. Higher values of the trail (obtained with high values of the caster angle as shown in Fig. 1-6) increase the stability of the motorcycle's rectilinear motion, but they drastically reduce maneuverability.

Consider, for example, "chopper" type motorcycles which became very popular following the success of the well-known film, “Easy Rider”. These motorcycles have caster angle values up to 40°, making them more adaptable to straight highways than to curving roads.

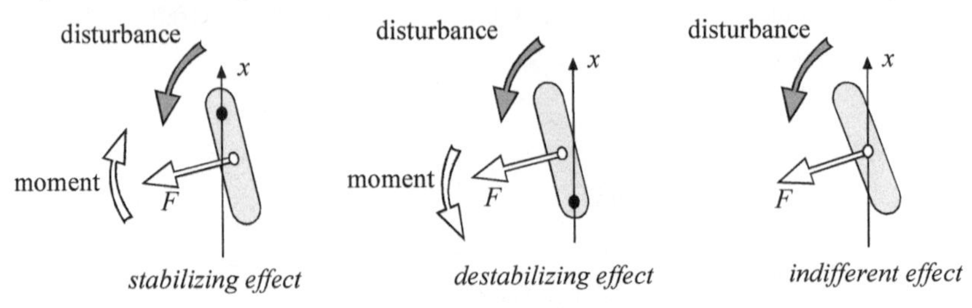

o contact point

• intersection of the steering head axis with the road plane

Fig 1-7 Summary of the effect of trail during forward movement.

During curvilinear motion, road gripping is assured by the lateral frictional forces, which are perpendicular to the line of intersection of the wheel plane with the road.

The front and rear lateral forces create moments around the steering head axis that are proportional respectively to distances a_n and b_n , which are related to the wheelbase and the trail by the equations:

$$a_n = a\cos\varepsilon \qquad\qquad b_n = (p+a)\cos\varepsilon$$

where a_n represents the normal front trail and b_n may be considered the normal trail of the rear wheel.

This simple consideration shows how the wheelbase and the trail are intimately connected to each other and should therefore be considered together. It is not entirely correct to define a trail as small or large without reference to the motorcycle’s wheelbase. As a comparison parameter, we could use the ratio between the front and rear normal trail:

$$R_n = a_n / b_n$$

In general the normal front trail is approximately 4-8% of the value of the rear one. The value of this ratio for racing motorcycles is approximately 6%; for sport and super sport motorcycles it is from 6 to 6.5%; and for touring motorcycles, which are more or less similar to sport motorcycles, it varies from 6 to 8%.

“Cruiser” motorcycles (heavy, slower motorcycles) are characterized by values of 5-6% and have trails that are modest in comparison with the wheelbase. This is probably due to the necessity of making the motorcycles maneuverable at low velocities. Since the load on the front wheels is high due to the weight of the motorcycle, the choice of a small trail lowers the value of the torque that the rider must apply to the handlebars to execute a given maneuver. In addition, it is worth pointing out that these motorcycles are normally used at rather low velocities, and they do not therefore need long trails, which, as previously noted, assures a high directional stability at high velocities.

This ratio is also low for scooters since they are used (or should be used) at low velocities and therefore maneuverability has a higher priority than directional stability.

Strictly speaking, the ratio should take into account the distribution of the load on the wheels. A motorcycle that has a heavy load on the front wheel needs a shorter trail. In fact, heavier loads on the front wheel generate greater lateral frictional forces in proportion to the lateral motion of the wheel. Therefore, for the same aligning torque acting around the axis of the steering head a smaller trail is sufficient.

The correct ratio on the basis of the load distribution, is expressed by the equation:

$$R_n = (a_n / b_n)(N_f / N_r)$$

where N_f is the load on the front wheel and N_r the load on the rear one.

1.4 Kinematics of the steering mechanism

It is clear that when turning the handlebars, keeping the motorcycle perfectly vertical, the steering head lowers and only begins to rise for very high values of the steering angle. We will demonstrate this statement by considering the following cases:

- steering mechanism with no fork offset, $d = 0$;
- steering mechanism with a non-zero fork offset, $d \neq 0$.

1.4.1 Steering mechanism with zero fork offset

In the case of the fork with no offset the center of the wheel is on the axis of the steering head. Let us add the following assumptions:

- the roll angle of the motorcycle is zero;
- the wheels have zero thickness.

As shown in Fig. 1-8, when the steering angle δ is zero, the wheel is perfectly vertical and lies in the xz plane.

The caster angle ε, the steering angle δ, the camber angle of the front wheel β, the kinematic steering angle Δ (projection of the angle of rotation δ onto the road plane) and the angle α are related to each other through the following trigonometric equations:

$$\tan\alpha = \tan\varepsilon \cdot \cos\delta \qquad \tan\Delta = \tan\delta \cdot \cos\varepsilon \qquad \sin\beta = \sin\alpha \cdot \sin\delta$$

It is possible to derive $\sin\alpha$ and $\cos\alpha$ as functions of δ and ε from the preceding equations:

$$\sin\alpha = \frac{\cos\delta \ \sin\varepsilon}{\sqrt{1 - \sin^2\delta \ \sin^2\varepsilon}} \qquad \cos\alpha = \frac{\cos\varepsilon}{\sqrt{1 - \sin^2\delta \ \sin^2\varepsilon}}$$

We now assume that the wheel center (point O) can neither rise nor fall. The δ rotation of the front wheel causes it to incline with respect to the vertical position and to detach itself from the horizontal plane xy. The distance OD of the wheel center from the road plane is greater than the radius of the wheel OP.

Actually, the wheel is not raised from the ground but rather lowered. Supposing that we keep the axis of the steering head immobile, the center of the wheel moves along the steering head axis to the point O_1. Consequently, the contact point P_1 moves forward, as shown in Fig. 1-8. In the final position the distance O_1P_1 is obviously equal to the radius of the wheel OP.

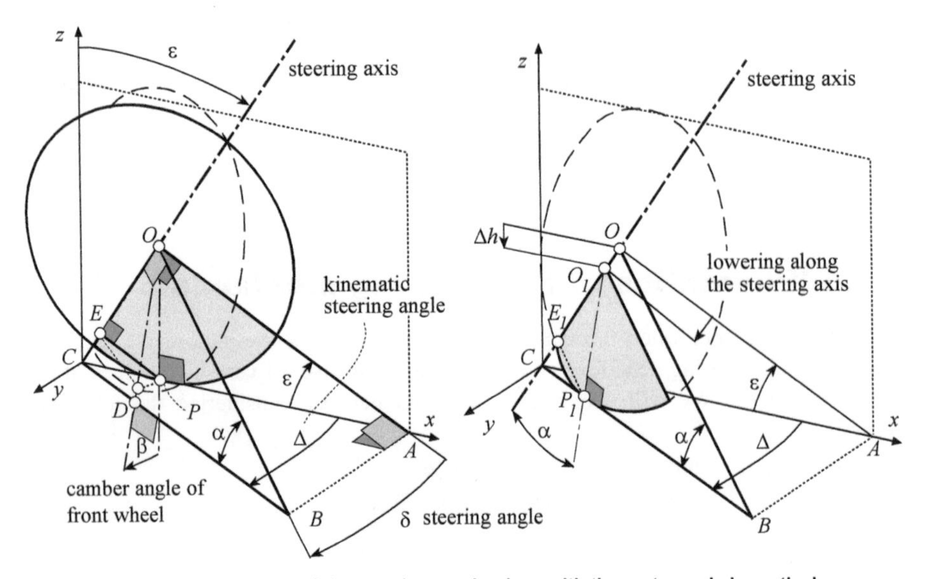

Fig. 1-8 Geometry of the steering mechanism, with the motorcycle in vertical position and no fork offset.

When the steering angle is zero (Fig. 1-8, left), the normal trail and the trail measured on the road plane are:

$$a_n = EP = R_f \sin\varepsilon \qquad\qquad a = CP = R_f \tan\varepsilon$$

Here R_f indicates the radius of the front wheel. When the steering angle δ is not zero, the normal trail $a_n = P_1E_1 = R_f \sin\alpha$ becomes,

$$a_n = R_f \frac{\cos\delta \cdot \sin\varepsilon}{\sqrt{1-(\sin\delta \cdot \sin\varepsilon)^2}}$$

The trail measured on the road plane is related to the normal trail and steering angle δ by the equation:

$$a = \frac{a_n}{\cos\alpha} = R_f \cdot \tan\varepsilon \cdot \cos\delta$$

The vertical displacement of the wheel center is given by the difference:

$$\Delta h = (OC - O_1C)\cos\varepsilon = \left(\frac{R_f}{\cos\varepsilon} - \frac{R_f}{\cos\alpha}\right)\cos\varepsilon$$

Expressing the angle α in terms of δ and ε, we have:

$$\Delta h = \left(1 - \sqrt{1 - \sin^2\delta\ \sin^2\varepsilon}\right)R_f$$

1.4.2 Steering with non-zero fork offset

Let us now consider the effect of offset d, i.e. the distance between the center of the wheel and the steering head axis. The considerations that have allowed us to express the lowering of the steering head as a function of the angles δ and ε in the case of zero-offset remain valid. However, the zero-offset formula must be corrected since the offset causes the center O of the wheel to move to O^*, as is shown in Fig. 1-9.

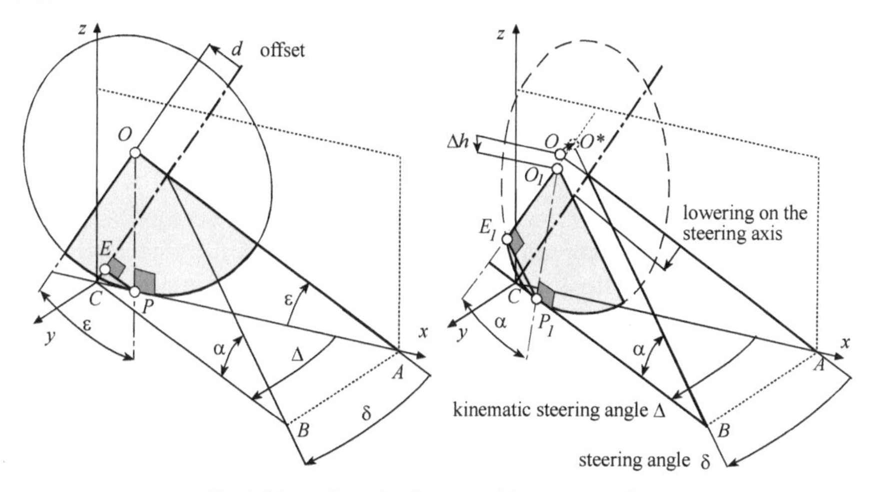

Fig. 1-9 Lowering wheel center with non-zero offset.

With a zero steering angle the trail is:

$$a_n = EP = R_f \sin\varepsilon - d \qquad a = CP = R_f \tan\varepsilon - \frac{d}{\cos\varepsilon}$$

With a non-zero steering angle δ the trail is:

$$a_n = R_f \frac{\cos\delta \cdot \sin\varepsilon}{\sqrt{1 - (\sin\delta \cdot \sin\varepsilon)^2}} - d \qquad a = R_f \cdot \tan\varepsilon \cdot \cos\delta - \frac{\sqrt{1 - (\sin\delta \cdot \sin\varepsilon)^2}}{\cos\varepsilon} d$$

The presence of the fork offset leads to a reduction, $d \cdot \sin\varepsilon \cdot (1 - \cos\delta)$, in the lowering of the wheel. This value must be subtracted from the lowering of the front axle, calculated without the offset.

$$\Delta h = R_f\left(1 - \sqrt{1 - \sin^2\delta\ \sin^2\varepsilon}\right) - d \cdot \sin\varepsilon \cdot (1 - \cos\delta)$$

In conclusion, with offset, there is less lowering of the front wheel center than with zero offset.

Example 1

Let us consider a motorcycle characterized by the following steering parameters:
- radius of the front wheel: $R_f = 0.3$ m;
- offset: $d = 0.05$ m;
- caster angle: $\varepsilon = 27°$.

Now calculate the effects of two different steering angles, 9° and 45°, on steering head lowering with and without fork offset.
With a steering angle $\delta = 9°$, the lowering is:
- with zero offset: $\Delta h = 0.75$ mm;
- with offset: $\Delta h = 0.478$ mm.

By increasing the steering angle to $\delta = 45°$ the lowering is:
- with zero offset: $\Delta h = 1.59$ mm;
- with offset: $\Delta h = 0.92$ mm.

The example shows that ignoring the offset causes a significant error in calculating the lowering of the front wheel center. It must be pointed out that the range of steering is generally less than $\pm 35°$.

Example 2

Let us consider two motorcycles in a vertical position, with the same mechanical trail ($a = 101$ mm), the same radius ($R_f = 0.3$ m) and different caster angles ($\varepsilon_1 =$ 27°, $\varepsilon_2 = 20°$).
If the steering angle is changed from $\delta = 0°$ to $\delta = 9°$, calculate the lowering of the front wheel center for each of the caster angle:
- with $\varepsilon_1 = 27°$ and $\delta = 9°$: $\Delta h = 0.50$ mm;
- with $\varepsilon_2 = 20°$ and $\delta = 9°$: $\Delta h = 0.40$ mm.

Lower caster angles reduce the lowering of the front wheel center.
If the steering angle δ is equal to 9°, calculate the change in trail for each caster angle.
The trail is reduced from the 101 mm to:
- with $\varepsilon_1 = 27°$ and $\delta = 9°$: $a = 99.5$ mm;
- with $\varepsilon_2 = 20°$ and $\delta = 9°$: $a = 99.8$ mm.

The value of the trail slightly depends on the steering angle.

* * *

The previous considerations have allowed us to find analytical equations that express the lowering of the steering head and the values of the trail in terms of the angles δ and ε and with the limiting hypotheses of zero roll angle and zero wheel thickness. In the following section, a more complicated kinematic model is used taking into account both the roll angle and the radius of the front tire cross section.

1.5 Roll motion and steering

Not only is the kinematics of a two-wheeled vehicle significantly more complex than of a four-wheeled vehicle, but it also presents some unique aspects.

For example, let us consider a motorcycle in rectilinear motion at velocity V, which at a certain point enters into a curve. The motorcycle passes from a vertical position, in which the steering angle was zero, to an tilted position with a roll angle φ. In order to stay balanced, the handlebar's angle of rotation will deviate from zero depending on the radius of the curve and the velocity.

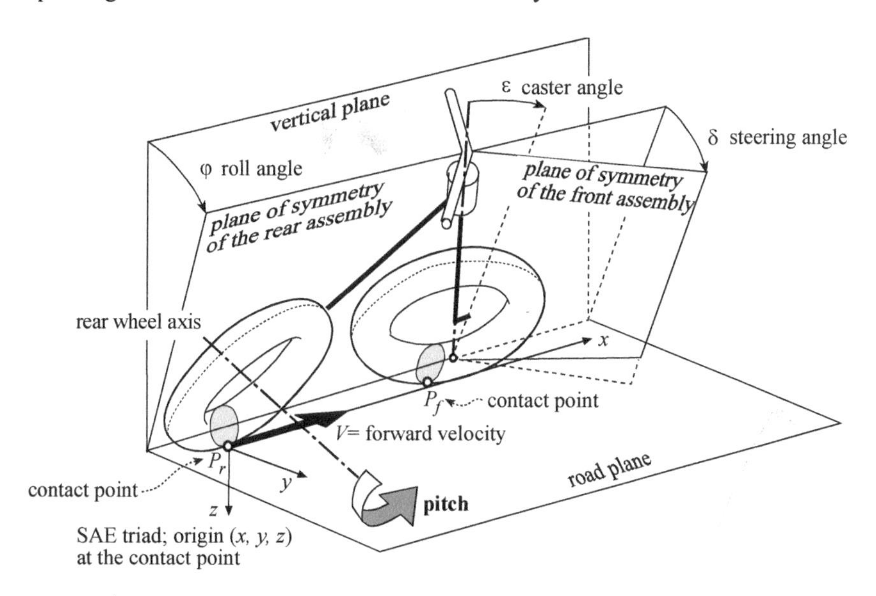

Fig. 1-10 Motorcycle in a curve.

We have seen that the rotation of the steering, considering zero wheel thickness, generates a small lowering of the steering head, which causes a small forward rotation of the rear frame around the axis of the rear wheel (pitch rotation).

We will now see how, in reality, following the roll motion, the contact point of the rear wheel with the road plane is displaced. Two triads can be defined as follows:

- a mobile triad (P_r, x, y, z), defined as specified by the Society of Automotive Engineers (SAE). The origin is established at the contact point P_r of the rear wheel with the road plane. The axis x is horizontal and parallel to the rear wheel plane. The z axis is vertical and directed downward while the y axis lies on the road plane. The road surface is therefore represented by the plane $z = 0$;
- a triad fixed to the rear frame (A_r, X_r, Y_r, Z_r) which is superimposed on the SAE triad when the motorcycle is perfectly vertical and the steering angle δ zero.

Let us now suppose that only the rear wheel is tilted at the roll angle φ. Consequently the triad fixed to the rear axle (A_r, X_r, Y_r, Z_r) rotates at the same angle around the x-axis. Therefore, the triad's origin A_r is translated with respect to P_r, as illustrated in Fig. 1-11b.

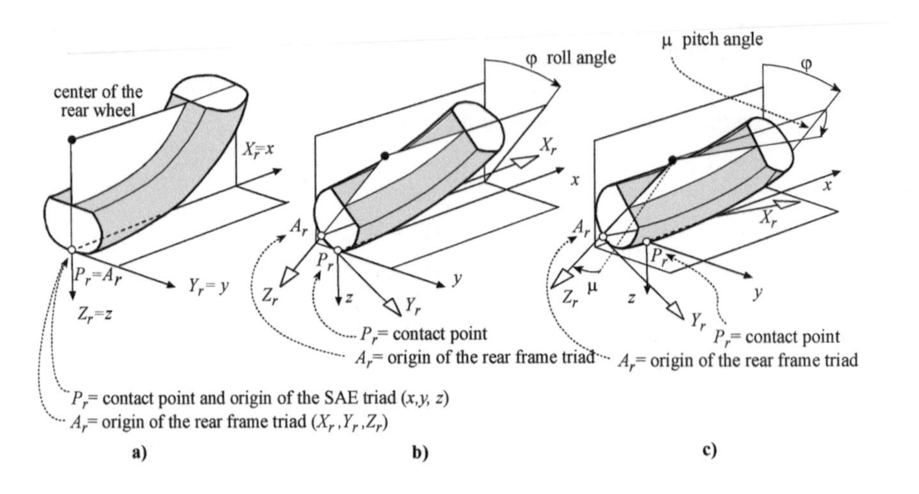

Fig 1-11 Rear wheel in a curve: displacement of the contact point.

It can be seen that the lowering of the steering head causes a small pitching rotation of the rear frame or, in other words, another rotation of the triad fixed to the rear frame, as shown in Fig. 1-11c. It is important to point out that the origin A_r of the triad is fixed to the rear frame and not to the rear wheel: A_r coincides with P_r only when the roll angle φ and the pitch angle μ are both zero.

The behavior of the front wheel is even more complicated, since, in addition to the rolling and pitching motion, the front wheel is also subject to rotation around the axis of the steering head. The change from a vertical to a tilted position was assumed to be a pure roll motion as if the slippage between the tire and the road plane were zero. In reality, the behavior is more complex. In order to produce the lateral reaction forces on the curve, a lateral slippage, which is expressed in terms of the sideslip angle λ, might be necessary. (The next chapter, on tires, will show that slippage can be either positive or negative depending on the value of the force generated by the camber angle of the wheel.)

Fig. 1-12 illustrates the cases of pure rolling motion and of motion with lateral slippage. The absence of slippage means that the velocity vector of the forward motion of the wheel's contact point lies in a plane parallel to the wheel itself, even when the motorcycle is traveling in a curve.

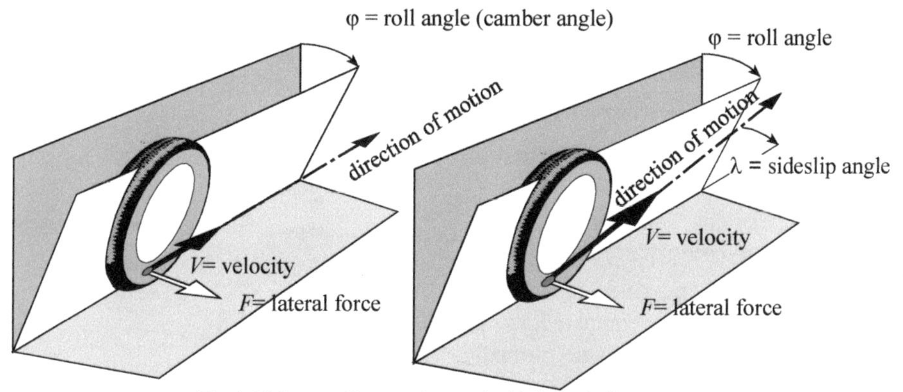

Fig. 1-12 Pure rolling motion and with lateral slippage.

1.6 Motorcycle pitch

We have shown that when a motorcycle is perfectly vertical ($\varphi = 0$), the rotation of the handlebars causes a lowering of the front wheel center and, therefore, a rotation of the rear frame around the rear wheel axis. In other words, the movement of the handlebars causes a pitching motion. Now we would like to study motorcycle pitch in a more general case, considering a roll angle φ other than zero and taking into account the size of the tire cross sections (see Fig. 1-10).

The pitch angle of the frame, indicated by μ, is assumed to be positive in a counterclockwise direction. Therefore, lowering the front wheel center leads to a negative value of the pitch angle. A kinematic analysis of motorcycles allows us to determine a non-linear equation, which connects the unknown pitch angle μ to a series of known quantities: the roll angle φ, the steering angle δ, the wheelbase p, the radii of the cross sections of the tires, t_f and t_r, the radii of the torus center circles ρ_r and ρ_f, and the caster angle ε.

$$\mu = \frac{(c_1 + c_2)\cos\varphi + c_3 \sin\varphi + t_f - t_r}{(c_4 + c_5)\cos\varphi}$$

where:

- $c_1 = d\sin\varepsilon\left(1-\cos\delta\right) + t_r - t_f$
- $c_2 = \rho_f\left[\cos\varepsilon\,\cos(\beta' - \varepsilon) - \cos\delta\,\sin\varepsilon\,\sin(\beta' - \varepsilon) - 1\right]$
- $c_3 = d\ \sin\delta + \rho_f \sin\delta\sin(\beta' - \varepsilon)$
- $c_4 = p - d\ \cos\varepsilon\left(1-\cos\delta\right)$
- $c_5 = \rho_f\left[\sin\varepsilon\cos(\beta' - \varepsilon) + \cos\delta\cos\varepsilon\sin(\beta' - \varepsilon)\right]$

$$\beta' = \varepsilon + \arctan\left(\frac{\sin\delta\ \tan\varphi - \sin\varepsilon\ \cos\delta}{\cos\varepsilon}\right)$$

The physical meaning of the angle β' is shown later in Fig. 1-22 of Section 1.7.2.

The previous equation was determined by ignoring the pitch angle μ with respect to ε, since its value is only a few degrees compared to the caster angle ε, whose value normally varies from 20° to 35°.

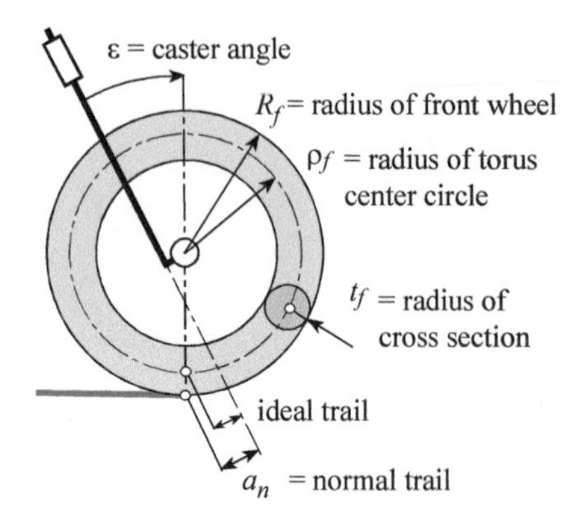

Fig. 1-13 Ideal trail.

Once the value of the pitch is known, it is easy to calculate the resulting lowering of the front wheel center, which is measured in the plane of the motorcycle. A good approximation of the lowering can be derived from the product of the pitch angle and the wheelbase.

The preceding equations can be significantly simplified if we consider small rotations of the steering angle δ ($\sin\delta \cong \delta$). The expression for the pitch then becomes:

$$\mu = -\frac{a_n - t_f \sin\varepsilon}{p}\delta \cdot \tan\varphi - \frac{t_r - t_f}{p}\left(\frac{1}{\cos\varphi} - 1\right)$$

The pitch is proportional to the geometric parameter ($a_n - t_f \sin\varepsilon$), which corresponds to an ideal normal trail, measured in correspondence with the circular axis of the torus center circle, as shown in Fig. 1-13. The pitch also depends on the difference between the radii of the tire sections ($t_r - t_f$): the need to mount a larger tire on the rear to improve the adherence during thrusting, increases the effect of lowering the steering head. It is worth noting that the second term does not depend on the steering angle δ, but only on the roll angle φ.

If we also ignore tire thickness, i.e. if we consider zero thickness wheels ($t_r = t_f = 0$), we obtain the simple equation:

$$\mu = -\frac{a_n}{p}\delta \cdot \tan\varphi$$

This latter expression shows that the normal trail is the parameter that has the greatest influence on the pitching motion.

1.6.1 Pitch in terms of steering and roll angles

Figure 1-14 shows the effect of the steering angle δ and roll angle φ on the pitch angle μ. It is important to stress that negative values of the pitch angle μ correspond to downward rotations of the vehicle around the rear wheel axle. Therefore, a negative value of the pitch angle μ causes a motorcycle's center of gravity to lower.

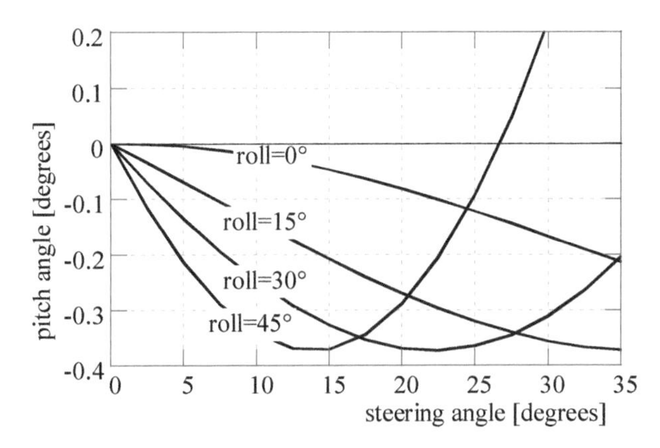

Fig. 1-14 The pitch angle μ as a function of steering angle δ at different roll angles φ.

[p = 1.4 m, a_n = 0.1 m, ε = 30°, $R_r = R_f$ = 0.36 m, $t_r = t_f$ = 0.06 m]

When the roll angle φ values are not high (0° and 15° in Fig. 1-14), an increase in the steering angle δ leads to a continuous lowering of the motorcycle's center of gravity G. Since the lowering corresponds to a reduction in potential energy, the increase in the steering angle takes place naturally, even without applying torque to the handlebars.

It is easy to verify this behavior, especially with a light two-wheeled vehicle, such as a bicycle. When the bicycle is tilted, the roll angle imposed determines the angle at which the handlebars naturally rotate. For high values of the roll angle (30° and 45° in Fig. 1-14), the variation in the pitch angle μ in terms of the steering angle δ stops decreasing and presents a minimum. At this point, a limiting value of the steering angle δ is reached, beyond which the pitch slope reverses its sign.

Let us now consider the minimum condition of the pitch angle μ. This corresponds to the minimum potential energy (center of gravity at its lowest point). From a physical point of view, this means that, if a determined roll angle φ is imposed and no external torques are applied to the handlebars, the front frame tends to rotate naturally towards the value of the steering angle δ which corresponds to the minimum value of the pitch angle μ.

In conclusion, as the roll angle φ gradually increases, the minimum value of the pitch angle μ corresponds to a lower steering angle δ.

1.6.2 Pitch as a function of the caster angle

Figure 1-15 shows the influence of the steering angle δ and steering head angle ε, on the pitch angle μ, for a fixed value of the roll angle φ.

The pitch angle μ becomes more negative as the steering angle δ increases. The influence of the caster angle is modest.

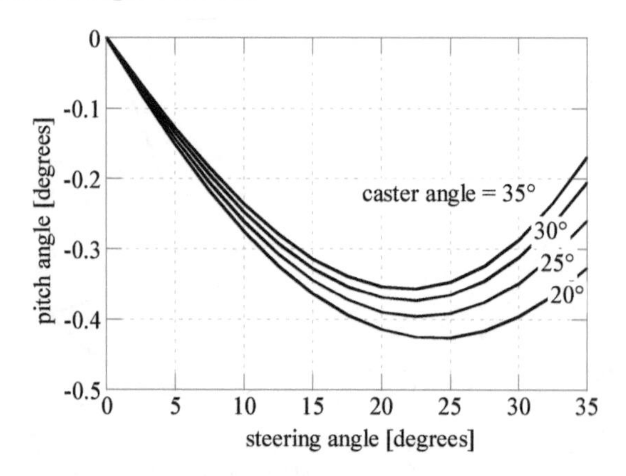

Fig. 1-15 The pitch angle μ versus the steering angle δ for various values of the caster angle ε and with a roll angle equal to 30°.

1.6.3 Pitch as a function of the normal trail

Figure 1-16 shows that the normal trail is the parameter which most influences motorcycle pitch. For example, when the steering angle δ is 10° and the offset is modified to obtain a 20% variation in the normal trail, the variation in the pitch angle is approximately 35%.

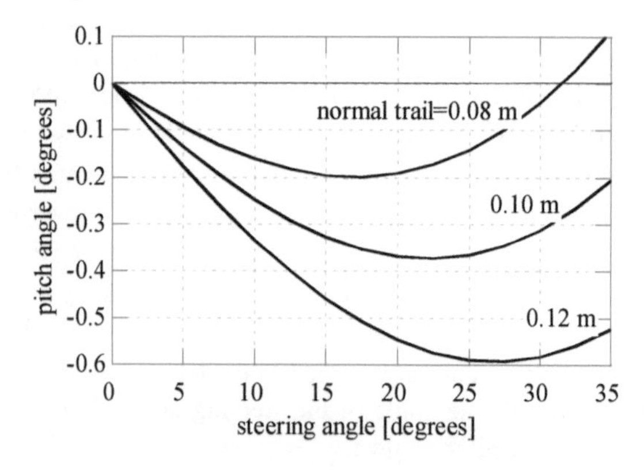

Fig. 1-16 The pitch angle μ as a function of the steering angle δ for various values of normal trail.

As is shown in Fig. 1-16, when the normal trail increases, the minimum condition of the pitch angle μ corresponds to increasing values of the steering angle. This is the opposite of what happens when the roll angle increases, as is clear by comparing Fig. 1-16 with Fig. 1-14.

1.7 The rear wheel contact point

1.7.1 The effect of camber and tire cross section

Let us consider a motorcycle that is initially in a vertical position. The cross section of the rear tire is larger than that of the front. The rear frame tilts, assuming that there is lateral roll without slippage on the road plane, as illustrated in Fig. 1-17b.

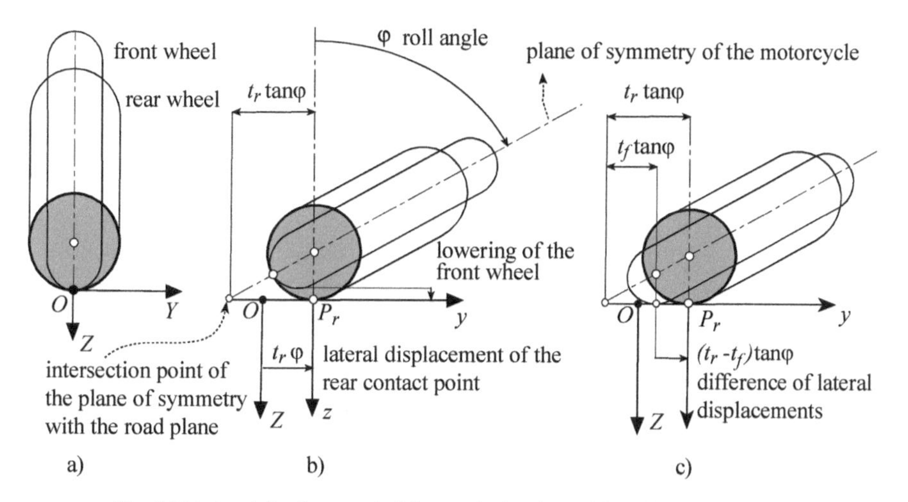

Fig. 1-17 Lateral displacement of the contact points without lateral slippage.

The contact point of the rear tire moves laterally, in the y direction, over a distance $t_r \cdot \varphi$, which is proportional to the radius of the tire cross section and the roll angle of the rear frame.

Let us suppose that the roll motion of the rear frame takes place while the steering angle is kept at zero, and that there is no pitch of the motorcycle around the axis of the rear wheel. Since the front wheel has a smaller section than the rear one, the front wheel would be raised from the road plane following the roll motion. However, contact of the front wheel with the road is assured by the simultaneous pitch rotation of the entire motorcycle around the axis of the rear wheel.

Once the roll and pitch rotations have taken place, the front wheel contact point moves to the left of the rear wheel contact point by the quantity $\left(t_r - t_f\right)\tan\varphi$, as is shown in Fig. 1-17c. It is clear that if the tires have equal sections, the lateral displacement of the two contact points have the same value.

1.7.2 The combined effect of roll and steering

The rotation of the handlebar generates lateral and longitudinal displacements of the front wheel's contact point.

Let us consider a motorcycle that is initially in a vertical position. The motorcycle is tilted through the roll angle φ and then the handlebars rotated through angle δ. Following this maneuver, the front wheel's contact point P_f moves away from the plane of the rear frame.

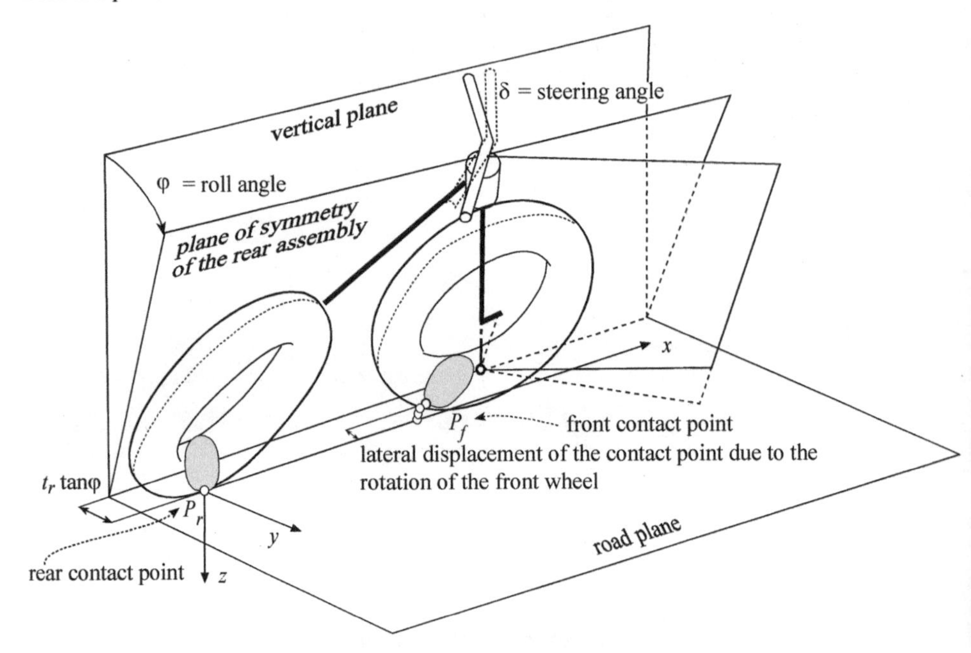

Fig. 1-18 Lateral displacement of the contact point P_f.

The coordinates of the point P_f in the SAE reference system, are expressed in the following equations:

$$x_{P_f} = (c_1 + c_2)\sin\mu + (c_4 + c_5)\cos\mu$$

$$y_{P_f} = \left[-(c_1 + c_2)\cos\mu + (c_4 + c_5)\sin\mu\right]\sin\varphi + c_3\cos\varphi - (t_f - t_r)$$

The quantities $c_1 \ldots c_5$ have been defined previously in the section 1.6.

Figure 1-19 shows the lateral and longitudinal displacements of the front contact point for four values of the roll angle and corresponding steering rotation.

The point moves forward as the steering angle δ and preset roll angle φ increase. Figure 1-19 shows how the x coordinate of P_f increases from the initial value, which is equal to the wheelbase.

The point moves initially to the left (the y coordinate of P_f is initially negative) and then returns to the right, passing over the x axis (the y coordinate of P_f changes its sign). At the point where the pass-over takes place, the steering angle δ decreases as the preset roll angle φ increases.

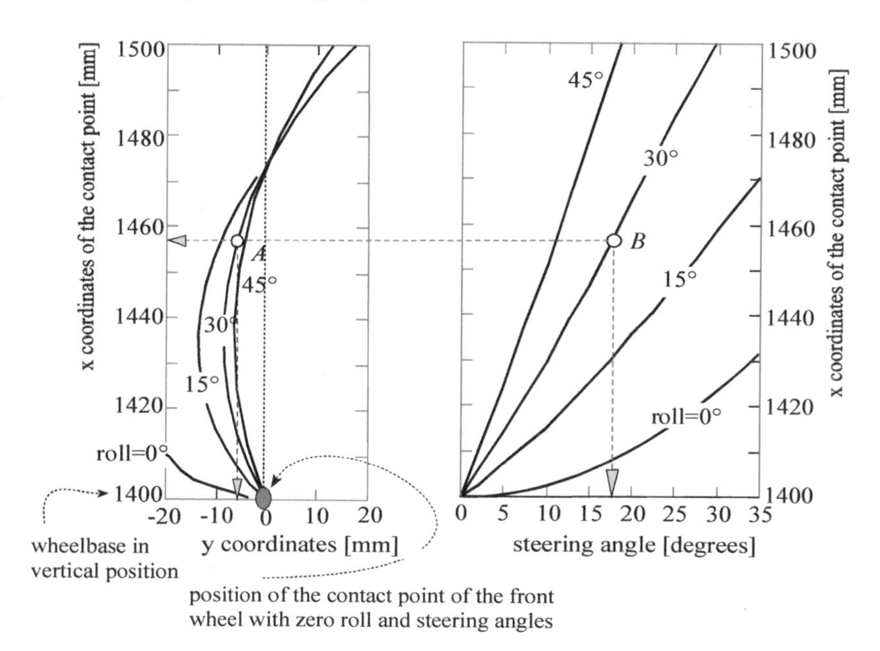

Fig. 1-19 Position of the front contact point P_f.

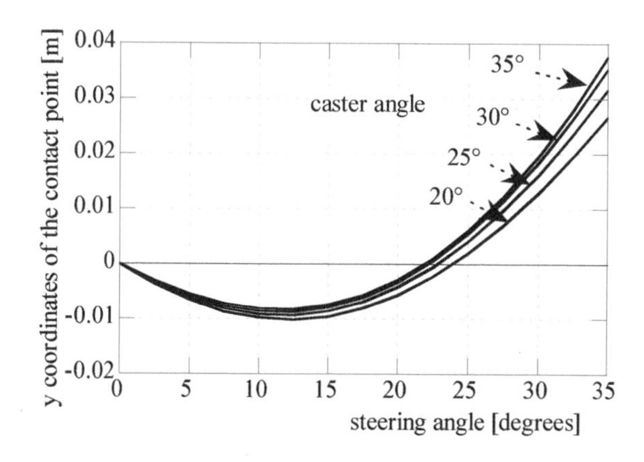

Fig. 1-20 Lateral position of the front contact point P_f as a function of the steering angle δ, for roll angle φ = 30°, and for various values of the caster angle ε.

The lateral displacement of the contact point P_f is not greatly affected by the caster angle ε, as is seen in Fig. 1-20, while it is very sensitive to the value of the normal trail a_n, as can be observed in Fig. 1-21.

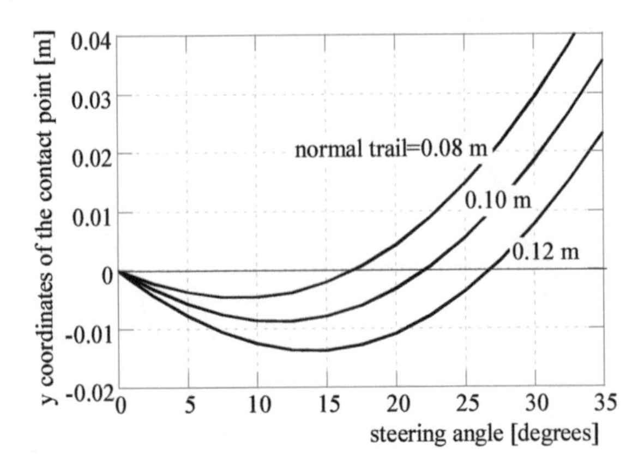

Fig. 1-21 Lateral position of P_f versus the steering angle δ, for angle φ=30° and for various values of the normal trail a_n [ε = 30°].

It is interesting to study the displacement of the front contact point in a triad fixed to the front frame. For this purpose, we consider a zero thickness wheel (Fig. 1-22). When the motorcycle is perfectly vertical (the roll and steering angles being zero), the contact point is located at A, as is shown in Fig. 1-22. While increasing the roll and the steering angles the contact point P_f moves along the arc AC up to its limiting position C,. The point P_f reaches the point C only when the roll angle φ is equal to 90°, i.e., if the motorcycle is horizontal.

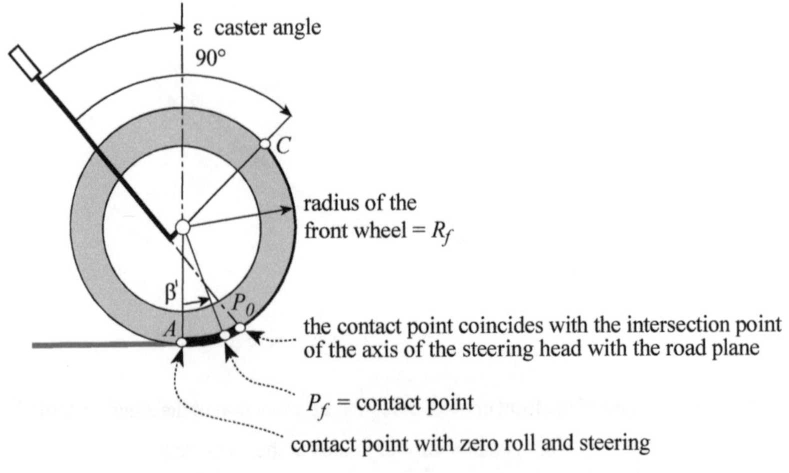

Fig. 1-22 The geometry of steering (zero thickness wheel).

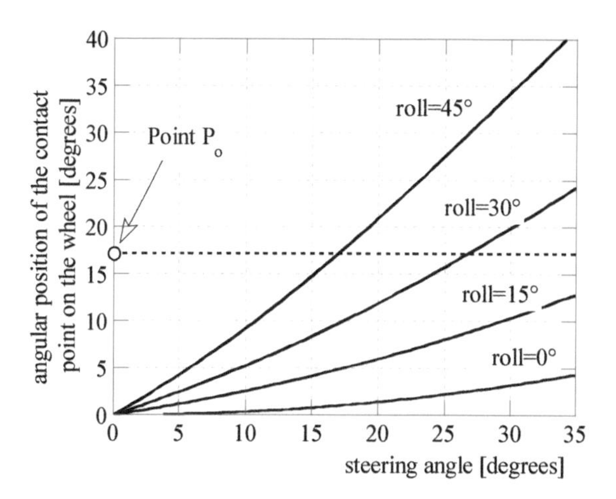

Fig. 1-23 Angular position β' of the front contact point of a zero thickness wheel versus the steering angle δ, for various values of the roll angle φ.

As is clear, the front contact point never reaches point C with the values of the roll and steering angles normally used in driving. In fact the contact point P_f moves within the arc AP_o, P_o being the intersection point of the steering axis with the profile of the wheel.

The front contact point P_f reaches point P_o depending on the steering and roll angles, as is shown in Fig. 1-22. When carrying out a steering maneuver to the right with a set roll angle, the contact point P_f moves forward along the arc AP_o, while its trace moves to the left and forward on the road surface. The effective trail is zero when the contact point is exactly at P_o. Further increases in the roll and steering angles move the contact point P_f towards the position C, while its trace on the road plane moves to the right and the trail becomes negative.

The preceding equations, that give the position of the front contact point, can be significantly simplified by assuming the rotations of the steering angle δ to be small ($\sin\delta \cong \delta$). The expression for the contact point P_f then becomes:

$$x_{P_f} = p\left[1 + \frac{\rho_f}{p}\cos\varepsilon \cdot \tan\varphi \cdot \delta + \frac{(R_r - R_f)}{p}\mu\right]$$

$$y_{P_f} = -\frac{a_n - t_f \sin\varepsilon}{\cos\varphi}\delta - (t_f - t_r)\tan\varphi$$

If we also ignore tire thickness, these equations are further simplified:

$$x_{P_f} = p + R_f \cos\varepsilon \cdot \tan\varphi \cdot \delta \qquad y_{P_f} = -\frac{a_n}{\cos\varphi}\delta$$

1.7.3 The influence of contact point lateral displacement on roll motion

It is clear from the preceding section that the leftward displacement of the contact point P_f, following a steering maneuver to the right, favors roll. This statement can be explained by Fig. 1-24, which represents the motorcycle, schematized as a rigid body of mass m, in equilibrium on a curve with a roll angle φ equal to 30°.

Assuming that we maintain a constant roll angle, the front contact point P_f moves to the outside of the curve as the steering angle δ increases. Therefore, the weight moment increases with the increase in the steering angle δ. This moment tends to tilt the motorcycle even more. The increase in the weight arm, as shown in Fig. 1-24, is proportional to the leftward lateral displacement of the front contact point. The lateral displacement Δy begins to decrease when a certain steering angle δ is reached.

The contact point reaches its maximum external displacement at a certain steering angle δ.This value of δ does not correspond to the δ value that minimizes the pitch angle μ. For example, with a roll angle φ equal to 30°, the maximum lateral displacement Δy occurs with a steering angle δ equal to 12.5°, while the pitch angle μ is at a minimum when δ is equal to 22.5°, as shown in Figs. 1-19 and 1-14 respectively.

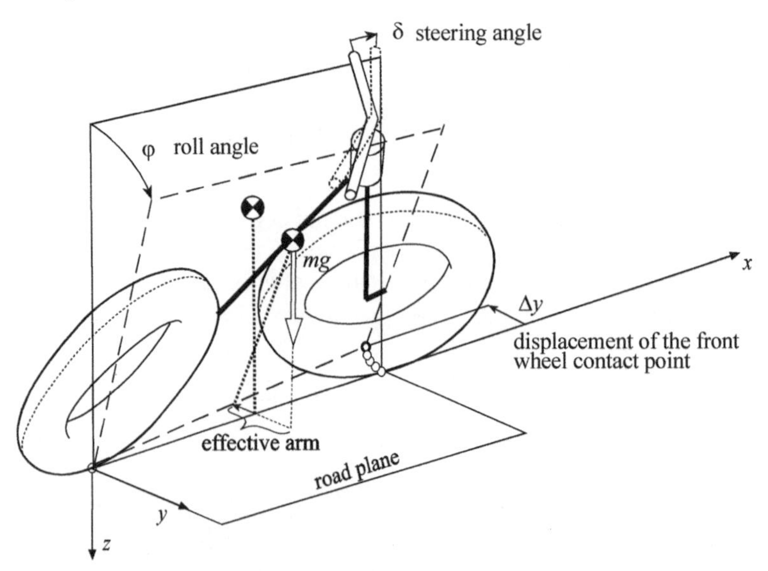

Fig. 1-24 Lateral displacement of the front contact point.

1.8 Front wheel camber angle

The camber angle β of the front wheel is different from the roll angle φ of the rear frame, when the steering angle δ is other than zero. As has already been shown, the front and rear frame roll angles coincide only for zero steering angle.

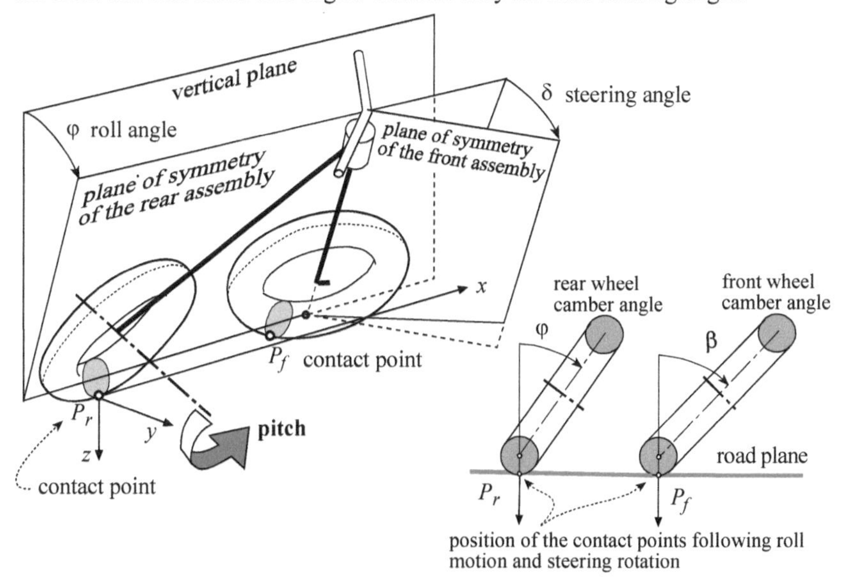

Fig. 1-25 Pitch of the motorcycle and camber angles of the front and rear wheels.

The camber angle of the front wheel β depends on the rear frame roll angle φ, the steering angle δ, the caster angle ε and the pitch angle μ:

$$\beta = \arctan\left(\frac{\tan\varphi\ \cos\delta + \sin\delta\ \sin(\varepsilon+\mu)}{\cos(\varepsilon+\mu)}\cos(\beta' - \varepsilon)\right)$$

The front frame is always more tilted with respect to the rear frame when steering angle is other than zero (same sign as roll angle). As the steering angle δ increases, so does the camber angle β.

If the pitch μ is ignored with respect to the caster angle ε, we obtain:

$$\beta = \arcsin(\cos\delta\sin\varphi + \cos\varphi\sin\delta\cdot\sin\varepsilon)$$

If the steering and the roll angles are small enough the front camber angle can be approximated as:

$$\beta = \varphi + \delta\cdot\sin\varepsilon$$

The equation shows that, for roll and steering angles "in phase", e.g. with the roll angle to the right and the handlebars also turned to the right, the front frame roll angle is always greater than the rear frame roll angle. This aspect is important because the tire lateral force, as will be seen in the next chapter, depends heavily on the camber angle.

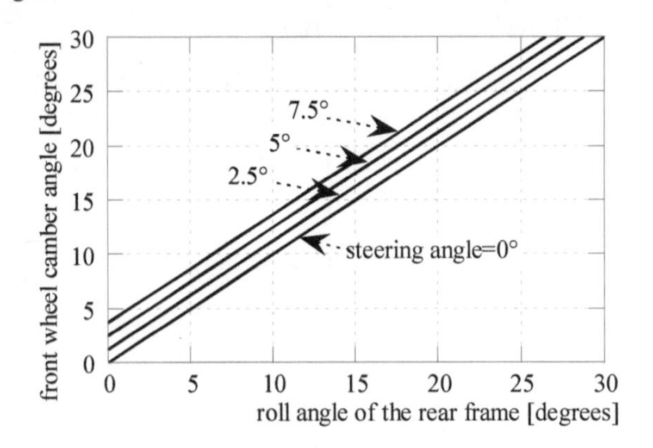

Fig. 1-26 Front wheel camber angle versus the rear frame roll angle φ for various values of the steering angle.

1.9 The kinematic steering angle

The kinematic steering angle Δ depends on the rear frame roll angle φ, steering angle δ, caster angle ε and pitch angle μ :

$$\Delta = \arctan\left(\frac{\sin\delta \cos(\varepsilon+\mu)}{\cos\varphi \cos\delta - \sin\varphi \sin\delta \sin(\varepsilon+\mu)}\right)$$

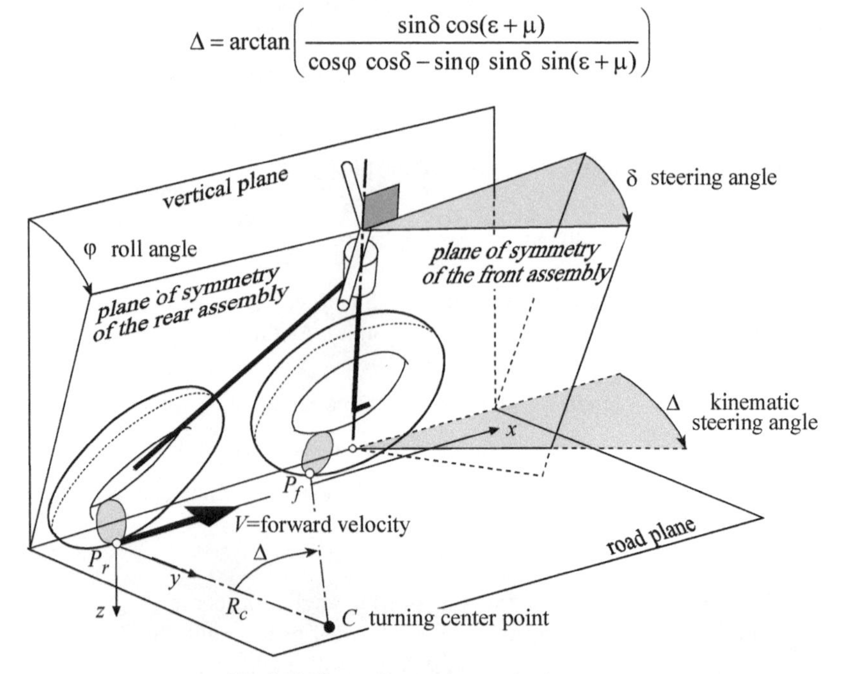

Fig. 1-27 Kinematic steering angle Δ.

From a strictly geometric point of view, the steering angle δ is the angle between the rear and front wheel planes, while the kinematic steering angle Δ represents the intersection of this actual angle with the road plane $z = 0$.

Figure 1-28 shows the variation in the kinematic steering angle Δ as a function of the steering angle δ for four different values of the roll angle φ. The dotted line represents the condition $\Delta = \delta$. Therefore, it appears immediately evident that there is a transition value of the roll angle, below which the kinematic steering angle Δ remains lower than the set value δ, and above which Δ is more than the set value δ. In the specific case examined, the transition value is approximately 27.5°.

In Fig. 1-29 the variation in the kinematic steering angle Δ is shown, this time in terms of the roll angle φ for four typical values of the steering angle δ. The horizontal dotted lines represent the condition $\Delta = \delta$ for each of the four set values of δ. Clearly, it can be observed that for lower values of the roll angle φ (25° to 30°) the steering mechanism is "attenuated" ($\Delta < \delta$). In this case, the steering mechanism is less sensitive to the rotation of the handlebars and the motorcycle can be more easily steered. The rider experiences the same ease of steering offered by wide handlebars even if the ones being used are not. On the other hand, for higher values of the roll angle φ, the steering mechanism is "amplified" ($\Delta > \delta$) making the motorcycle more sensitive to changes in direction.

The kinematic steering angle Δ also depends on the geometry of the steering mechanism. Figure 1-30 is carried out with the roll angle set to φ=30°. The figure, in which the dotted line represents the condition $\Delta = \delta$, shows that decreasing the caster angles make the steering mechanism more sensitive ($\Delta > \delta$). This sensitivity is practically independent of the value of the normal trail. In fact, it is well known that small caster angles are needed for motorcycles to be very sensitive to rapid steering and that high caster angles values make steering more controllable.

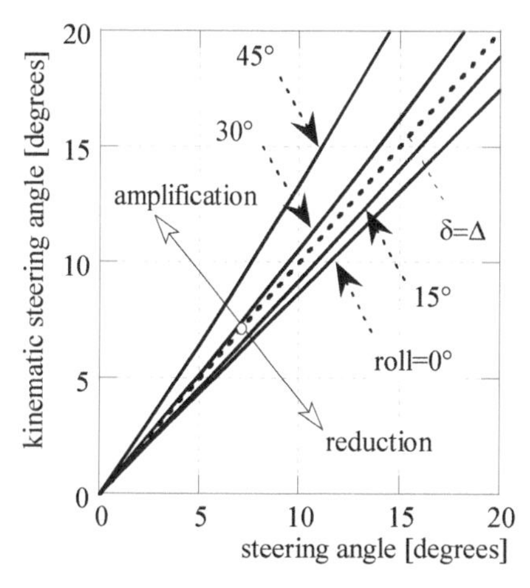

Fig. 1-28 Kinematic steering angle Δ as a function of the steering angle δ for different values of the roll angle φ [ε =30°].

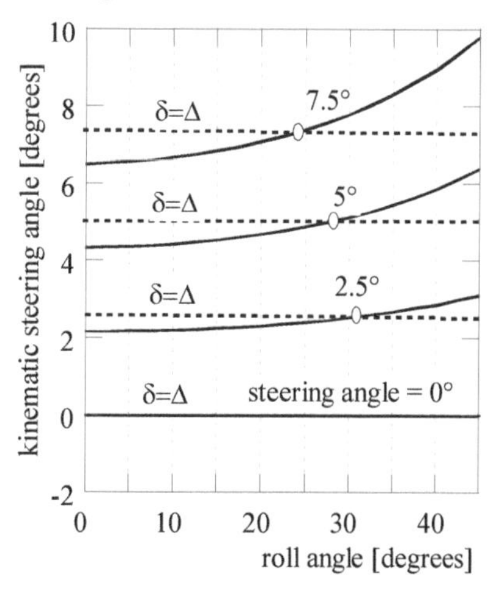

Fig. 1-29 Kinematic steering angle Δ as a function of the roll angle φ for various values of the steering angle δ [ε =30°].

If we ignore the pitch μ with respect to the caster angle ε and the term $\sin\varphi \sin\delta \sin\varepsilon$ with respect $\cos\varphi \cos\delta$, the approximate equation for the kinematic steering angle then becomes:

$$\Delta = \arctan\left(\frac{\cos\varepsilon}{\cos\varphi}\tan\delta\right)$$

This equation can also be obtained on the basis of simple geometric considerations.

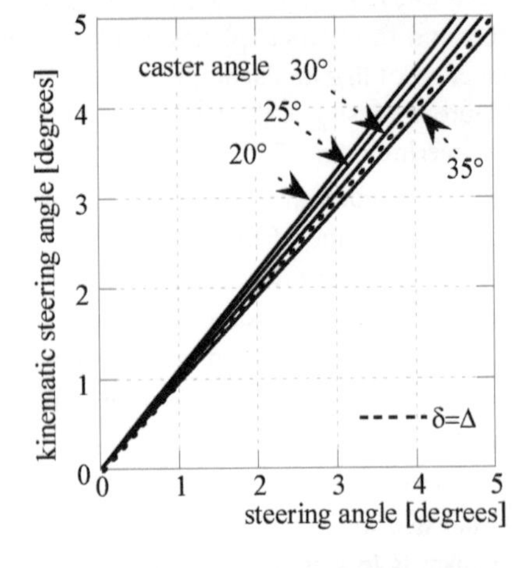

Fig. 1-30 Kinematic steering angle versus the steering angle δ, for various values of the caster angle ε [φ=30°].

Let's consider a motorcycle in a vertical position and suppose that the handlebar rotates, while locking the contact point P_f of the front wheel. The rear contact point P_r moves back slightly, while the intersection point of the steering head axis with the road plane moves laterally. This movement describes an approximately circular trajectory, as shown in Fig. 1-31.

The rear frame rotation angle δ_p depends on the steering angle δ_n as the following equation shows:

$$(p+a)\tan\delta_p = a\tan(\delta_p+\delta_n)$$

The steering angle δ (which by definition is measured in a plane orthogonal to the steering head axis) is related to the angle δ_n (which is measured in a plane orthogonal to the motorcycle plane), by the expression (see also section 1.4):

$$\tan\delta_n = \tan\delta \cdot \cos\varepsilon$$

Assuming that the rotations are small, the following simplified equation is obtained:

$$\delta_p \cong \frac{a}{p}\delta\cos\varepsilon$$

Therefore, the displacement of the rear plane from the front contact point, keeping zero roll angle, is:

$$y_{P_f} \cong a\delta\cos\varepsilon = a_n\delta$$

The displacement y_{P_f} is proportional to the value of the trail and decreases with an increase in the caster angle as demonstrated previously on page 24.

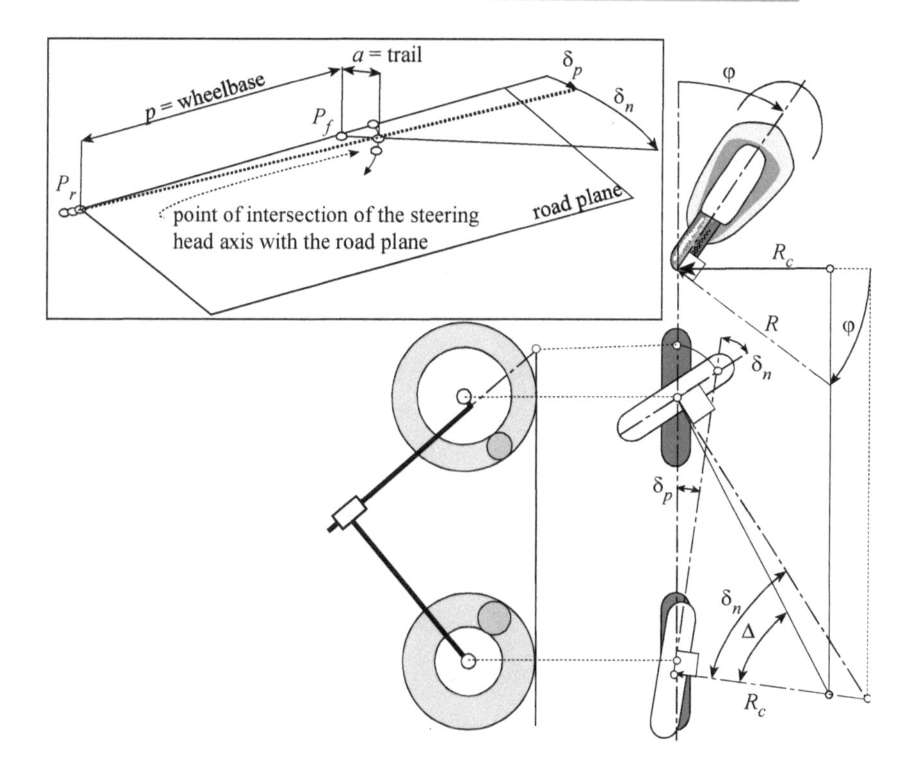

Fig. 1-31 Steering geometry.

Now that the steering angle δ has been fixed, let us tilt the motorcycle through a set roll angle φ, as in Fig. 1-31 and 1-32. The kinematic steering angle Δ is represented by the angle formed by the direction of forward motion of the front and rear wheels. The approximate equation of this angle is given by the ratio of the wheelbase and the radius of curvature:

$$\tan\Delta \cong \frac{p}{R_c}$$

The steering angle δ_n, measured in the plane normal to the rear frame plane, is:

$$\tan\delta_n = \frac{p}{R}$$

The kinematic steering angle Δ , in terms of the roll angle φ, caster angle ε and steering angle δ, is then:

$$\tan\Delta = \frac{\tan\delta_n}{\cos\varphi} = \frac{\cos\varepsilon}{\cos\varphi}\tan\delta$$

which was shown previously on page 28.

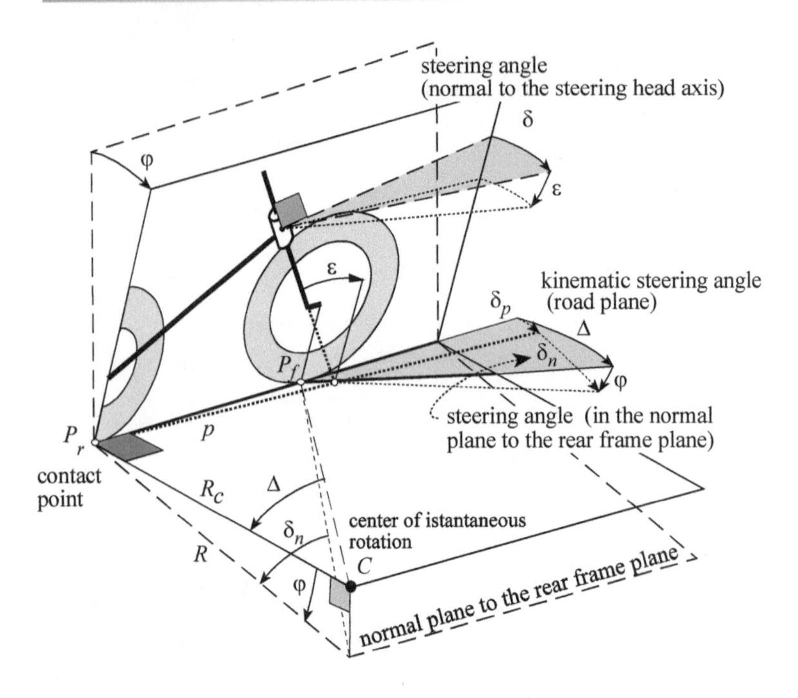

Fig. 1-32 Kinematic steering angle.

On the basis of this equation we can draw the following conclusions:

- only when the roll angle φ is equal to the caster angle ε, can the kinematic steering angle Δ be equal to the rotation angle of the handlebars δ;
- an attenuation, $\Delta < \delta$, occurs for low values of the roll angle, while an amplification, $\Delta > \delta$, occurs for large roll angles:
- with high values of the caster angle ε (like choppers), a greater rotation of the handlebar is needed to produce the same value of the kinematic steering angle.

1.10 The path curvature

The kinematic study of the path traced by a motorcycle is carried out assuming that there is no lateral slippage between the wheels and the road plane ("kinematic steering"). The curvature C of the path (the inverse of the path radius) depends on the position of the front contact point P_f and the kinematic steering angle Δ:

$$C = \frac{\tan\Delta}{x_{P_f} + y_{P_f}\tan\Delta} \cong \frac{\tan\Delta}{p}$$

For small steering angles ($\sin\delta \cong \delta$), the curvature C can be expressed in terms of the roll angle φ and the steering angle δ:

$$C = \frac{\cos\varepsilon}{p\cos\varphi\left(1 + \dfrac{R_f \cos\varepsilon}{p}\delta\tan\varphi\right)}\tan\delta$$

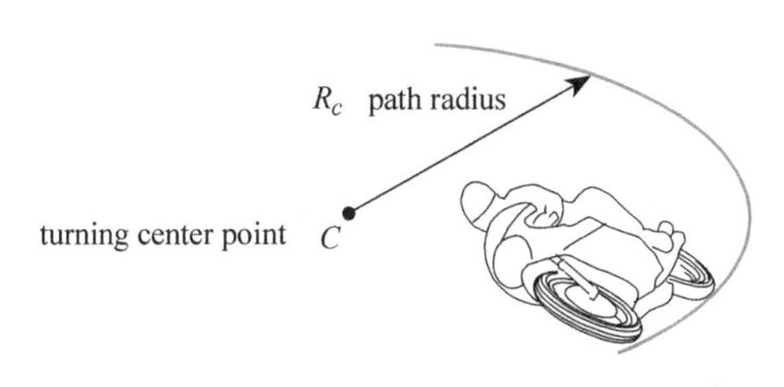

Fig. 1-33 Radius of path.

Since the displacement of the contact point P_f of the front wheel is small with respect to the wheelbase, the curvature can be computed with the following simplified formula:

$$C = \frac{1}{R_c} \cong \frac{\tan\Delta}{p} = \frac{\cos\varepsilon}{p\cos\varphi}\delta$$

It can be observed that the path's radius is directly proportional to the wheelbase. Fig. 1-34 shows how the curvature C varies with the steering angle δ, for various values of the roll angle φ. The maximum error using the approximate formula is equal to about 2%.

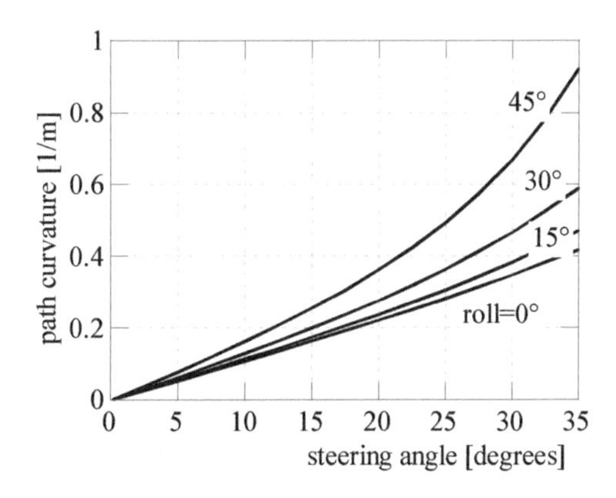

Fig. 1-34 Curvature C versus the steering angle δ for various values of the roll angle φ.

1.11 The effective trail in a curve

The trail is the distance between the front contact point and the intersection point of the steering head axis with the road plane. On the other hand, the normal trail is the perpendicular distance between the front contact point and the steering head axis (Fig. 1-35).

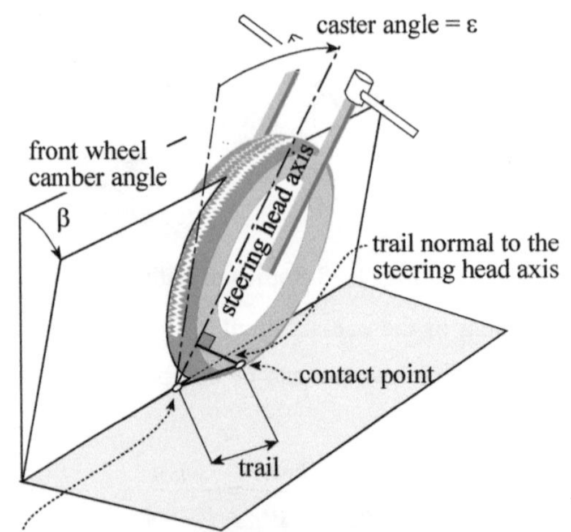

Fig. 1-35 Effective trail with the front wheel tilted and steered.

In cornering conditions the normal and mechanical trail depend on the wheelbase, the caster angle, the offset of the front wheel, the geometrical properties of the tires, the position of the front contact point and the pitch angle.

The normal trail in cornering condition a_n^* is:

$$a_n^* = \sqrt{a_1^2 + a_2^2 + a_3^2}$$

$$a_1 = -(y_{Pf} \cos\varphi + t_r \sin\varphi)\cos(\varepsilon + \mu)$$

$$a_2 = \left[x_{Pf} \cos(\varepsilon + \mu) - p\cos\varepsilon + R_r \sin\varepsilon - a_n\right]\cos\varphi - \left[\rho_r \cos\varphi + t_r\right]\sin(\varepsilon + \mu)$$

$$a_3 = y_{Pf} \sin(\varepsilon + \mu) + \left[x_{Pf} \cos(\varepsilon + \mu) - p\cos\varepsilon + R_r \sin\varepsilon - a_n - \rho_r \sin(\varepsilon + \mu)\right]\sin\varphi$$

The mechanical trail in cornering condition a^* is:

$$a^* = \sqrt{(b_1 - x_{Pf})^2 + (b_2 - y_{Pf})^2}$$

$$b_1 = \frac{(a_n + p\cos\varepsilon - R_r \sin\varepsilon)\cos\varphi + (\rho_r \cos\varphi + t_r)\sin(\varepsilon + \mu)}{\cos\varphi\cos(\varepsilon + \mu)}, \qquad b_2 = -t_r \tan\varphi$$

It is worth highlighting that the trail depends on the geometry of the front tire because the pitch angle and the front contact point position depend on ρ_f and t_f.

The importance of the normal trail derives from the fact that the moments generated by the tire reaction forces (vertical load and lateral force) acting around the steering head axis are proportional to the value of the normal trail.

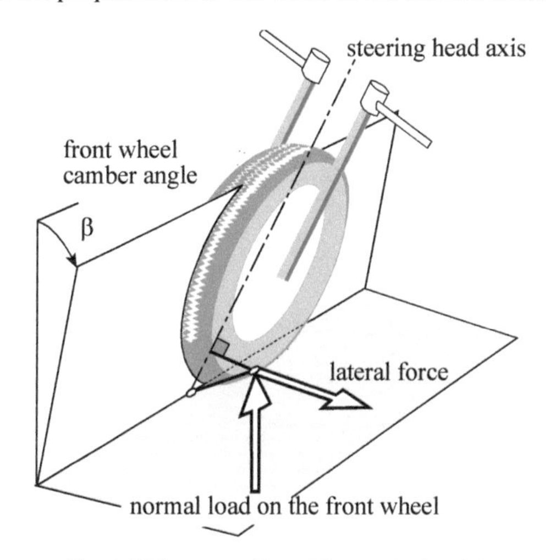

Fig. 1-36 Forces acting at the contact point.

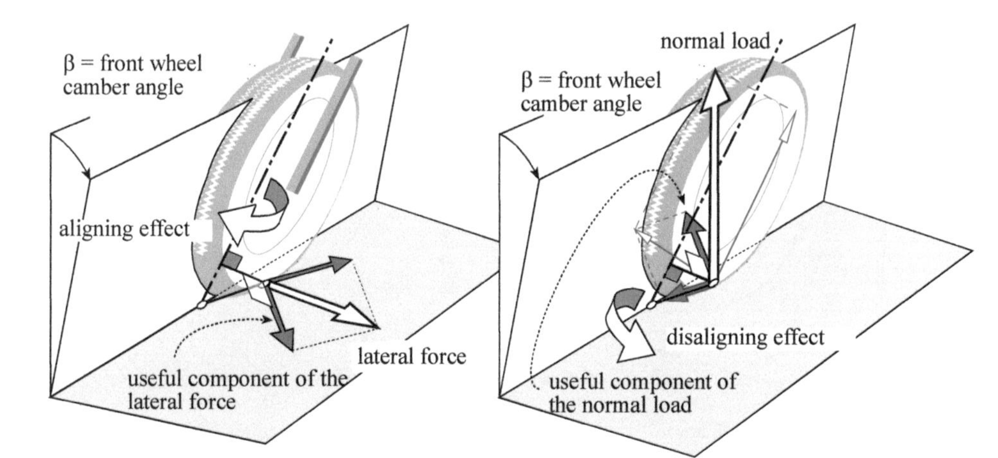

Fig. 1-37 Components of the reaction forces that generate a moment around the steering head axis.

Let's consider the lateral force and normal load applied at the front contact point (Fig. 1-36). Each force can be split into components that act perpendicular to the steering axis and normal trail (therefore in a position to produce a moment around the axis) and components parallel to or intersecting the steering axis (which do not produce a moment). This is shown schematically in Figs. 1-36 and 1-37.

The normal trail represents the arm of the useful components. The useful component of the lateral force, tends to align the wheel to the forward velocity, while the useful component of the vertical load, has a misaligning effect, i.e., it tends to cause the wheel to rotate towards the inside of the curve. The values of the moments around the steering head, generated by these two useful components, are important for the equilibrium of the front section (around the steering head axis). The torque the rider must apply to maintain equilibrium depends on them.

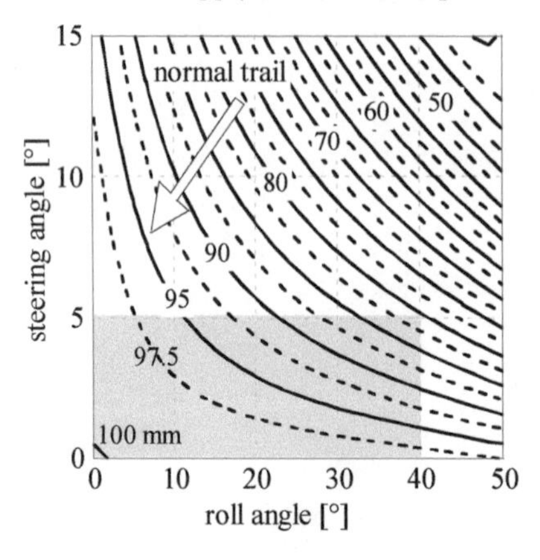

Fig. 1-38 Normal trail [radius of front cross section t_f = 50 mm].

Fig. 1-39 Normal trail [radius of front cross section t_f = 80 mm].

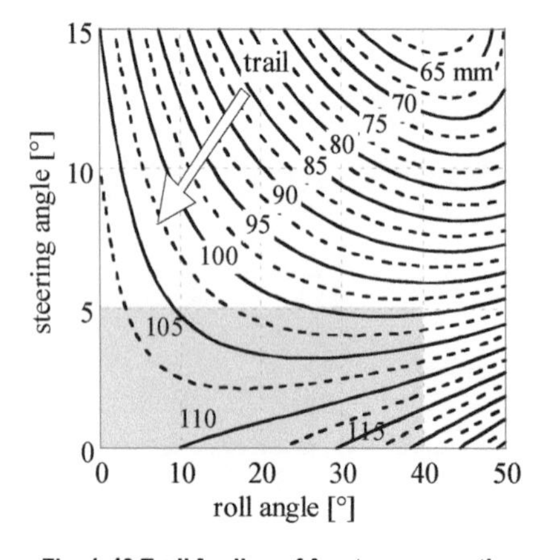

Fig. 1-40 Trail [radius of front cross section t_f = 50mm].

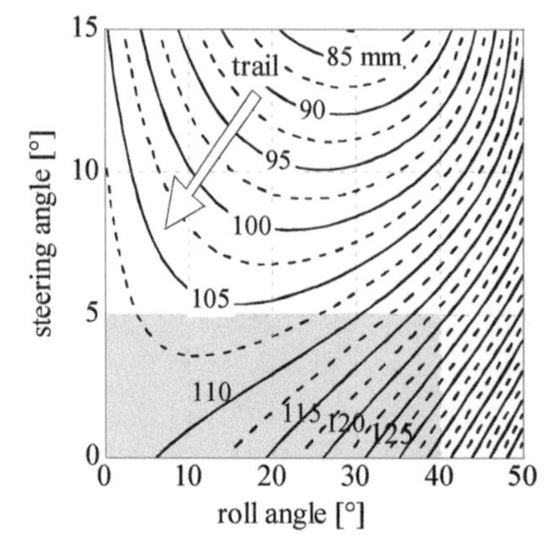

Fig. 1-41 Trail [radius of front cross section t_f = 80 mm].

We will now try to understand whether and how the normal trail varies when the motorcycle is in a curve. Figure 1-38 shows the normal trail as a function of the roll angle and the steering angle. It can be noted that the normal trail diminishes when the roll angle increases and even more so when the steering angle increases. However, if we consider steering angles below 5° and roll angles below 40°, the trail variations remain below 20%.

Fig. 1-39 shows that an increase in the radius of the front tire cross section from 50 mm to 80 mm further reduces these variations. With a steering angle of 5° and a roll angle of 40° the reduction of the trail goes from 20% to 10%.

Changing the type of front tire can cause a variation in the cross section radius. Therefore, the normal effective trail in turning, i.e. the arm of the reactive forces, varies. Since the rider "feels" the behavior of the front section through the torque applied on the handlebars, it is clear that a variation in the cross section radius can produce a different feeling. If the caster angle is varied, very similar graphs are obtained. However, the caster angle value influences the variation in the normal trail to a lesser extent.

In conclusion, it can be stated that when comparing different motorcycles it is important to refer to the normal trail since it has a precise physical meaning. In Fig. 1-40 and Fig. 1-41 the trail is presented as a function of the roll angle and steering angle. It can be observed that an increase in the roll angle, with small values of the steering angle, produces an increase in the trail. This is different from what happens with the normal trail.

To summarize we can say that:

- both trail and normal trail diminish with an increase in the steering angle δ,
- the value of the trail, whether normal or measured on the road plane, also depends on the roll angle φ,
- a reduction in the trail, with the increase in the steering angle δ, is attenuated as the front tire cross section radius and the external tire radius increase.

1.12 The effect of tire size on the rear frame yaw

We would now like to study another particular aspect of motorcycles which occurs when the motorcycle tilts to the side: the yawing effect caused by different cross section sizes of the tires.

Consider a motorcycle initially in a vertical position and with a zero steering angle (Fig. 1-42a). As has already been stated, according to the SAE reference system, the triad's origin is in the rear wheel contact point P_r and the x-axis represents the motorcycle's forward motion.

Let's suppose that the motorcycle tilts while the steering angle is held at zero, as shown in Fig. 1-42b. If the tires' cross-sections have the same radii ($t_r = t_f$), the intersection of the rear frame plane with the road plane coincides with the direction of the forward motion.

In this case, the rear plane does not yaw, but rather moves laterally due to the lateral rolling of the tires. The lateral displacement, $t_r\varphi = t_f\varphi$, is equal to the product of the roll angle φ times the radius of the tire's cross-section.

If the radii of the cross sections have different values ($t_r > t_f$), also shown in Fig. 1-42b, the tilting motion, with the steering angle fixed at zero, produces a rotation ψ of the rear frame plane, i.e., a yaw motion whose value is given by:

$$\psi = \frac{(t_r - t_f)(\tan\varphi - \varphi)}{p}$$

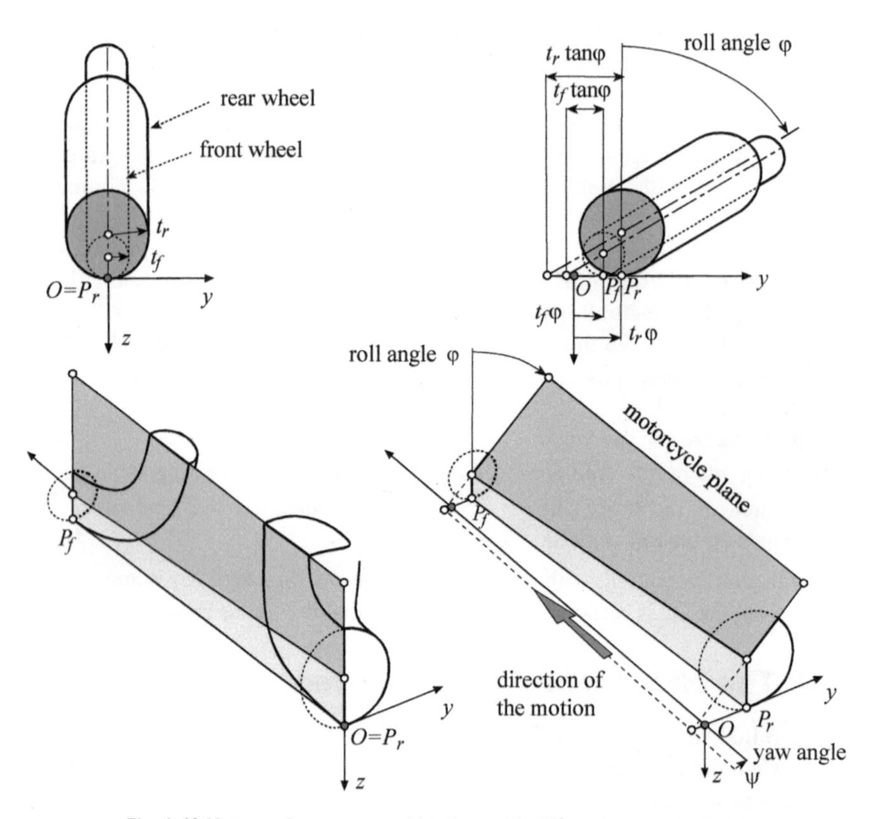

Fig. 1-42 Motorcycle yaw caused by tires with different cross sections.

Example 3

Consider a motorcycle with the following characteristics: wheelbase p = 1400 mm, the radii of the tire cross sections are t_r = 100 mm and t_f = 40 mm.

If the motorcycle changes from straight running (vertical motorcycle, φ = 0) to turning (roll angle $\varphi = 45°$), calculate the yaw angle of the rear frame.

The yaw angle of the rear frame plane, due to the difference in the radii of the cross sections, is equal to ψ = 0.53°.

Moto Guzzi 500 cc of 1924

2 Motorcycle Tires

The tire is one of the motorcycle's most important components. Its fundamental characteristic is its deformability, which allows contact between the wheel and the road to be maintained even when small obstacles are encountered.

In addition to improving the comfort of the ride, the tire improves adherence, an important characteristic both for the transfer of large driving and braking forces to the ground, and for the generation of lateral forces. The performance of a motorcycle is largely influenced by the characteristics of its tires. In order to understand their importance, one must consider that control of the vehicle's equilibrium and motion occurs through the generation of longitudinal and lateral forces acting between the contact patches of the tires with the road plane. The forces originate as a result of action taken by the rider through the steering mechanism, the accelerator and the braking system.

2.1 Contact forces between the tire and the road

From the dynamic view of the motorcycle, it is fundamental to portray the overall behavior of the tire in various conditions of use through a model capable of representing the forces and moments of contact in terms of forward velocity, camber angle, longitudinal slip, lateral slip and load acting on the tire itself.

From a macroscopic viewpoint, the interaction of the tire with the road can be represented by a system composed of three forces and three moments, as in Fig. 2-1:

- a longitudinal force acting along the axis parallel to the intersection of the wheel plane with the road plane, and passing through the contact point (assumed positive if driving and negative if braking), in x direction;

- a vertical force orthogonal to the road plane (a vertical load that acts on the wheel, assumed positive in an upward direction), along the z axis;
- a lateral force, in the road plane, orthogonal to the longitudinal force, in y direction;
- an overturning moment around the x-axis,
- a rolling resistance moment around the y-axis,
- a yawing moment around the z-axis.

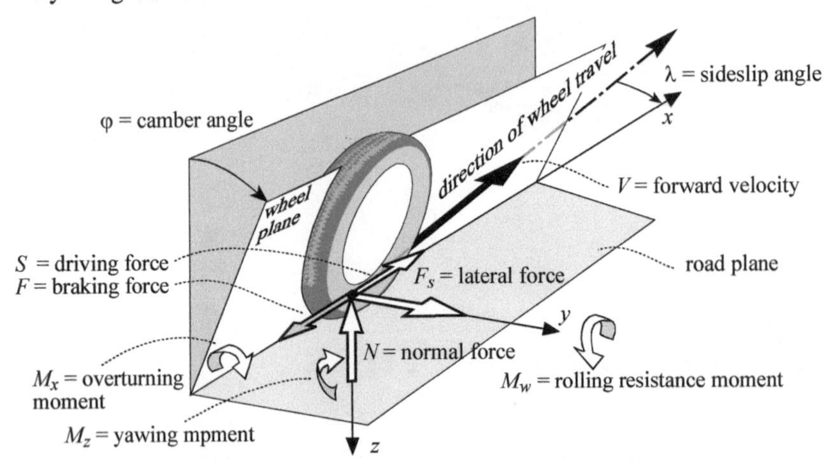

Fig. 2-1 Forces and torques of contact between the tire and the road plane.

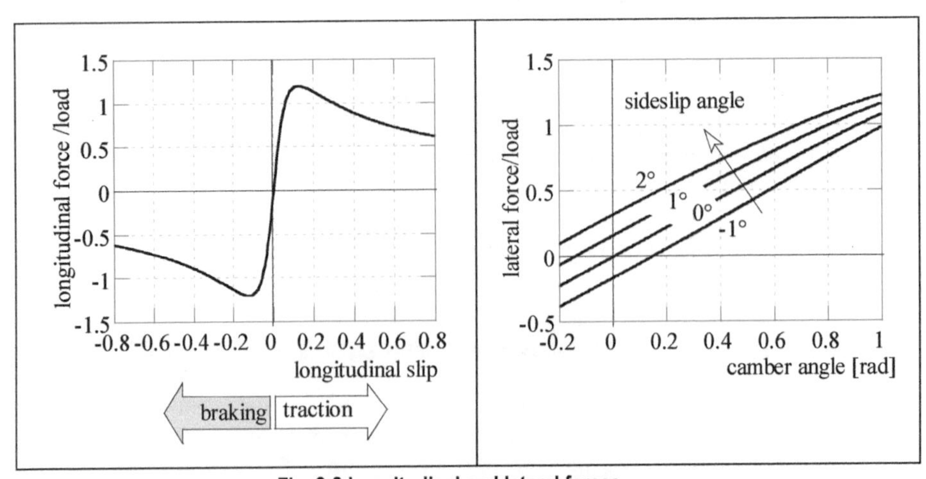

Fig. 2-2 Longitudinal and lateral forces.

- In Fig. 2-2 typical longitudinal and lateral forces have been depicted in the condition of pure slip. Pure slip represents the situation when either longitudinal or lateral slip occurs in isolation. The longitudinal force depends on the longitudinal slip and shows a clear peak while the lateral force is a function both of the camber angle and of the sideslip angle. Curves which exhibit a shape like the forces depicted in Fig. 2-2 can be represented by a mathematical formula named the "Magic Formula".

2.2 The "Magic Formula" for representing experimental results

The model proposed by Pacejka (1993) is very much in use. The approach is substantially empirical and the results reproduce the real behavior of the tire very well. The entire model revolves around what is called the "magic formula," that is, a single expression that can be used to represent the longitudinal driving or braking force, the lateral force or the moment around the z axis. The expression is as follows:

$$Y(x) = y(x) + S_v$$
$$y(x) = D \cdot \sin\left\{C \cdot \arctan\left[Bx - E\left(Bx - \arctan Bx\right)\right]\right\}$$
$$X = x + S_h$$

where B, C, D and E are four parameters, S_v indicates the translation of the curve along the y axis, and S_h indicates the translation of the curve along the x axis.

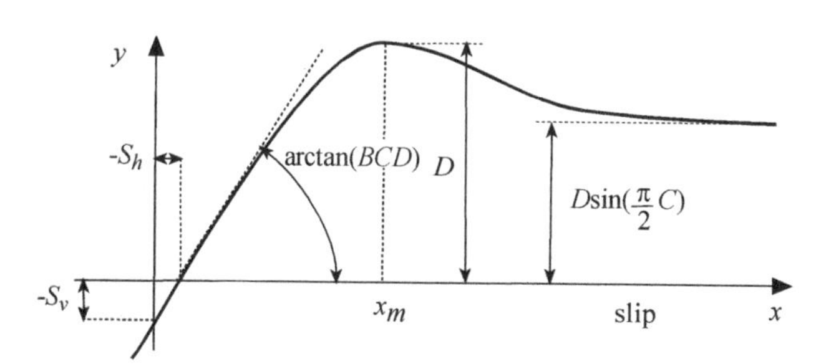

Fig. 2-3 Meaning of the parameters in the "magic formula".

The magnitude y can represent either the longitudinal thrust or the lateral force, while x represents the corresponding slip quantity. Figure 2-3 reproduces the typical variation of the Pacejka curve and is effective in visualizing the meaning of the four parameters appearing there.

- Parameter D represents the peak value (only with $E < 1$ and $C \geq 1$) and depends on the vertical load.
- Parameter C controls the asymptotic value assumed by the curve as the slip tends to infinity, and in this way determines the resulting form of the curve.
- Parameter B determines the slope of the curve from the origin.
- Parameter E characterizes the curvature near the peak, and at the same time determines the position of the peak itself.

It can be shown that the gradient at the origin is given by the product BCD.

2.3 Rolling resistance

Consider a wheel that rotates without slippage on a flat surface. The rolling radius is defined by the ratio of the forward velocity to its angular speed:

$$R_0 = \frac{V}{\omega}$$

The effective rolling radius in free motion is, as shown in Fig. 2-4, smaller than the radius of the unloaded tire because of the deformation of the tire. Its value depends on the type of tire, its radial stiffness, the load, the inflation pressure and the forward velocity. It can be demonstrated that its value in free motion is smaller than that of the radius of the unloaded tire but greater than the distance from the center of the tire to the road plane. An approximate value is given by the equation:

$$R_0 = R - (R - h)/3$$

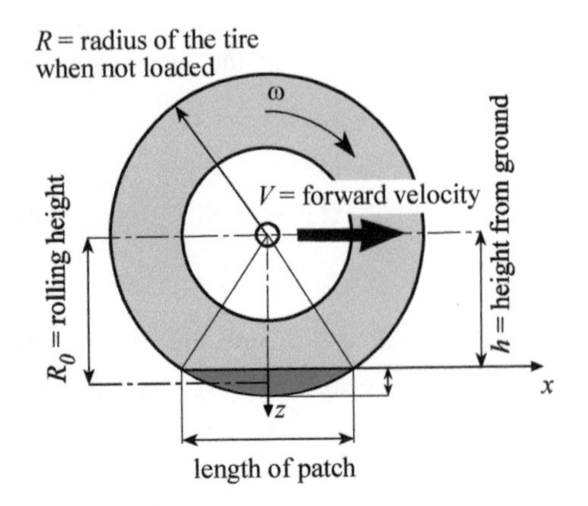

Fig. 2-4 Effective rolling radius of the tire.

During the tire's rolling, the portion of the circumference that passes over the track undergoes a deflection. In the contact area stresses are generated, which are both normal (due to the load) and shear due to the difference in length of the arc of circumference and its tread chord (that represents the length of the contact tread). Because of the hysteresis of the tire material, part of the energy that was spent in deforming the tire carcass is not restored in the following phase of relaxation, or is restored late. This causes a change in the distribution of the contact pressures, which therefore are not symmetric, but are higher in the areas in front of the wheel's axis.

As shown in Fig. 2-5, the resultant of the normal contact pressures is displaced forward with respect to the center of the wheel by the distance d. The forward displacement is called the rolling friction parameter. Hence, to move the wheel with constant forward velocity it is necessary to overcome a rolling resistance moment equal to:

$$M_W = d\,N$$

The resistance to rolling is expressed via a resistance force that opposes the forward motion, and whose value is given by the product of the rolling resistance coefficient f_W and the vertical load.

$$F_W = f_W N = \frac{d}{R} N$$

In addition to the type of tire (either radial or bias-ply), its dimensions, the characteristics of the tire, the temperature and the conditions of use the rolling resistance coefficient depends principally on the forward velocity and on the inflation pressure. The rolling resistance coefficient increases with the camber angle. Typical values are on the order of 0.02.

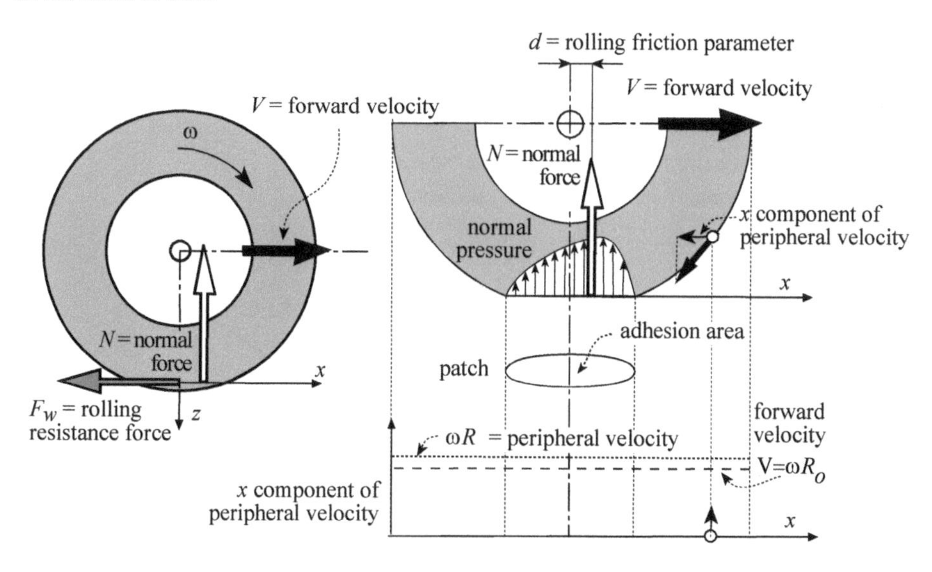

Fig. 2-5 Description of contact pressures and forces acting on a rolling wheel.

Kevin Cooper (see [J. Bradley, 1996]) has proposed the following empirical formula for calculating losses through resistance due to the rolling of the motorcycle tires. The formula takes inflation pressure and forward velocity into account:

$$f_W = 0.0085 + \frac{0.018}{p} + \frac{1.59 * 10^{-6}}{p} V^2 \quad \text{for velocities below 165 km/h}$$

$$f_W = \frac{0.018}{p} + \frac{2.91 * 10^{-6}}{p} V^2 \quad \text{for velocities above 165 km/h}$$

Velocity is expressed in kilometers per hour and the tire pressure p in bar (1 bar ≅ 1 atm). Figure 2-6 shows the variation of the rolling resistance coefficient versus the variation of velocity at certain values of tire pressure. It can be observed that an increase in pressure diminishes the resistance to rolling.

The power that is dissipated because of the rolling resistance force, is given by the product of the resistance force and the forward velocity:

$$P = \left(2.36 * 10^{-6} V + \frac{4.88 * 10^{-6}}{p} V + \frac{4.41 * 10^{-10}}{p} V^3 \right) N \quad \text{for velocities} < 165 \text{ km/h}$$

$$P = \left(\frac{4.88 * 10^{-6}}{p} V + \frac{8.09 * 10^{-10}}{p} V^3 \right) N \quad \text{for velocities} > 165 \text{ km/h}$$

Here N represents the load on the wheel (expressed in Newtons); the power dissipated P is expressed in kilowatts.

To summarize it may be said that the rolling resistance force depends on:

- inflation pressure
- the deformation of the tire (in view of the hysteresis of the material),
- the relative slip between the tire and the road,
- the aerodynamic resistance due to the ventilation.

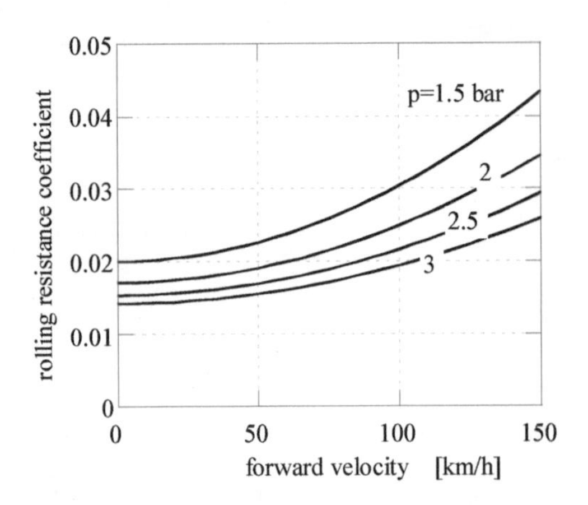

Fig. 2.6 The rolling resistance coefficient versus the forward velocity for various values of tire pressure.

Of these three causes, the first one is by far the most important. Losses through ventilation are caused by the interaction between the wheel and the circulating air, which in turn depends on the form of the wheel itself (arm or spokes), the profile of the tire and the rotational velocity.

Example 1

Consider a motorcycle with a mass of 200 kg and two different velocities: 100 km/h and 250 km/h. Assuming the tire pressure is 2.25 bar, determine the power dissipated to overcome rolling resistance.

The power dissipated in order to overcome the rolling resistance forces at a velocity of 100 km/h is only 1.1 kW, while at a velocity of 250 km/h the power rises to 12 kW.

2.4 Longitudinal force (driving-braking)

The presence of driving or braking forces generates further longitudinal shear stresses along the area of contact. The circumferential stress, in the case of driving force, compresses the fibers in the contact area (Fig. 2-7); in the case of braking forces, the fibers are engaged in tension (Fig. 2-8).

The forward velocity of the contact point is therefore less, in the case of traction, than the tire's peripheral velocity. Alternatively, in the case of braking, it is greater than the tire's peripheral velocity. This is expressed by the longitudinal slip, defined by the ratio between the slip velocity $(V-\omega R)$ and the forward velocity V:

$$\kappa = -\frac{V-\omega R}{V}$$

The longitudinal slip is positive in the case of traction and negative in the case of braking. In the latter case, longitudinal shear stresses have the opposite sign of the forward velocity.

In the case of driving wheel, some longitudinal shear stresses are generated in the contact area having the same sign as the forward velocity and therefore the tire tread in the contact patch is compressed. In the first part of the patch the contact is one of adhesion, but in the second part the contact occurs with sliding (Fig. 2.7).

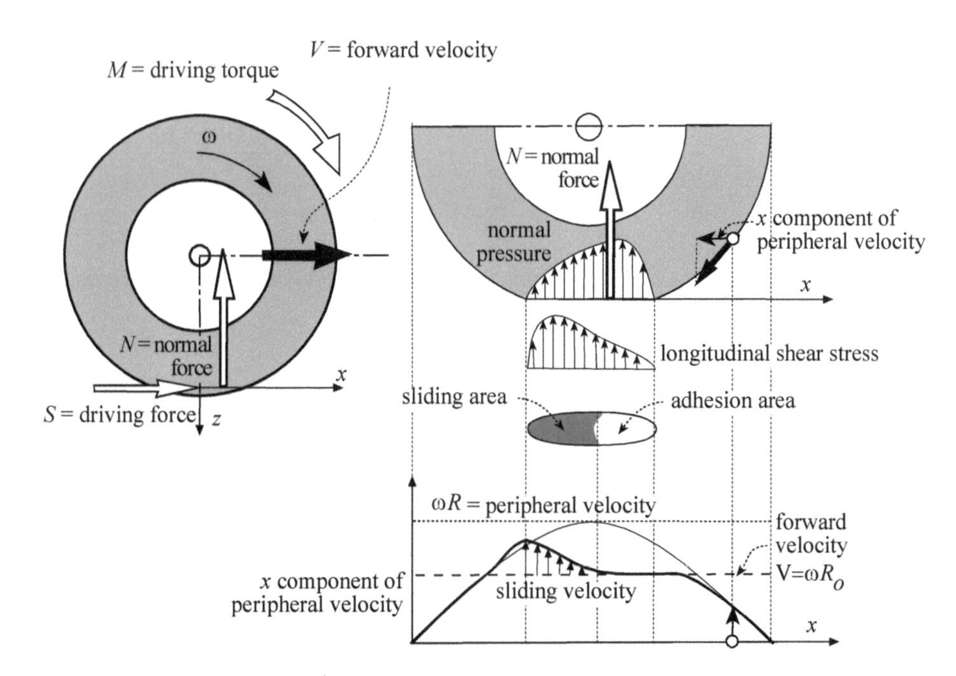

Fig. 2-7 Longitudinal shear stress in the contact area and forces acting on a driving wheel.

In braking, the instantaneous rolling radius, which in conditions of pure rolling is less than the peripheral radius of the wheel, increases with an increase of the braking force until it becomes greater than the wheel's radius (in a sudden stop that includes locking the wheel, this radius is infinite). In the first part of the patch, the contact is one of adhesion. At a certain point, the difference between the forward velocity and the peripheral velocity produces shear stresses greater than those that can be generated in conditions of adhesion, and for this reason a sliding zone is generated. The length of the sliding zone is approximately proportional to the braking force (Fig. 2.8).

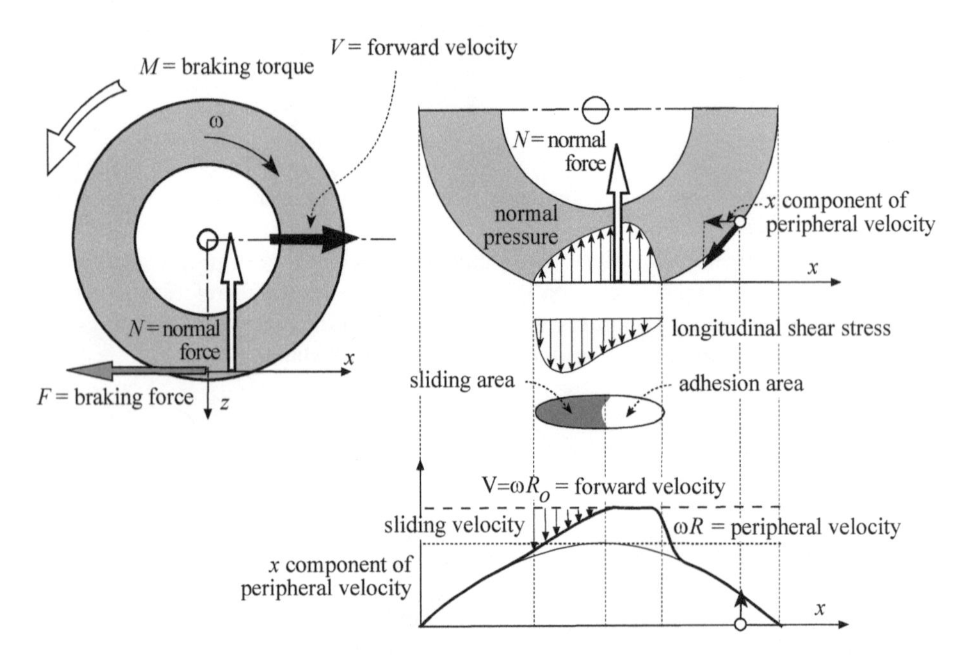

Fig. 2-8 Longitudinal shear stress in the contact area and forces acting on a braking wheel.

2.4.1 Non-linear model

The longitudinal force of both traction and braking is proportional in a first approximation to the load applied; the ratio μ between the longitudinal force and the load (normalized longitudinal force) is called longitudinal braking/driving force coefficient.

The longitudinal force at nominal load N can be described by means of the Magic Formula:

$$\left.\begin{array}{l} F(braking) \\ S(thrusting) \end{array}\right\} = D_\kappa \cdot \sin\left\{C_\kappa \cdot \arctan\left[B_\kappa \kappa - E_\kappa\left(B_\kappa \kappa - \arctan B_\kappa \kappa\right)\right]\right\} N$$

The coefficient $D_\kappa = \mu_p$ represents the peak of the braking/driving force coefficient while the product $D_\kappa C_\kappa B_\kappa$ is the longitudinal slip stiffness.

Fig. 2-9 shows in qualitative terms the ratio of the longitudinal force to the normal load, versus variation of the value of the longitudinal slip. The maximum value (braking/driving traction coefficient) depends strongly on road conditions.

2.4.2 Linear model

The force, in braking and thrusting phases respectively, can be expressed by a linear equation, such as:

$$F = K_\kappa \kappa = (k_\kappa N)\kappa$$
$$S = K_\kappa \kappa = (k_\kappa N)\kappa$$

where,

$$K_\kappa = \left.\frac{dF}{d\kappa}\right|_{\kappa=0}$$
$$K_\kappa = \left.\frac{dS}{d\kappa}\right|_{\kappa=0}$$

indicates the dimensional stiffness (N) of longitudinal slip, and

$$k_\kappa = \frac{1}{N} K_\kappa$$

the non-dimensional longitudinal slip stiffness.

The order of magnitude of the value of longitudinal stiffness k_κ (gradient of the curve of zero slippage) ranges from 12-30 (non-dimensional value).

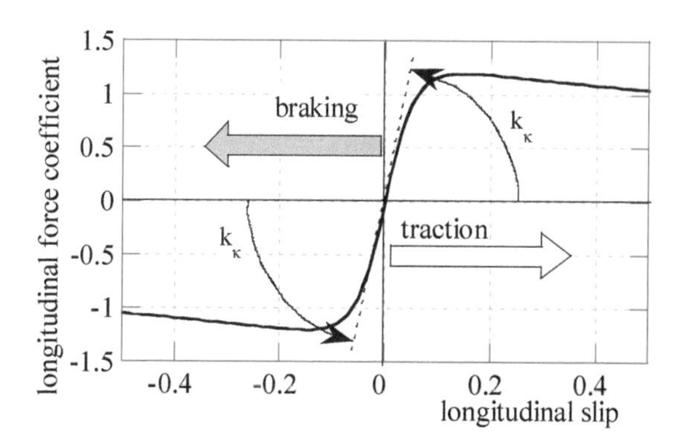

Fig. 2-9 Qualitative variation of the braking/driving force coefficient versus slip.

Example 2

Suppose that a vehicle of 81 kW power (110 HP) attains a maximum velocity of 270 km/h (75 m/s). Determine the driving thrust required to attain this velocity. Next assuming maximum thrust with the entire load on the motorcycle on the rear wheel, determine the driving force coefficient.

The necessary driving thrust is equal to the ratio between power and velocity:

$$S = 81*1000/75=1079.5 \text{ N}$$

The driving force coefficient is equal to:

$$\mu = \frac{S}{N_r} = \frac{1079.5}{1500} = 0.72$$

The slip necessary to produce this normalized longitudinal force can be determined if we know the variation of the longitudinal force coefficient in terms of the slip. The value of the necessary slip depends on the type of tire. In the two curves given in Fig. 2-10, the value of the friction coefficient 0.72 is obtained, with 3.6% slippage in the case of tire A and 8% with tire B. It is clear that tire B is subject to more rapid wear than tire A, because of the greater longitudinal slip necessary for generating the same thrust force.

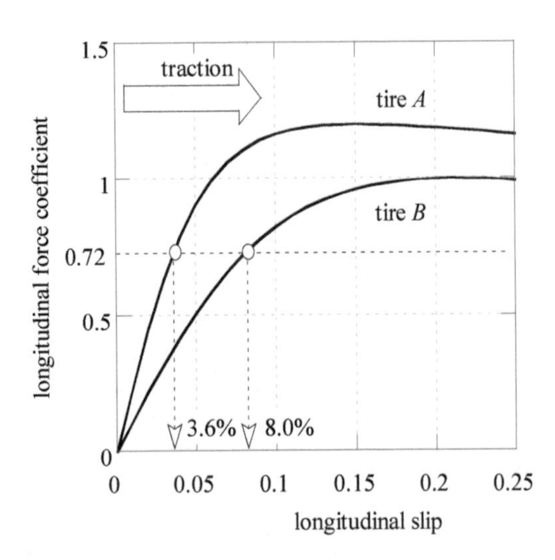

Fig. 2-10 Longitudinal force coefficient versus longitudinal slip for two different tires.

2.5 Lateral force

The lateral force, which the tire exerts on the ground, depends on both the sideslip angle λ and the camber angle φ. The sideslip angle is defined as the angle measured in the road plane between the direction of travel and the intersection of the wheel plane with the road plane, as can be seen in Fig. 2-11. Sideslip forces depend on tire carcass distortion while camber forces depend primarily on geometry.

The tire is deformed on contact with the ground, producing a patch of variable shape and dimensions according to the characteristics of the tire, the roll angle, the sideslip angle, as well as external factors such as the load, the inflation pressure, etc. Any presence of lateral forces and braking or driving torques introduces further deformations to the contact patch. In general, the patch is not symmetrical with respect to the x and y-axes.

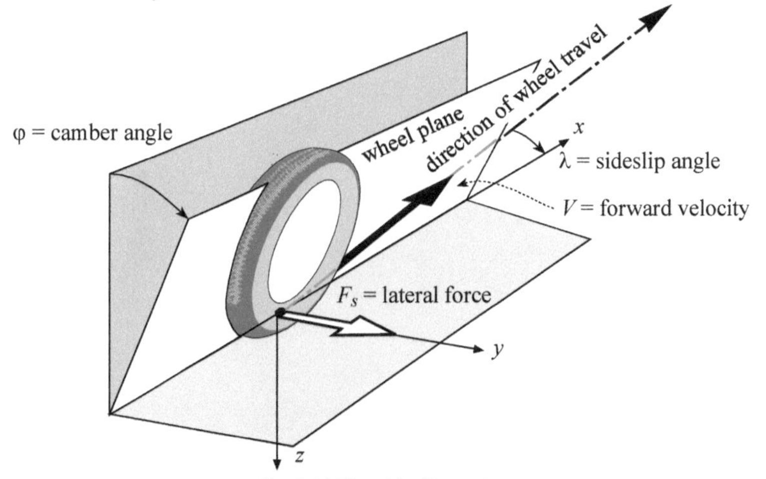

Fig. 2-11 The sideslip angle.

2.5.1 Lateral force generated by the camber angle

First let us consider the case of a tire inclined to a set camber angle, which moves forward in the direction of its plane and has a zero sideslip angle (Fig. 2-12). In the case of an undeformable tire carcass, the patch is dot-shaped and the generic point P, situated on the external surface of the torus of the wheel, describes a circular trajectory in space whose projection on the road plane is a curve in the form of an ellipse. It therefore touches the road at the single contact point A. There is no lateral deformation of the tire; therefore, it generates no camber force.

In the case in which the tire's carcass is deformable, the contact zone is extended, and point P at the moment when it enters the area of contact with the ground is obliged to abandon the theoretical elliptical trajectory and to move along a rectilinear trajectory in the direction of the wheel's forward motion; this direction is indicated with the line $a-a$ in Fig. 2-12.

We can imagine that the deformation of the tire carcass PP'' will occur in two distinct phases: first, the vertical load generates the vertical deformation PP', then the lateral force of the camber thrust generates the deformation $P'P''$. The lateral force due to camber is important, especially at small slip angles.

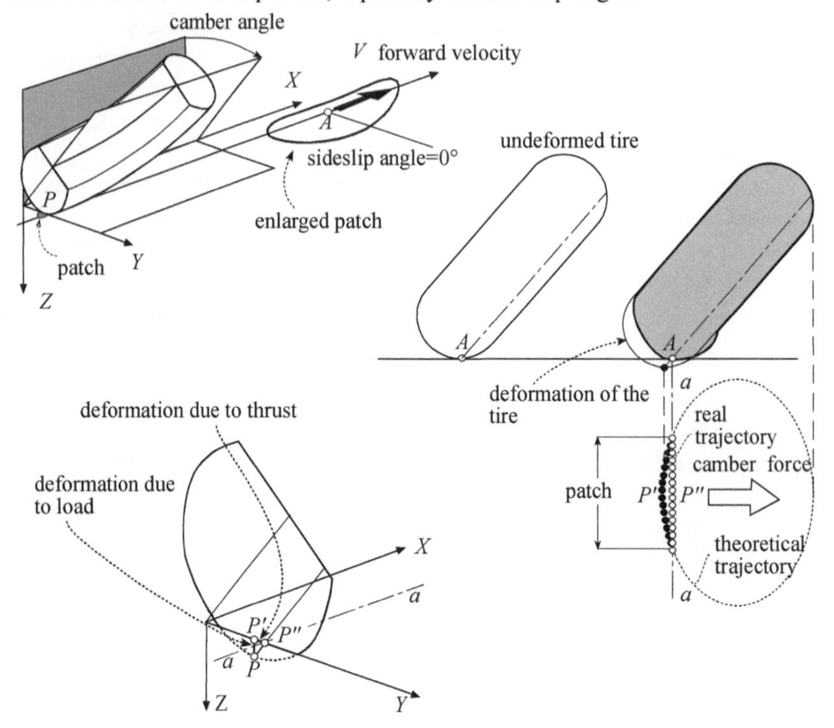

Fig. 2-12 Origin of the camber thrust.

2.5.2 Lateral force generated by lateral slip

Consider a wheel that rotates and at the same time slips laterally. In this case the form of the contact patch is distorted as shown in Fig. 2-13.

Consider a point P situated on a tread that reaches contact with the ground at point A. When point P moves to a determined point indicated with B, it describes a rectilinear trajectory. Its velocity has the direction of the forward velocity V. When it reaches point B, the elastic restoring shear stress, due to the deformation of the carcass and of the rubber elements in the tire tread, becomes greater than the adhesion forces and therefore become such as to make it deviate in the opposite direction, causing it to slide on the ground until the trailing edge C.

Two zones are therefore to be distinguished in the contact area:

- a front zone where adhesion takes place;
- a rear zone in which there is sliding.

The sliding zone is more extended the greater the slip angle is. Once a limiting value of the lateral force has been reached, the entire contact zone becomes a sliding area.

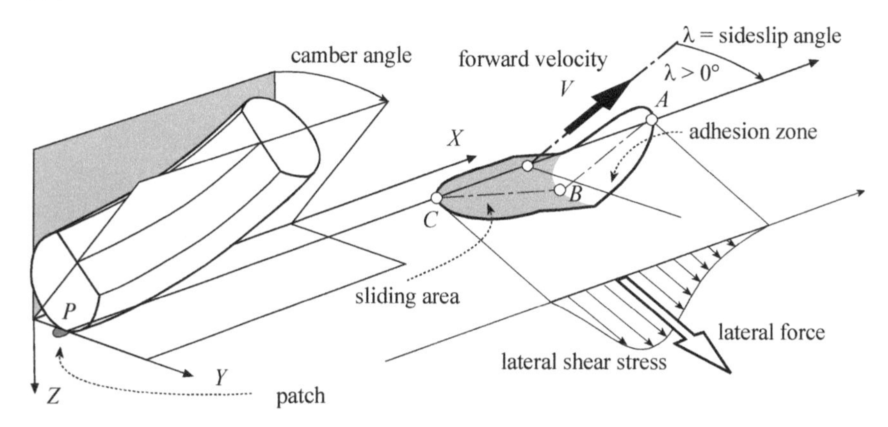

Fig. 2-13 The patch of a motorcycle tire in the presence of lateral slip.

2.5.3 Non-linear model

Figure 2-14 shows in qualitative terms, the normalized lateral force versus the slip angle and versus the camber angle. The maximum value of the force that can be obtained, given a certain tire, is strongly dependent on road conditions.

The forces were measured by means the rotating disk test machine described in [Cossalter et al., 2003] and shown in Fig. 2-15. The disk rotates around a vertical axis and is equipped with a safety walk track. The wheel under testing rolls on the track and is placed in position by an articulated arm that makes it possible to set the camber and sideslip angles at assigned values.

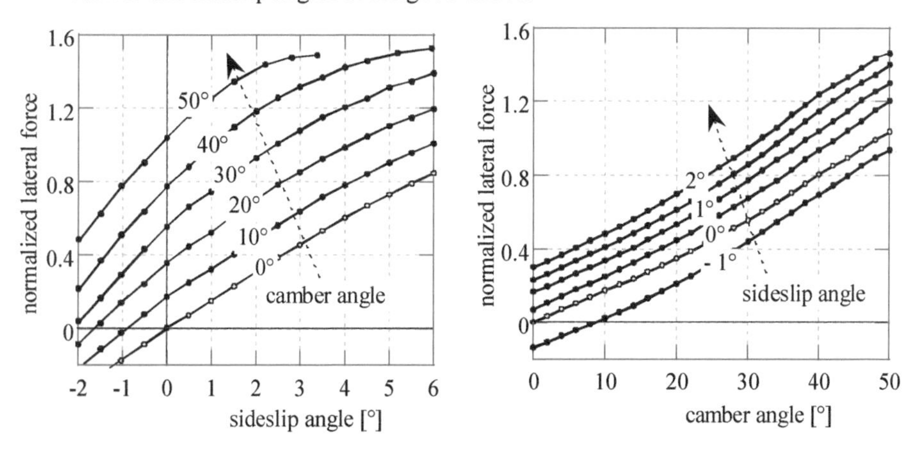

Fig. 2.14 Measured values of lateral force as a function of the sideslip angle λ and for various values of the camber angle φ (left) and as a function of the camber angle φ for various values of the sideslip angle λ (right) [front tire 120/70/17].

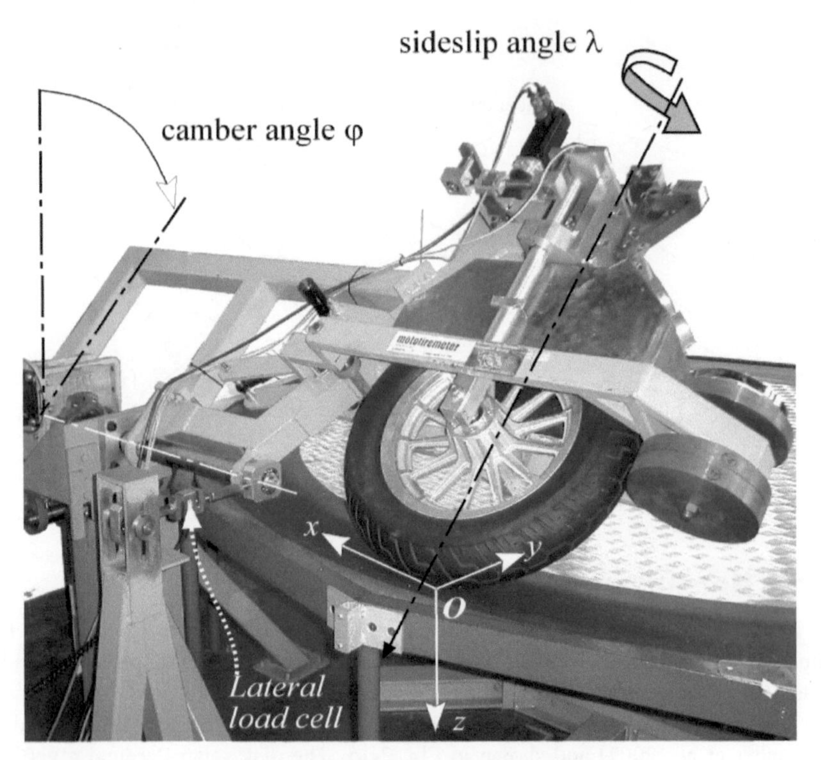

Fig. 2.15 The rotating disk tire test machine.

As shown in the figure the lateral force is a function of the vertical load, sideslip and camber angle. The coupling between the cornering and the camber components can be expressed with the equivalent force approach by means of the following expression:

$$F_s = D_s N\Big[\sin\big\{C_\lambda \cdot \arctan\big[B_\lambda \lambda - E_\lambda\big(B_\lambda \lambda - \arctan B_\lambda \lambda\big)\big]\big\} + \\ + \sin\big\{C_\varphi \cdot \arctan\big[B_\varphi \varphi - E_\varphi\big(B_\lambda \varphi - \arctan B_\lambda \varphi\big)\big]\big\}\Big]$$

This approach is the most recent of Pacejka's formulations for motorcycle tires [Pacejka, 2005].

$D_s = \mu_y$ is the peak of the lateral force coefficient, $D_\lambda C_\lambda B_\lambda = k_\lambda$ is the cornering stiffness coefficient and $D_\varphi C_\varphi B_\varphi = k_\varphi$ is the camber stiffness coefficient.

2.5.4 Linear model

The contact forces between the tire and the road plane depend on the slip angle and the camber angle. It can be seen that for small slip angles, the dependence on the slip angle is nearly linear while the camber component is almost a linear function of the camber angle.

The lateral force for small slip angle and limited camber angle can be expressed by means of the linear expression:

$$F = K_\lambda \cdot \lambda + K_\varphi \cdot \varphi = (k_\lambda \cdot \lambda + k_\varphi \cdot \varphi)N$$

Figure 2-16 shows, on the left, the typical variation of the normalized lateral force with respect to the vertical force, versus the slip angle for various values of the camber angle and, on the right, versus the camber angle for various values of the slip angle.

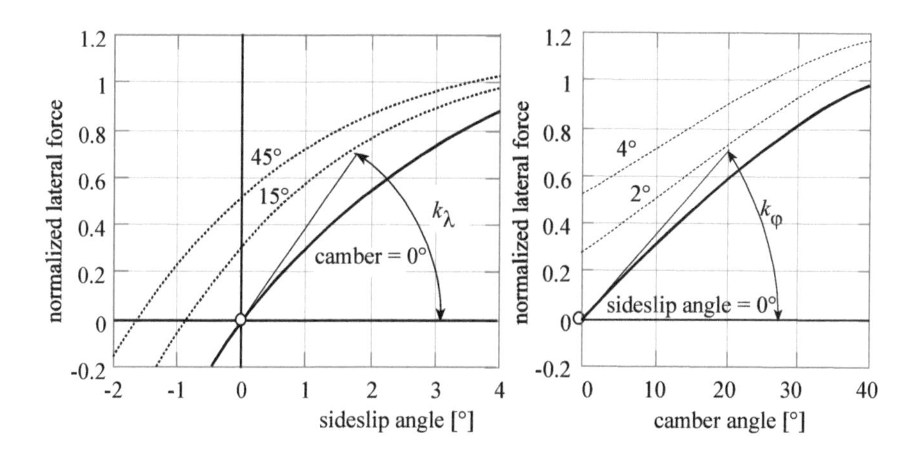

Fig. 2-16 Geometric interpretation of the cornering k_λ and camber k_φ stiffness.

The characteristics of the tire, as far as the lateral force is concerned, are defined by the cornering and camber dimensional stiffnesses (N/rad):

$$K_\lambda = \left.\frac{dF_s}{d\lambda}\right|_{\lambda=0,\varphi=0} \qquad K_\varphi = \left.\frac{dF_s}{d\varphi}\right|_{\lambda=0,\varphi=0}$$

$$k_\lambda = \frac{1}{N} K_\lambda \qquad k_\varphi = \frac{1}{N} K_\varphi$$

The cornering stiffness coefficient k_λ varies with the variation of the characteristics of the tires. Its field of variability ranges from approximately 10 rad^{-1} up to values of 25 rad^{-1}.

The camber stiffness coefficient k_φ is of the order of magnitude of 0.7 to 1.5 rad^{-1}.

The ratio between the maximum lateral force and the vertical load, can reach values of 1.3 to 1.6 when the road surface is clean and dry.

2.5.5 Lateral force needed for motorcycle equilibrium

Consider a motorcycle in a curve in steady state. The equilibrium of the moments of the forces acting on the center of mass shows that the normalized lateral force necessary to assure the motorcycle's equilibrium is equal to the tangent of the roll angle, as represented in Fig. 2-17.

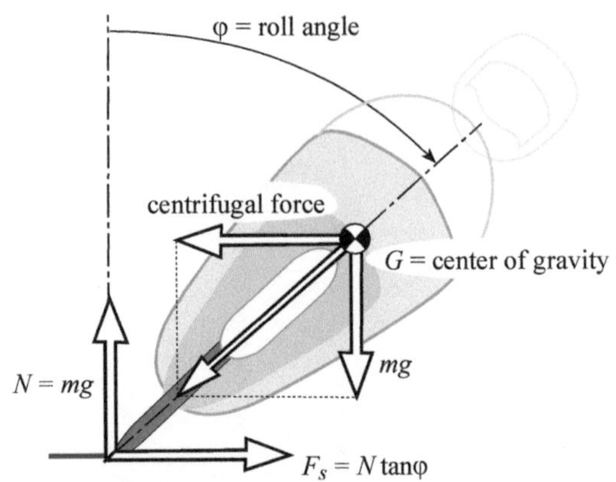

Fig. 2-17 Equilibrium of the motorcycle in a curve.

Consider Fig. 2-18. The right graph shows, for a certain type of tire, the component of lateral thrust due to camber alone (the straight line) and the lateral force necessary for the equilibrium of the motorcycle on a curve (dotted line). The forces are normalized with respect to the vertical load N. The graph on the left represents the normalized lateral force versus the lateral slip angle.

It can be observed that the straight line, which approximates the variation of the camber thrust, intersects the curve $\tan\varphi$ in correspondence with a roll angle of 28°. This means that within the range 0 to 28°, the lateral force needed for equilibrium is less than the thrust force generated by camber alone. Since the lateral force generated must be exactly equal to that needed for equilibrium, the diminution of the lateral force is obtained through a negative sideslip angle. That is, the wheel presents a lateral velocity component towards the interior of the curve.

Figure 2-18 shows, for example, that in the case of a camber angle of 10°, the equality of the force generated with that needed is obtained when there is a negative slip angle of 0.3° (Point A). With a camber angle of 28° the lateral slip is zero (Point B). For values of the camber angle greater than 28° the lateral force produced by camber alone is not sufficient for the equilibrium of the motorcycle and therefore the increase in the lateral force is obtained with the lateral slip of the tire (positive slip).

This behavior is a characteristic of motorcycle tires in which the lateral force generated is, up to determined roll angles, almost entirely due to the camber component. Since this component appears more rapidly with respect to the component due to slip, it plays a fundamental role in safety. The camber component appears more rapidly because it depends on the carcass deformation while the cornering component depends on the slip angle which needs some time to occur.

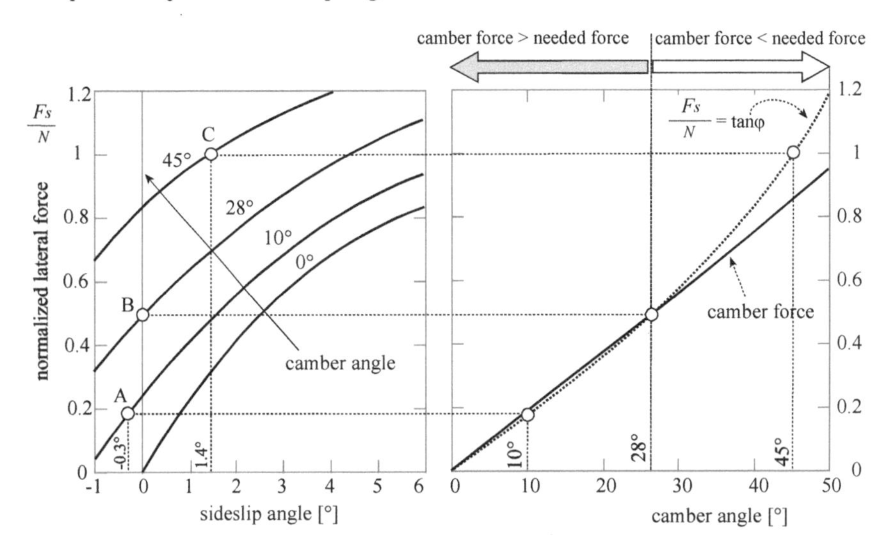

Fig. 2-18 Components of the lateral force generated by camber and slip. Tire A.

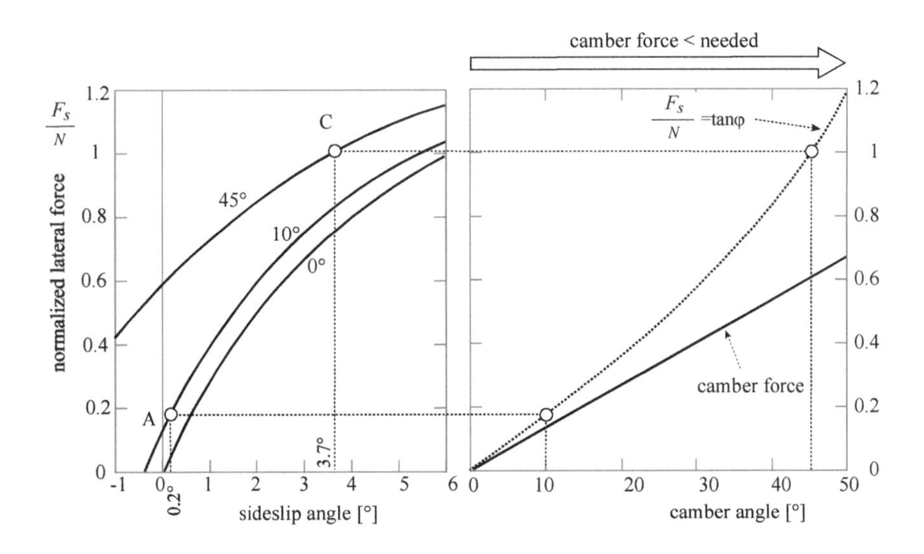

Fig. 2-19 Components of lateral force generated by camber and slip. Tire B.

Now consider the graph in Fig. 2-19, which refers to a different type of tire. In this case the camber thrust is always inferior to the lateral force needed for equilibrium. This means that it is always necessary to have lateral slip in order to generate the additional lateral force required for equilibrium. The lateral forces, the way they are produced and their dependence on the camber angle and the slip angle, play a fundamental role in the motorcycle's under-steering or over-steering behavior.

If the generation of lateral front force requires a slip angle larger than that needed for the generation of lateral rear force, the motorcycle will tend, as the roll angle increases, to skid more with the front wheel. This behavior causes the vehicle to under-steer. On the other hand, if the slip in the rear wheel is greater than that of the front one, the behavior will be over-steering. Neutral behavior occurs when the slip angles are equal. On the basis of these considerations, the tire's ideal behavior occurs when the slip angle is zero, that is, when the lateral force necessary for equilibrium is produced by camber alone.

2.5.6 Dependence of lateral force on load, pressure, temperature

The tire's capability for lateral grip is well represented by the plots of lateral force versus camber angle and sideslip angle, as depicted in Fig. 2.20 for several very different types of front and rear tires.

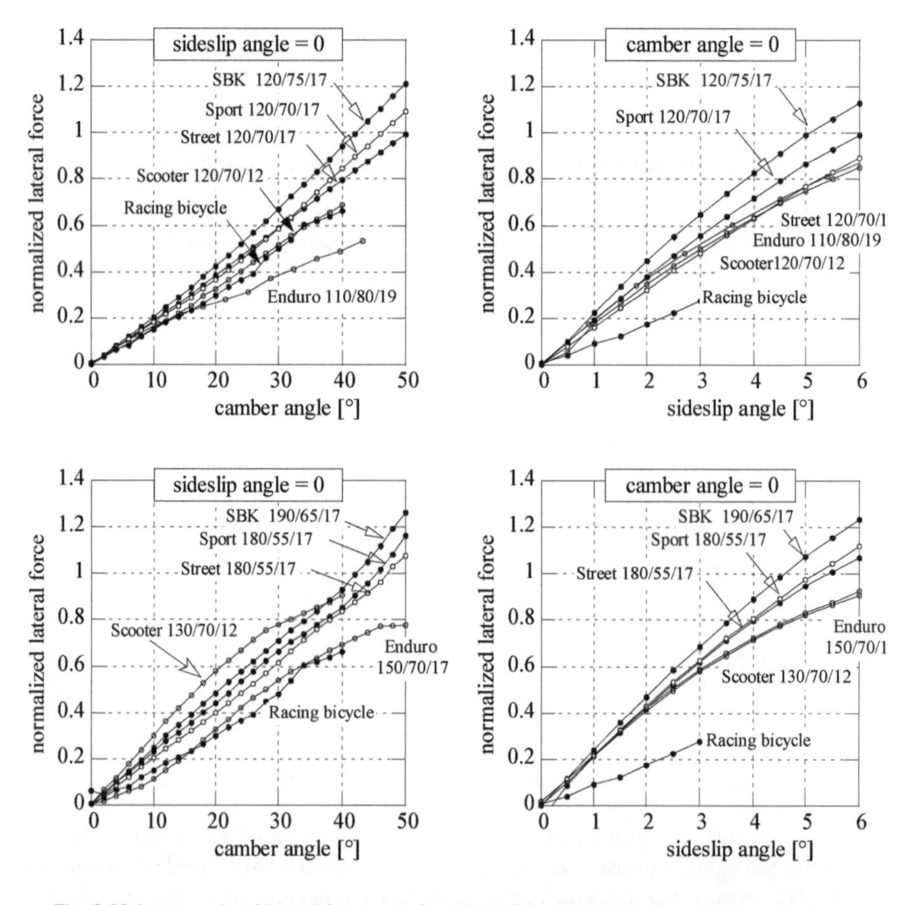

Fig. 2-20 An example of lateral force as a function of the camber angle φ (left) and of the sideslip angle λ (right) for different front (top) and rear (bottom) tires.

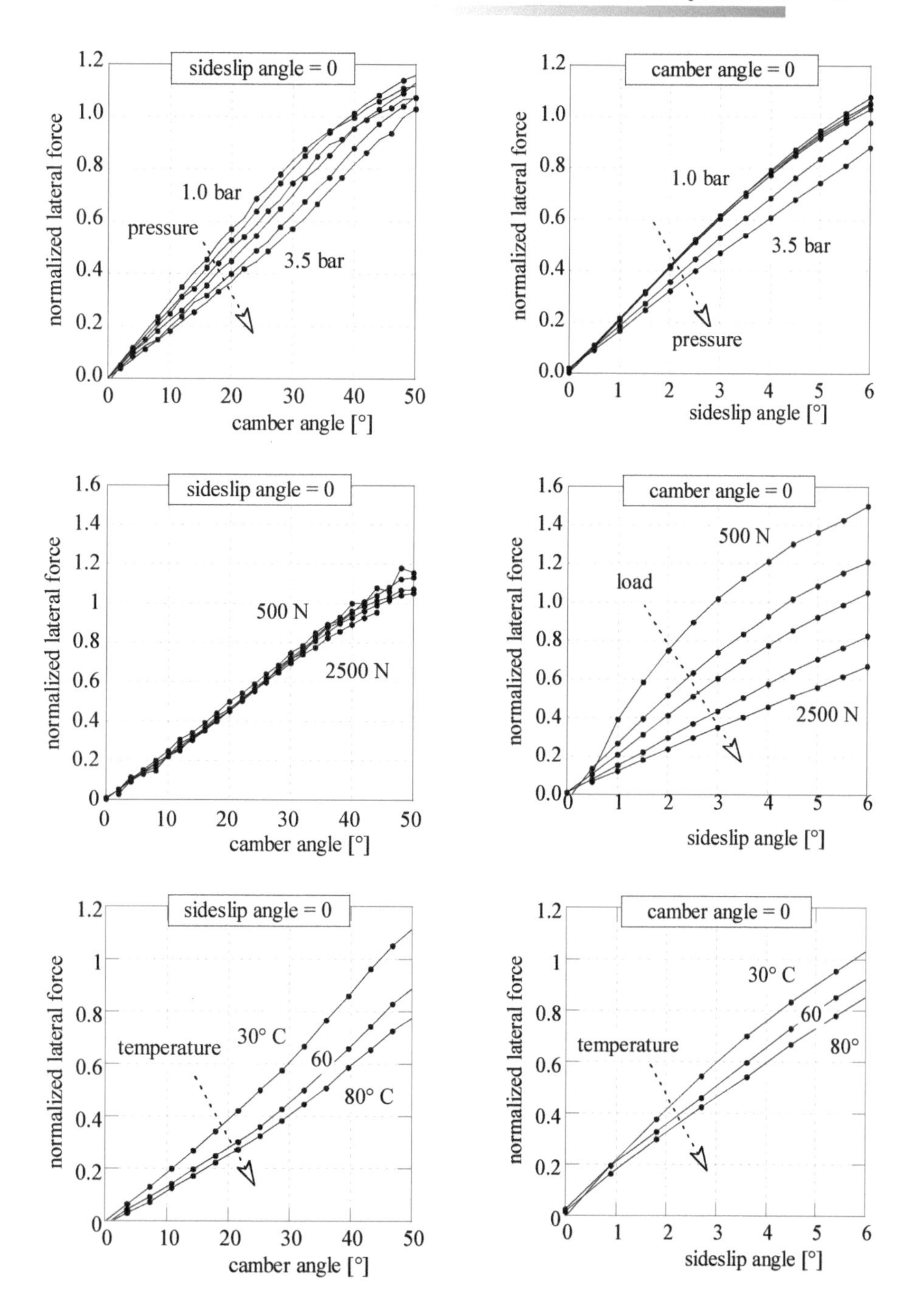

Fig. 2-21 An example of lateral force as a function of the camber angle φ (left) and sideslip angle λ (right) for various values of pressure, load and temperature.

Figure 2-21 shows as an example the measured normalized lateral force versus both the slip angle and the camber angle for various values of inflation pressure, of vertical load and temperature of the carcass.

Cornering stiffness is not to be confused with the tire's lateral or radial stiffness. Lateral stiffness is the ratio between applied lateral force and the resulting lateral deformation of the tire's carcass. It depends on the tire's construction characteristics. The order of magnitude of its value is of 100 to 200 kN/m. The radial stiffness of the tire is the relation between the vertical load and the vertical deformation and has values in the range of 100 to 350 kN/m. Inflation pressure and forward velocity influence both of these structural stiffnesses.

An increase in vertical load decreases the cornering stiffness coefficient whereas the camber stiffness coefficient is almost not influenced. An increase in inflation pressure, decreases the cornering stiffness coefficient and, to a lesser extent, decreases the camber stiffness coefficient. Tires with larger sections, or greater cross section radii, usually have a larger cornering stiffness coefficient. An increase in temperature decreases both cornering and camber stiffness coefficients but increases the maximum value of the ratio between the maximum lateral force and the vertical load.

2.5.7 Lateral force in transient state

We have stated that the tire lateral force does not arise instantaneously. To appear, the wheel needs to roll a certain distance, which depends on the tire's cornering characteristics and lateral stiffness.

Suppose that the motorcycle is initially in a state of vertical equilibrium. The roll and slip angles are zero with corresponding zero values of the lateral contact forces. If we instantaneously assign non-zero values to the roll and slip angles, the lateral contact forces increase exponentially, from zero to the steady state value corresponding to the assigned roll and sideslip angles; the contact forces therefore follow, with a delay, the variation of the angles on which they depend. This is due to the fact that the carcass distortion takes some time to establish itself. The component due to camber, depending primarily on tire geometry, has a lesser delay than that of the component due to lateral slip.

The tire's behavior in transient state can be represented by the model (Fig 2-22), which is composed of a spring (with stiffness k_S expressed in N/m) and damper in series (with damping coefficient c expressed in kg/s). The spring k_S represents the tire's lateral stiffness and depends mainly on the form and characteristics of the tire's carcass while the damper c describes the behavior of the tire under conditions of lateral slip. If we ignore the inertia of the tire's carcass, the cornering force is equal and opposite to the elastic force generated by the deformation of the tire:

$$F_S = c\dot{y} = c\lambda' V = K_\lambda \lambda' = -k_S\left(y - y_i\right)$$

Here $\lambda' = \dot{y}/V$ represents the transient slip angle, that is the slip angle of the contact patch (point P), y the lateral displacement of the contact patch and y_i the displacement imposed on the wheel.

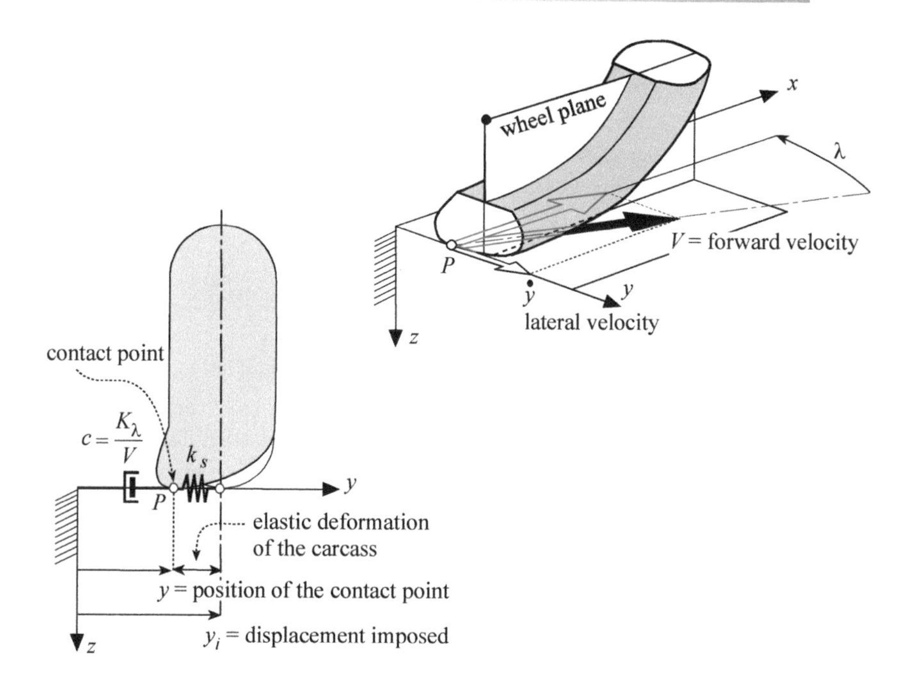

Fig. 2 -22 Spring and damper connected in series to represent the lateral behavior of the tire.

Taking into account the definition of the slip angle and the apparent damping coefficient $c = K_\lambda / V$, the equation can be expressed in the form:

$$\frac{K_\lambda}{V}\frac{dy}{dt} + k_s y = k_s y_i$$

The differential equation can also be expressed in terms of the lateral force produced. After some manipulation, the following expression, with the imposed wheel slip angle $\lambda = \dot{y}_i / V$, is obtained:

$$\frac{L}{V}\frac{dF_s}{dt} + F_s = K_\lambda \lambda$$

Here $L = K_\lambda / k_s$ represents the relaxation length.

Suppose that the wheel is suddenly subjected, at instant $t = 0$, to a lateral motion with constant slip angle λ_o. As a result the lateral force increases exponentially:

$$F_s = K_\lambda \lambda_o (1 - e^{-\frac{V k_s}{K_\lambda} t})$$

Keeping in mind that the distance x traversed by the wheel is given by the product of velocity and time, we have:

$$F_s = K_\lambda \lambda_o (1 - e^{-\frac{k_s}{K_\lambda} x})$$

Figure 2-23 shows the variation of the lateral force, normalized with respect to the value $K_\lambda \lambda_o$, which it assumes in a steady state, as a function of the distance x covered by the wheel. The tangent constructed through the origin is equal to the ratio k_s / K_λ. The inverse of the tangent is called relaxation length:

$$L = \frac{K_\lambda}{k_s}$$

The relaxation length represents the distance the wheel has to cover in order for the lateral force to reach 63% of the steady state force. Integrating the differential equation gives us the lateral force once we have assigned a temporal variation to the slip angle.

The values of the relaxation length of the cornering force ranges between 0.12-0.45 m. The small values correspond to low velocity (20 km/h), the higher values to very high velocity (250 km/h). It increases slightly with the load. It is interesting to highlight that the relaxation length is almost constant with respect to the ratio between the frequency ν of the sideslip angle oscillation and the forward velocity V. This ratio is called the path frequency and represents the number of cycles per meter of forward motion.

On the other hand the values of the relaxation length of the camber force in some experimental tests has been found to be almost negligible. However, further experimental results are needed to verify this behavior.

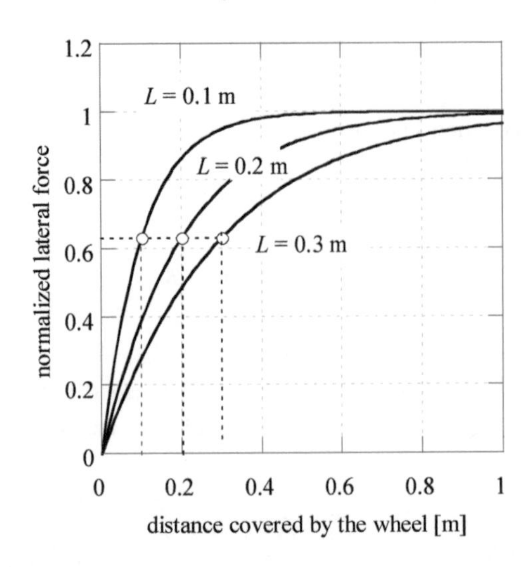

Fig. 2-23 Lateral force as a function of the distance covered by the wheel.

2.6 Moments acting between the tire and the road

2.6.1 Self-alignment moment

The distribution of the lateral shear stress generated by the lateral slip of the tire is not symmetric. The resulting force is therefore applied at a point situated at a certain distance from the center of the patch, a center which, in a first approximation, can be assumed to coincide with the theoretical contact point of the rigid toroid with the road plane. The distance a_t is designated the trail of the tire or pneumatic trail. It is clear from Fig. 2-24 that the lateral force generates a moment that tends to rotate the tire in such a way as to diminish the slip angle. For this reason this moment is called the self-aligning moment of the tire.

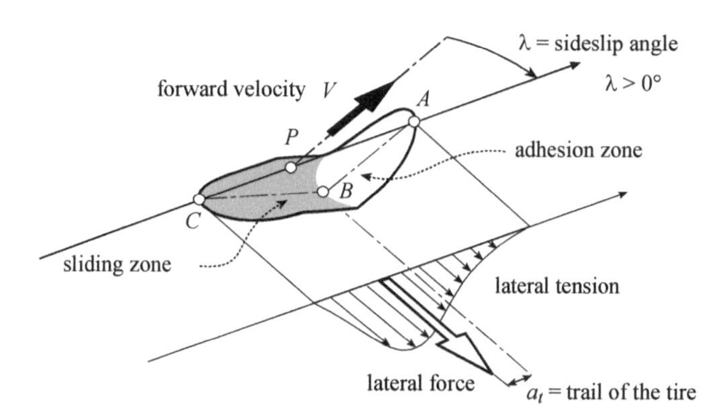

Fig. 2-24 Trail of the tire.

The self-aligning moment M_z is expressed as the product of the cornering force F_s and the trail of the tire a_t.

$$M_z = -a_t F_s$$

Experimental results show that the trail is at a maximum when the slip angle is zero; that it decreases with an increase in the slip angle until it reaches zero, and that it increases with increases in the vertical load. It can be approximately expressed in terms of the slip angle by the following linear equation:

$$a_t = a_{t_o}\left(1 - \left|\frac{\lambda}{\lambda_{max}}\right|\right) \qquad a_t = 0 \quad \text{if} \quad |\lambda| > \lambda_{max}$$

where a_{t_o} represents the maximum value of the tire trail (a_{t_o} ranges from 1.5 to 5 cm) and λ_{max} the slip angle at which the tire trail becomes zero.

Figure 2-25 represents the typical variation of the sideslip force, of the pneumatic trail and of the self-aligning moment as a function of the slip. When the slip angle reaches the value λ_{max} (about 15°) the moment is zero since the lateral force passes through the center of the patch because the sliding zone covers the whole patch.

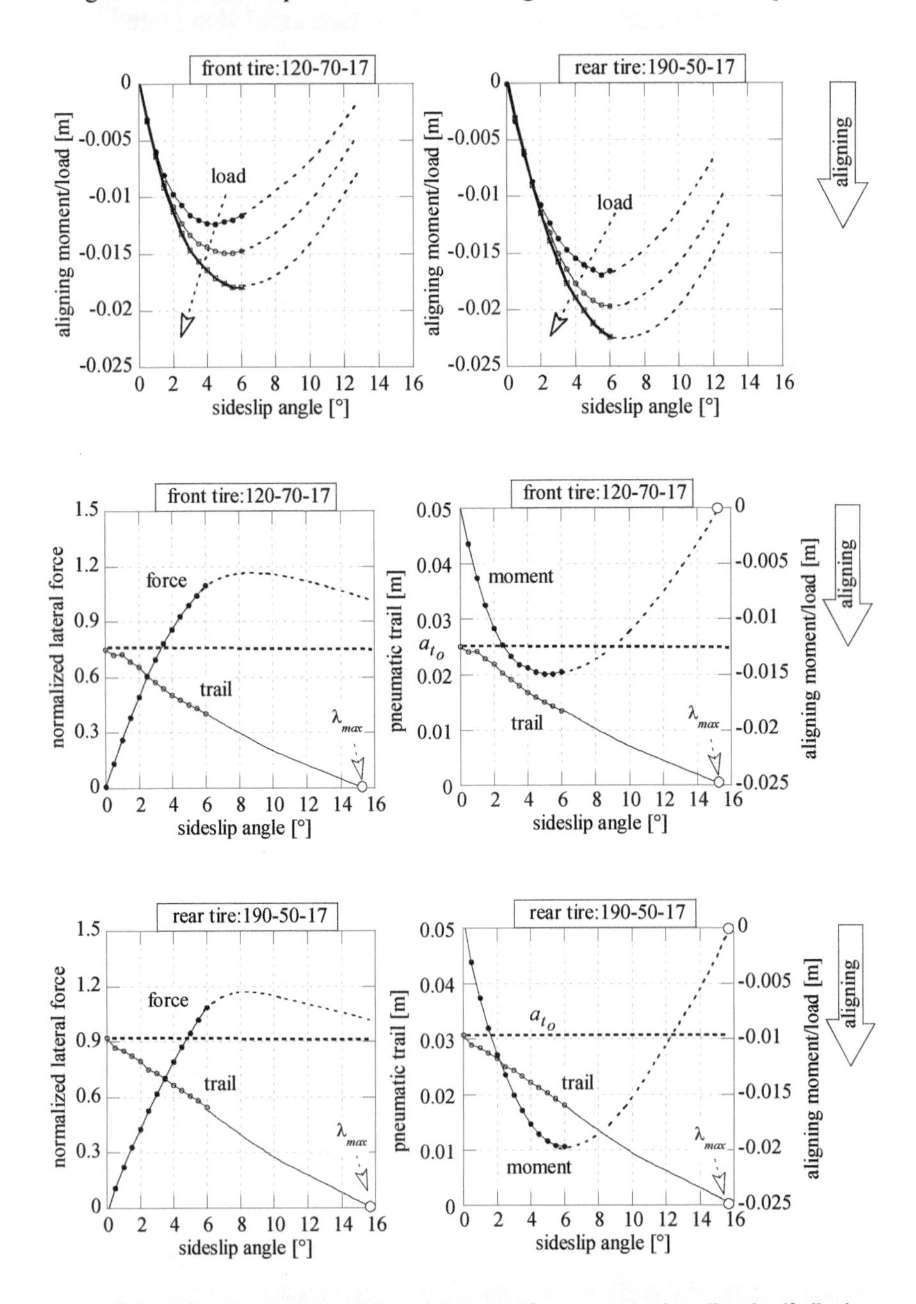

Fig. 2-25 Example of measured variation of the lateral force, pneumatic trail and self-aligning moment versus sideslip angle. Camber angle=0°, Front nominal load =1300 N, rear nominal load=1400 N.

2.6.2 Twisting moment

Consider an inclined wheel that rolls over the road plane with angular velocity ω about the wheel axis (Fig. 2-26).

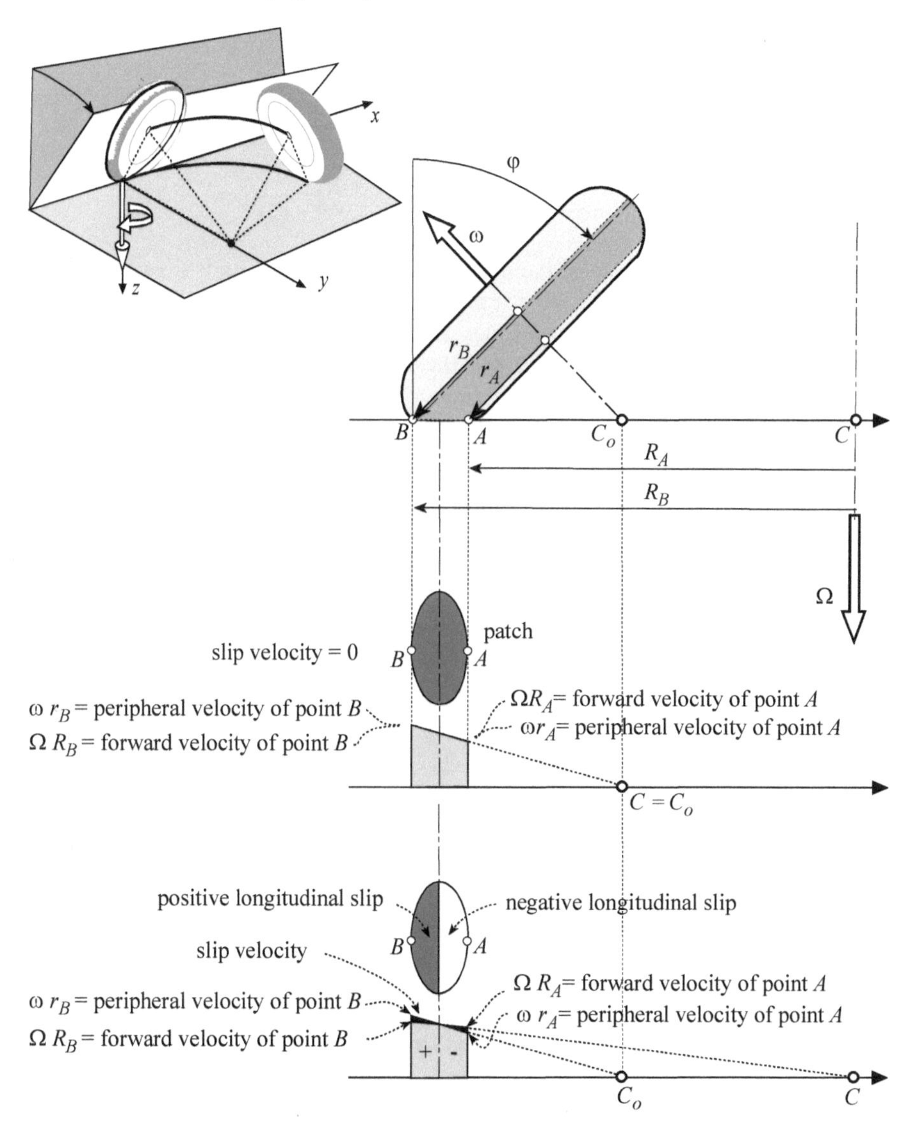

Fig. 2-26 Origin of the twisting moment.

We indicate with C_o the point of intersection of the wheel's axis with the road plane. If the turn center point C of the circular trajectory described by the wheel coincides with the point C_o, motion occurs without longitudinal slippage (under kinematic conditions). In fact, the peripheral velocities of the two points A and B

of the tire, which are part of the patch, are equal to the forward velocities due to the rotation of the wheel around the point C with angular velocity Ω.

$$\omega r_A = \Omega R_A \qquad \omega r_B = \Omega R_B$$

In reality, at free rolling the center of curvature C is always located externally with respect to the point C_o.

Suppose that at the midpoint of the patch the peripheral velocity is equal to the forward velocity:

$$\omega \frac{r_A + r_B}{2} = \Omega \frac{R_A + R_B}{2}$$

In the most external area of the patch the peripheral velocity is greater than the forward velocity, while in the interior area of the patch the contrary is true. Motion therefore occurs with slip, and two zones can be distinguished in the patch: one with positive longitudinal slip velocity, and the other with negative longitudinal slip velocity. Therefore, there are forward directed shear stress in the external zone and backward directed shear stress in the internal zone.

These shear stresses generate a twisting moment that tends to move the wheel along a trajectory with a smaller curvature radius, thereby acting to twist the wheel out of alignment. The twisting moment is approximately proportional to the camber angle. A typical variation is represented in Fig. 2-27.

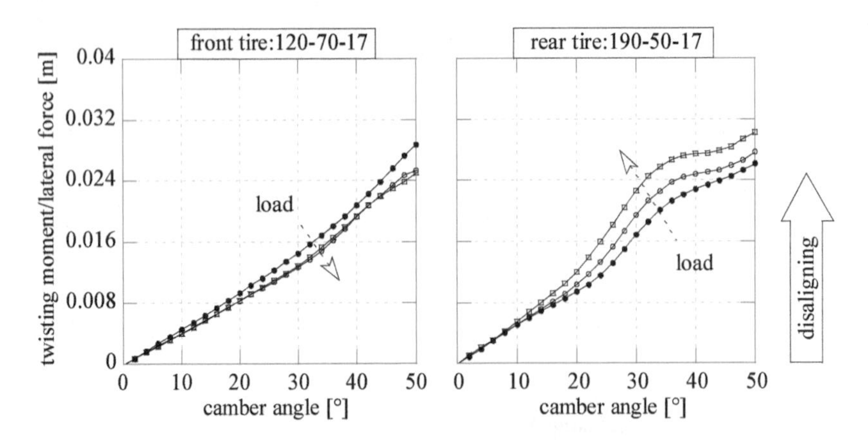

Fig. 2-27 Example of measured variation of twisting moment.

We have seen that two moments of opposite sign act on the tire: the self-aligning moment and the twisting moment. Their sum defines the yawing moment of the tire, whose qualitative variation against the sideslip angle, is shown in Fig. 2-28. The yawing moment M_z is zero when the slip and roll angles are zero; it increases with increases in the roll angle and has a minimum corresponding to a slip angle of $\lambda = 2°$ to $6°$.

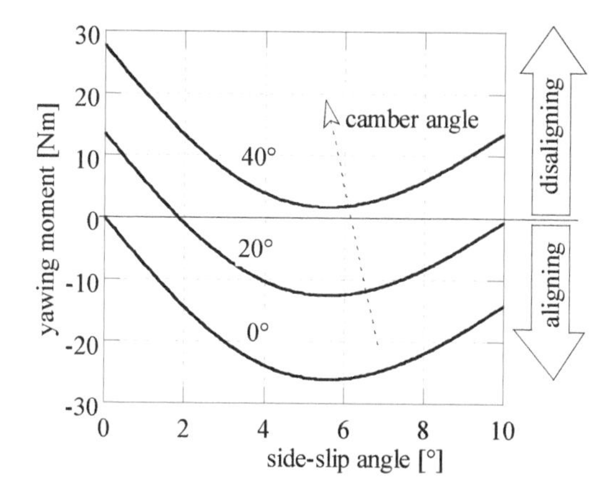

Fig. 2-28 Example of the yawing moment.

2.6.3 Torque generated by the driving or braking force

The driving force generates a moment that tends to align the plane of the tire in the direction of velocity, while the braking force generates a moment of opposite sign which therefore moves it out of alignment. The arm of the longitudinal force depends on the lateral deformation of the tire.

$$\text{driving:} \quad M_z = -s_p\, S = -\frac{F_s}{k_s}\, S$$

$$\text{braking:} \quad M_z = s_p\, F = \frac{F_s}{k_s}\, F$$

where k_s indicates the lateral stiffness of the tire's carcass.

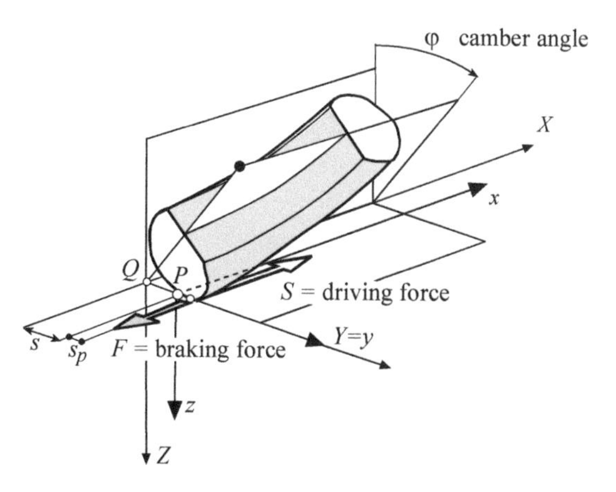

Fig. 2-29 Moments generated by driving or braking force.

With respect to point Q the arm of the longitudinal force also depends on the radius of the cross section and the camber angle.

$$\text{driving:} \quad M_Z = -(s + s_p)S = -(t \cdot \tan\varphi + \frac{F_s}{k_s})\, S$$

$$\text{braking:} \quad M_Z = (s + s_p)F = (t \cdot \tan\varphi + \frac{F_s}{k_s})\, F$$

where t indicates the radius of the cross section of the tire.

In general, the lateral deformation s_p, has a negligible value with respect to the lateral displacement s of the contact point of the tire.

2.7 Combined lateral and longitudinal forces: the friction ellipse

The longitudinal force F_x, either driving (positive value of F_x) or braking (negative value of F_x), is assumed to be assigned since it is controlled by the rider. The lateral force F_y that can be exercised is reduced by the simultaneous presence of the longitudinal force.

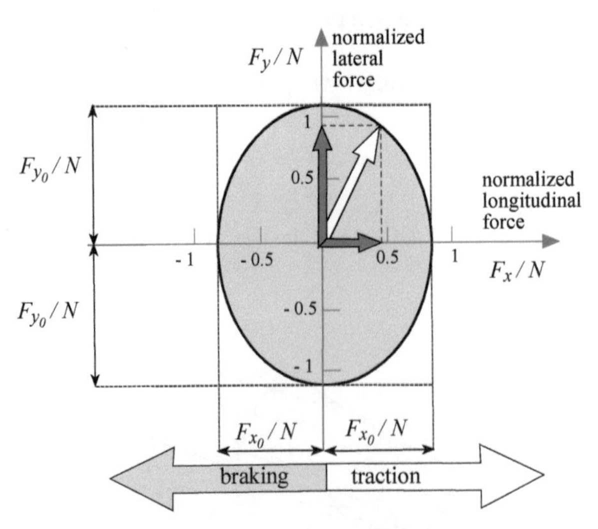

Fig. 2-30 The friction ellipse.

Their resultant must be within the friction ellipse that has the maximum values F_{x_o} longitudinal, and F_{y_o}, lateral respectively, when they act alone:

$$F_{x_o} = \mu_{x_p} N \qquad F_{y_o} = \mu_{y_p} N$$

where μ_{x_p} is the longitudinal traction coefficient and μ_{y_p} is the lateral traction coefficient. For this reason the formula that yields the lateral force is multiplied by a correction coefficient that depends on the longitudinal force applied:

$$\sqrt{1-\left(\frac{F_x}{F_{x_o}}\right)^2}$$

Figure 2-31 shows the variation of the normalized lateral force curves with the variation of the longitudinal force applied.

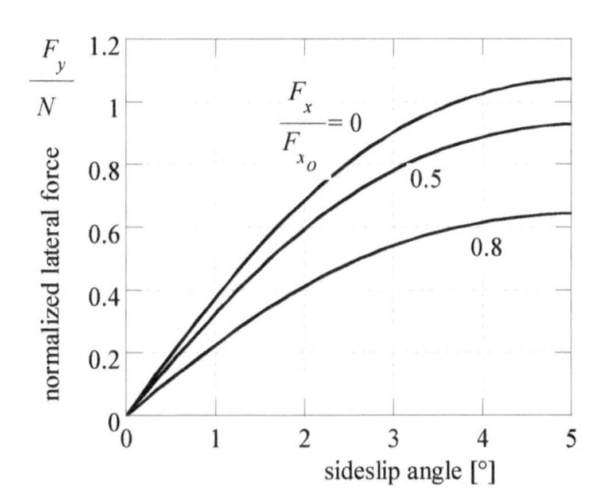

Fig. 2-31 Variation of the normalized lateral force for various values of the longitudinal force (roll angle = 0°).

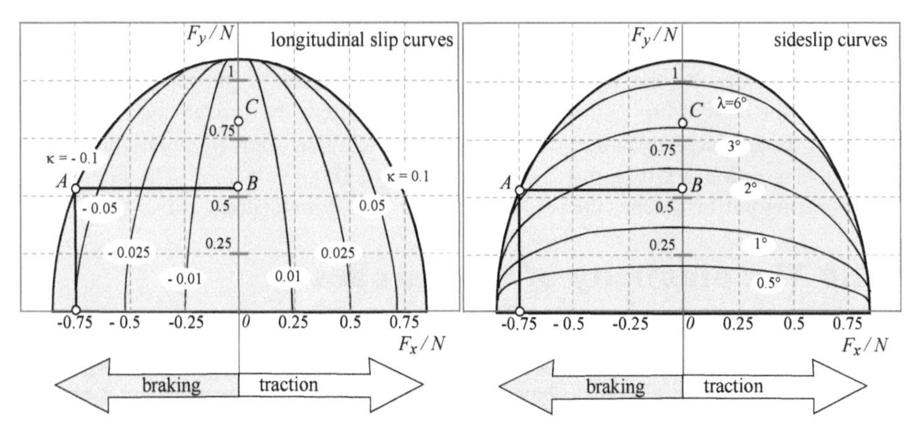

Fig. 2-32 Lateral and longitudinal forces for various values of longitudinal slip κ and sideslip λ (roll angle φ = 0°).

The interaction between the longitudinal and lateral forces can be shown by representing the constant lateral slip curves and constant longitudinal slip curves in a diagram that has as its ordinate the normalized lateral force F_y / N and as its abscissa the normalized longitudinal force F_x / N.

The curves for constant longitudinal slippage and constant lateral slip are represented in Fig. 2-32.

Example 3

Consider a motorcycle braking as it enters a curve. Suppose that the rear wheel has a normalized longitudinal force equal to 0.75 and a normalized lateral force equal to 0.53 (point A), which correspond to a sideslip angle of 3.5°.

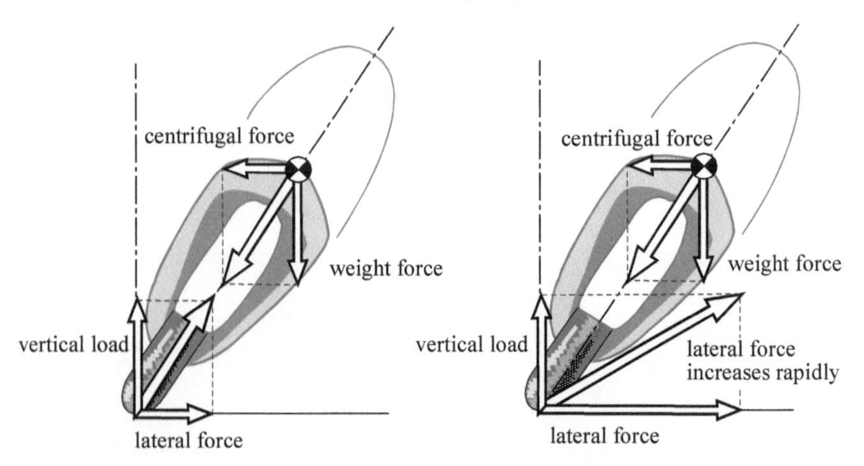

Fig. 2-33 Force acting on the motorcycle.

If the braking is suddenly stopped, the lateral force corresponding to a 3.5° slip angle is increased sharply, there is motion from point A to point C. The lateral force generated by the slip is now equal to 0.78, which is greater than the 0.53 needed for equilibrium. Since the motorcycle is tilted, the sudden increase in the lateral force generates an acceleration of the vehicle that tends to be returned to the vertical position and to project the rider upward (high-side fall). The lateral slip diminishes until it reaches the value needed for equilibrium (Point B).

2.8 The elasticity of the carcass

When lateral and vertical forces are applied to the tire, both lateral and radial elastic deformation of the carcass arise. Additionally the driving/braking force generates, in the longitudinal plane, a deformation that mainly consists in a relative rotation between the rim and the carcass. Because of tire deformation, the contact is no longer dot shaped, but involves a *contact patch* surface whose form depends on the camber angle, on the load and on the inflation pressure.

The length and width of the contact patch of motorcycle tires change in a rather regular manner with the vertical load and camber angle as long as the contact patch is not very large (large loads) and the camber angle does not approach 40°-45°. The effect of inflation pressure on contact patch is important if it is lower than the nominal value 2-2.5 bar

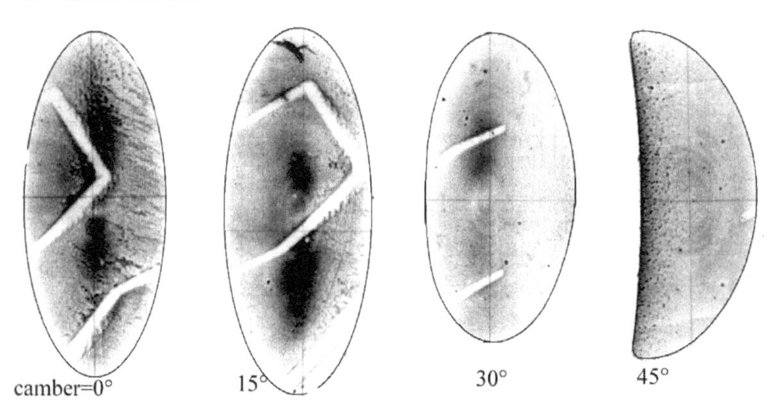

Fig. 2-34 The effect of camber angle on contact patch shape.
[Rear tire $Fz = 2000\ N$ and $p = 2\ bar$].

In the road plane the tire changes footprint and the contact point moves laterally depending on the geometry of the carcass as shown in Fig. 2-34.

In the presence of camber, a pure vertical load induced both horizontal and vertical deflection of the carcass. However, by expressing results with respect to the wheel cambered reference frame, the relationships between forces and deformations are simpler. In fact the elastic properties of the carcass can be effectively described by means of a pair of springs which act in the radial direction Z and the lateral direction Y, as shown in the Fig.2-35.

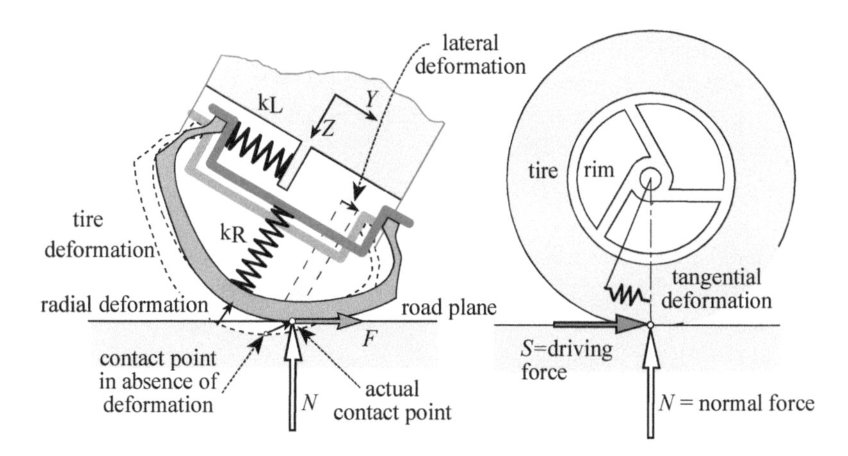

Fig. 2-35 Elasticity of the tire carcass.

Typical values of structural lateral stiffnesses ranges from 100 kN/m to 250 kN/m while radial stiffnesses range from 100 kN/m to 200 kN/m.

2.9 Model of the motorcycle tire

Nowadays multi-body codes make it possible to calculate the points of contact between the road and the motorcycle equipped with rigid or elastic toroidal tires. Hence in the tire model the forces can be applied in the area around the point of contact between the road and the toroidal tire.

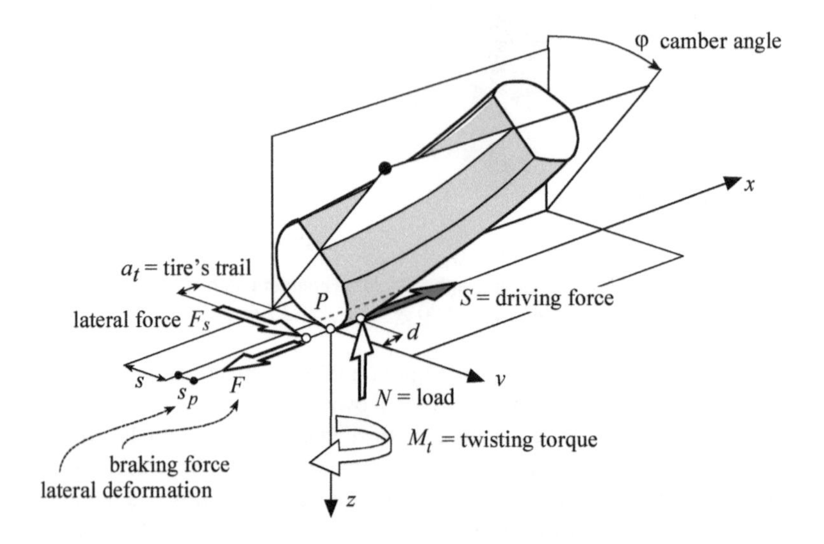

Fig. 2-36 Forces and moments acting on the tire.

The model of the motorcycle tire takes into consideration the forces acting on points near the theoretical contact point defined by the tire's geometry.
The forces under consideration are as follows:
Normal force. Normal force is applied at a point that precedes, by the distance d, the position of the theoretical contact point. The distance d depends on the rolling resistance coefficient and the tire's radius:

$$d = f_w R$$

Lateral force. Lateral force acts in the direction orthogonal to the intersection of the wheel plane with the road plane. The application point is displaced backwards with respect to the theoretical contact point by a distance a_t that represents the tire's trail, which varies with the sideslip angle.
Longitudinal force. The force is applied to a point displaced laterally from the theoretical contact point because of the tire's lateral deformability. The lateral displacement s_p, depending on the lateral stiffness of the tire, is generally negligible with respect to the geometric displacement s deriving from the roll inclination of the wheel.

The moments acting around the x, y and z-axes are generated by the forces described above and by the twisting moment.

Overturning moment. The overturning moment M_x is generated by the vertical load N whose arm is the lateral deformation s_p.

$$M_x = -s_p N$$

Rolling resistance moment. Rolling resistance moment is generated by the asymmetric distribution of normal stresses that causes a forward displacement of vertical load. The tire's rolling resistance moment is:

$$M_y = d\ N$$

Yawing moment. Yawing moment includes two contributions. The first term, due to the lateral force, tends to align the plane of the tire in the direction of velocity. The second term increases with the camber angle and works against alignment.

$$M_z = -a_t F_s + M_t$$

Fig. 2-37 shows a motorcycle tire modeled with virtual prototyping. The tire, launched on the road at a certain velocity and with an initial inclined orientation describes a trajectory that depends on the tire's characteristics and in particular:

- on the component of the lateral force due to camber (function of the camber angle),
- on the tire's trail (function of the lateral slip),
- on the twisting moment (function of the camber angle).

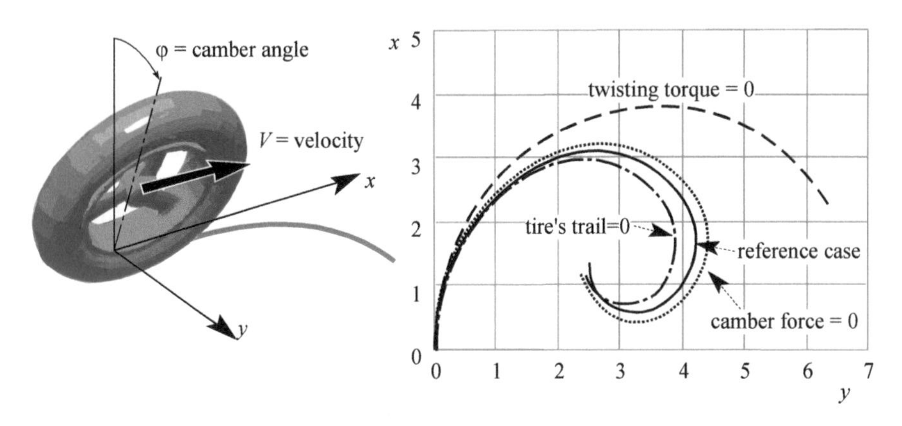

Fig. 2-37 Path of the different tires.

The graph in Fig. 2-37 shows the various trajectories described by the tire. If the component of the lateral force due to camber is zero, equilibrium is assured only by the component generated by lateral slip. Lateral slip is therefore always greater than that present in the reference case. Because of the greater lateral slip, the trajectory covered is more external than that of the reference case. The reduction of the tire's trail to zero also reduces to zero the self-aligning moment generated by the lateral force. The trajectory described is therefore more inner with respect to that of the reference case. The twisting moment, depending on the camber angle, has a significant influence on the tire's behavior. Since its effects work against alignment, that is, it tends to cause the tire to yaw more, its zeroing causes a large change in the trajectory covered. The tire moves along a path characterized by a notably larger curvature radius.

2.10 Vibration modes of the tires

The dynamic properties of tires have an important influence on several features of motorcycle behaviour such as comfort, shock-absorption and braking, which are related to in-plane dynamics, along with stability and handling, which are related to out-of-plane dynamics.

Motorcycle tires' modes of vibration can be divided into in-plane modes, out-of-plane modes and mixed modes. In-plane modes are characterised by radial and/or circumferential displacement of the points located in the symmetry plane of the wheel. Out-of-plane modes are dominated by lateral displacement of the points located in the symmetry plane of the wheel. Mixed modes exhibit combinations of radial, circumferential and lateral displacement.

In plane modes can be classified according to number n of circumferential waves: the $n=0$ mode does not exhibit any circumferential wave and is a breath mode; the $n=1$ mode exhibits one circumferential wave and is essentially a displacement of tire tread with respect to the rim; the $n=2$ mode exhibits two circumferential waves and the tread has an oval shape.

In-plane modes are the most excited modes, since in steady state conditions (rectilinear path) the resultant of tire forces (load N, braking force F and driving force S) stays approximately in the symmetry plane of the wheel. If a coordinate system fixed to the wheel is considered, the resultant of tire forces rotates around the wheel with angular velocity $\omega = V / R_o$, where R_o is the tire rolling radius and V is forward speed. The rotating force may excite in resonance conditions the circumferential mode that exhibits n circumferential waves if the following condition is satisfied:

$$\omega = 2\pi\nu_j / n$$

where ν_j is the natural frequency of the mode.

The presence of road unevenness and grooves on the tire surface are other sources of excitation in the high frequency range. Finally, the transient maneuvers (e.g. braking, changing lanes), which correspond to sudden variations in tire forces and torques, may excite both in-plane and out-of-plane modes.

Generally, the first natural frequencies of tires are in the range 100÷200 Hz and correspond to out-of-plane modes, then there is sometimes a band of frequency with mixed modes, when the frequency is higher than 300÷400 Hz, the modes with large in-plane displacement dominate.

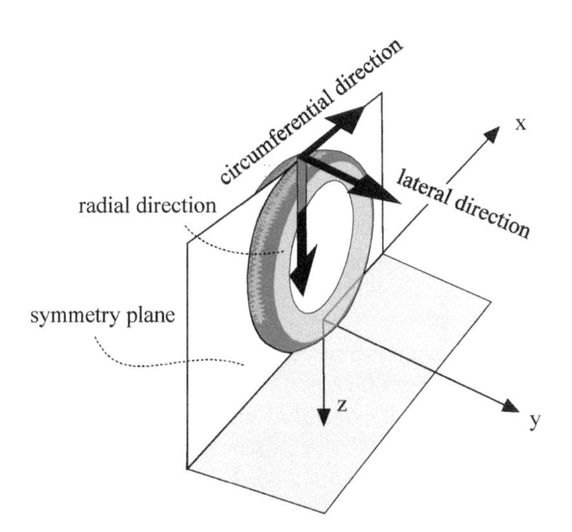

Fig. 2-38 Carcass deformations of the tire.

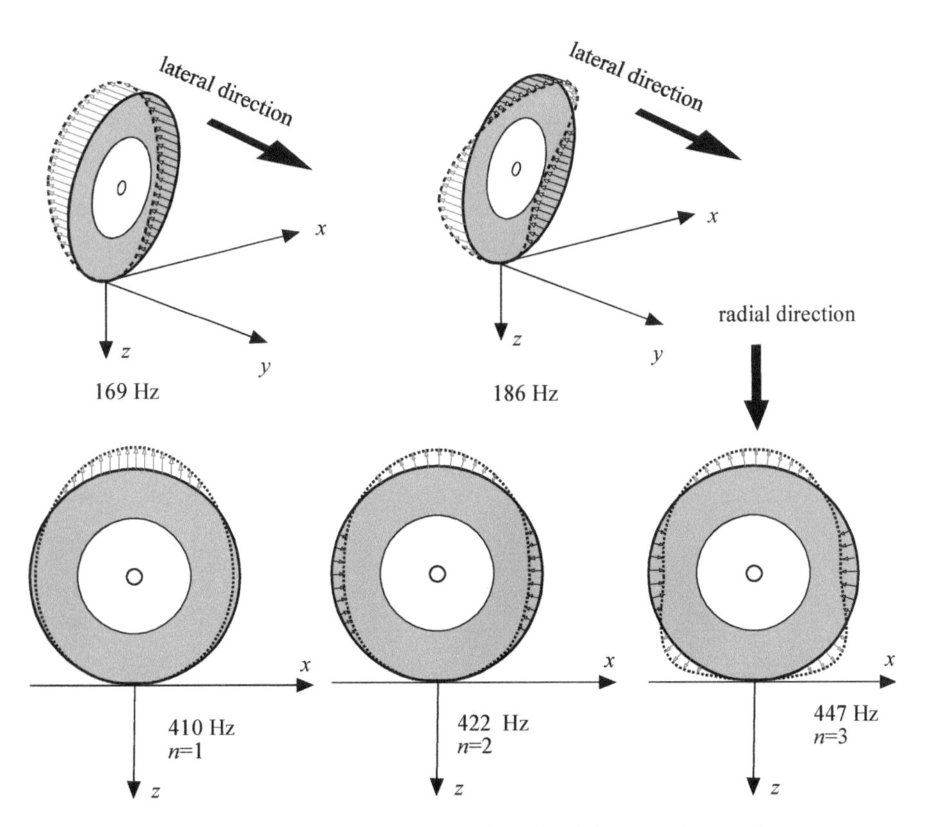

Fig. 2-39 Example of vibration modes of the tires (with ground contact).

Natural frequencies, loss factors and mode shapes strongly depend on tire size, construction (radial-ply, bias-ply) and material. The comparison between the modal properties of radial-ply and bias-ply motorcycle tires shows that natural frequencies of radial-ply tires are higher than the ones of the similar modes of bias-ply tires; the difference is large especially in the case of in-plane modes.

The presence of contact with the ground increases the complexity of modes and, between the modes that were measured in free conditions, new modes having intermediate shape and frequency appear. The range of natural frequencies and loss factors of the modes in contact with the ground are not very different from the ones measured in free conditions.

Figure 2-39 shows the natural frequencies and modes of a 120/65 R17 front tire inflated to 2.2 bars. This tire is a high performance radial-ply tire with 0° steel belts. The first out-of-plane mode is the lateral displacement of the tire tread with respect to the rim. Then there is a banana-shaped mode (with 1.5 waves in the lateral direction and minor displacements in the other directions). The first in-plane mode (410 Hz) is essentially an in-plane displacement of tire tread and derives from the n=1 mode measured in free condition (without contact).

The following modes (422 and 447 Hz) derive from the modes with two circumferential waves (n=2) and three circumferential waves (n=3) measured in free conditions. Because of the high circumferential rigidity of the 0° steel belt, the breath mode is not identified in the 0-500 Hz range of frequencies. The natural frequencies are higher than those of car tires. Loss factors of in-plane modes are similar to those measured in car tires. Loss factors of out-of-plane modes are higher than the ones of in-plane modes, because they are mainly influenced by the tire's side-walls.

Benelli 250 cc of 1938

3 Rectilinear Motion of Motorcycles

The behavior of motorcycles during rectilinear motion depends on the longitudinal forces exchanged between the tires and the road, the aerodynamic forces induced through this motion, and the slope of the road plane. The study of rectilinear motion highlights certain dynamic aspects that are also important for safety, such as the motorcycle's behavior during braking with possible forward overturning, and in acceleration, with possible wheeling.

3.1 Resistance forces acting on motorcycles

During steady state motion, the thrust produced by the engine is equated to the forces that oppose forward motion and depend essentially on three phenomena (Fig. 3-1):

- resistance to tire rolling;
- aerodynamic resistance to forward motion;
- the component of the weight force caused by the slope of the road plane.

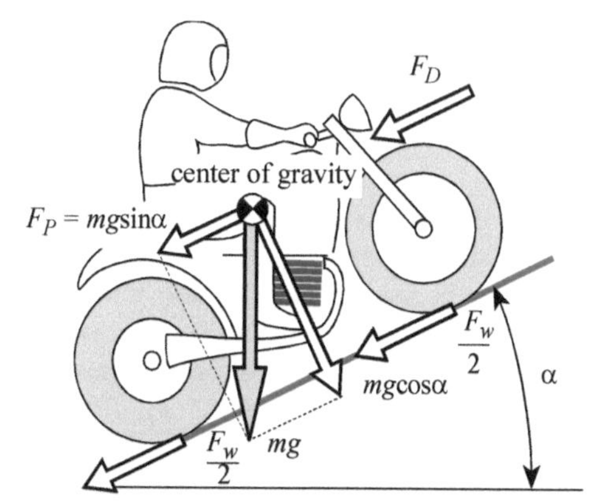

Fig. 3.1 Resistance forces acting on the motorcycle.

Resistance to tire rolling, F_w, was amply discussed in the previous chapter. It was seen that it could generally be considered equal to about

2% of the weight force.

3.1.1 Aerodynamic resistance forces

All the aerodynamic influences that act on the motorcycle can be represented by three forces, which are assumed to be applied on the center of gravity, and by three moments acting around the center of gravity axes x, y, z, as shown in Fig. 3-2:

- the drag force, in opposition to forward motion;
- the lift force that tends to raise the motorcycle;
- the lateral force that pushes the motorcycle sideways;
- the pitching moment;
- the yawing moment;
- the rolling moment.

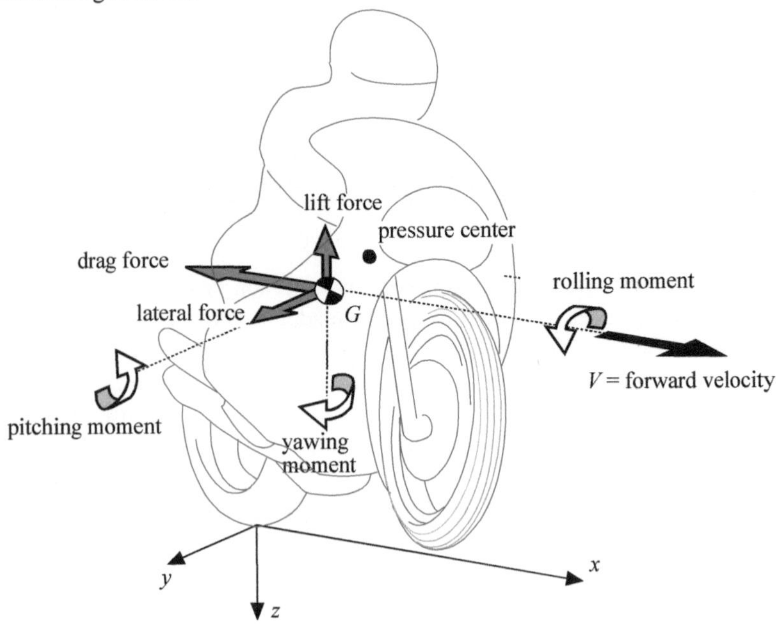

Fig. 3-2 Aerodynamic forces and moments.

The most important components are the drag and lift forces. They are applied at a point, called the pressure center, which does not coincide with the center of gravity, but rather is generally located above it. The resultant of the two aerodynamic forces therefore generates a pitching moment around the y-axis.

The drag force influences both the maximum attainable velocity and performance in acceleration. The drag force F_D is approximately proportional to the square of the motorcycle's forward velocity:

$$F_D = \frac{1}{2}\rho \cdot C_D \cdot A \cdot V^2$$

- ρ represents the density of the air (equal to 1.167 kg/m^3 at an atmospheric pressure of 987 mbar and a temperature of 20° C);
- A is the frontal area of the motorcycle;
- C_D represents the coefficient of aerodynamic resistance (drag coefficient);
- V is the forward velocity of the motorcycle.

The value of the coefficient C_D is strongly influenced by the shape of the motorcycle, in particular by the presence or lack of a fairing. In general, it can be stated that there is a significant increase in aerodynamic resistance when vortex wakes are formed and the boundary layer breaks from the surface of the fairing.

The interaction of the motorcycle with air also generates a lift force F_L proportional to the square of the velocity, which reduces the load on the front and in some cases the rear wheel:

$$F_L = \frac{1}{2}\rho \cdot C_L \cdot A \cdot V^2$$

where C_L represents the lift coefficient.

Motorcycle lift is dangerous since it reduces the load on the wheels and, thus, tire adherence. This is especially true regarding the front tire since the center of pressure is generally in front of and above the center of gravity. Typical motorcycles generate positive (upward) lift force, however, in order to counteract this phenomenon and increase load on the wheels, it would be necessary affix some sort of wing at the front of the motorcycle as in the case of racing cars. To lessen the undesired lift effects, modern fairings are designed to reduce lift to a minimum.

The aerodynamic characteristics of motorcycles are given by the drag area C_DA (drag coefficient times the frontal area) and by the lift area C_LA (lift coefficient times the frontal area).

The value of the product C_DA can vary from 0.18 m^2 for speed record contenders that are completely faired to 0.7 m^2 for motorcycles with no fairing and the rider in an erect position. A typical value for "super bike" motorcycles is 0.30 to 0.35 m^2, while "Grand Prix" motorcycles reach 0.22 m^2 or even smaller values. Touring and/or sporting motorcycles with a small front fairing have values around 0.4 to 0.5 m^2. The change from an erect to a prone riding position leads to a reduction in the value of the product C_DA that varies from 5 to 20%, depending on the type of motorcycle and the rider's body structure.

The resistance to forward motion is influenced in different ways by the various motorcycle components. For example, the following are the effects of some components on the product C_DA:

- front fairings produce an improvement ranging from 0.02 to 0.08 m^2;
- side fairings decrease the C_DA by a quantity of approximately 0.15 m^2;
- side mirrors increase the drag area from 0.012 to 0.025 m^2;
- the presence of a rear fairing improves it by a factor of 0.015 m^2;
- the saddlebags, if appropriately designed, improve it by 0.02 m^2;

- a lower spoiler improves it by a factor that varies from 0.01 to 0.02 m^2.

The frontal area A differs according to the type of motorcycle and is strongly influenced by the body of the rider and his/her position during travel. Reference values may vary from 0.6 to 0.9 m^2 for large displacement touring motorcycles, from 0.40 to 0.6 m^2 for sporting models and from 0.4 to 0.5 m^2 for Grand Prix motorcycles. Small displacement Grand Prix class motorcycles (125 cc) reach values around 0.32 m^2. If the frontal area A and the product $C_D A$ values are known, the resistance coefficient C_D, which is usually on the order of 0.4 to 0.5, can be evaluated. The product of the lift coefficient times the frontal area of the section $C_L A$ ranges from 0.06 to 0.12 m^2.

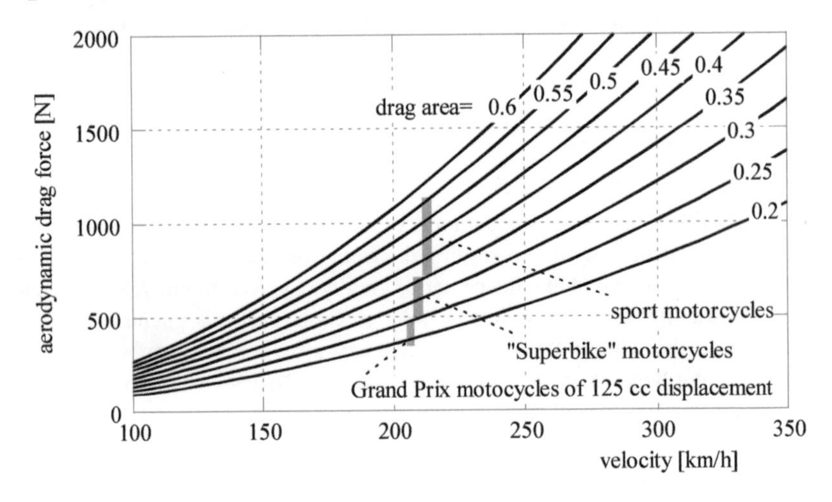

Fig. 3-3 Drag force versus velocity.

The pitching moment caused by the aforementioned forces can be dangerous, since it leads to a decrease in the load on the front wheel and an increase in the rear one. These variations can significantly modify the dynamic behavior of the motorcycle.

In rectilinear motion, if there is no crosswind, the x-z plane of the motorcycle with rider is the plane of symmetry and the forward velocity of the motorcycle lies in that plane. The lateral aerodynamic force and the rolling and yawing moments are zero. However, they are not zero if the rider moves from a symmetric position, if there is lateral wind, or if the sideslip angles of the tires are not zero. In particular, when the rider moves into the curve, displacing his or her body and knee towards the inside of the curve, an aerodynamic yawing moment is generated helping the motorcycle move into the curve. During the curve if the rider stays in this leaned position the lateral aerodynamic force persists.

Since the power dissipated by the aerodynamic forces depends on the cube of the velocity, a lot of power is needed to attain high velocities. Figure 3-3 shows the variation of the drag force against velocity for various values of the drag area.

The forces and aerodynamic moments can be measured in a wind tunnel by mounting the motorcycle on a force balance. The wind tunnel tests make it possible to identify the presence of vortices, if any, and the current lines surrounding the motorcycle. Smoke is used to visualize the current lines, as can be seen in Fig. 3-4.

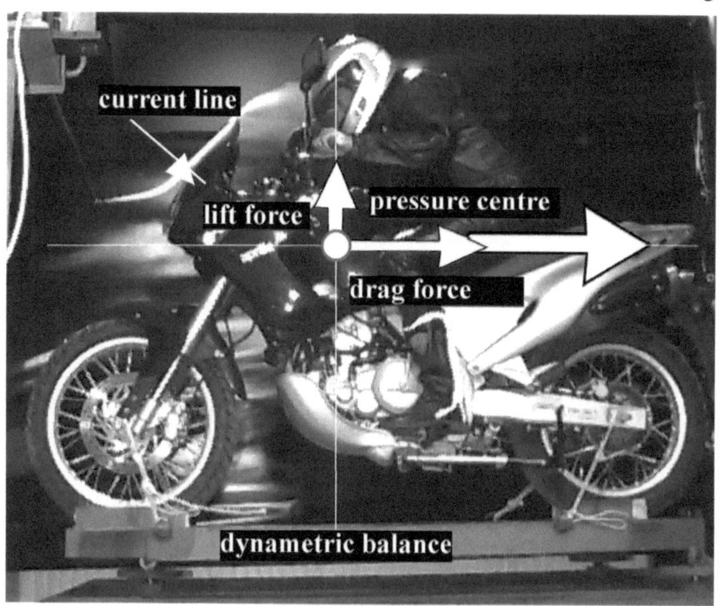

Fig. 3-4 Current lines surrounding the windscreen and rider (University of Perugia wind tunnel).

If no wind tunnel is available, the drag area $C_D A$ can be determined in the following ways. The motorcycle can be driven at its maximum velocity on a straight road recording the engine RPM (revolutions per minute) and the maximum velocity. The power, corresponding to the number of revolutions measured, is determined by the dynamometer curve. The product of the drag coefficient times the frontal area is:

$$C_D \cdot A = \frac{2}{\rho \cdot V^3} P$$

without considering the rolling resistance. There can be significant errors if the maximum velocity is not determined correctly or if the actual power of the motorcycle does not correspond to the dynamometer curve used for the calculation..

A second approach is as follows. The motorcycle can be driven on a flat road at a sustained velocity and then placed in neutral. The time Δt that the motorcycle needs to slow down from an initial velocity ($V_{initial}$) to a lower one (V_{end}) is measured and the drag area is given by:

$$C_D \cdot A = 2 \frac{1}{\rho \cdot \Delta t} m \left(\frac{1}{V_{end}} - \frac{1}{V_{initial}} \right)$$

There can be a certain operative difficulty trying to idle at a sustained velocity. Furthermore, the mass m should, strictly speaking, also take into account the rotating inertia.

Example 1

What power is required to push a sport motorcycle ($C_D A = 0.35$) to a velocity of 250 km/h and 275 km/h, taking into account both the rolling resistance force and the drag force?

The power on the wheel for a maximum velocity of 250 km/h is 71.1 kW. To increase the maximum velocity by 10% (275 km/h), a 32% increase in power is necessary (94.0 kW).

3.1.2 Resistant force caused by road slope

The resistant force F_P caused by the slope of the road plane is equal to the component of the weight force in the motorcycle's direction:

$$F_P = mg \cdot \sin\alpha$$

where α represents the slope of the road plane.

The graph in Fig. 3-5 displays the curves of different power levels at the wheel versus the velocity and the road slope.

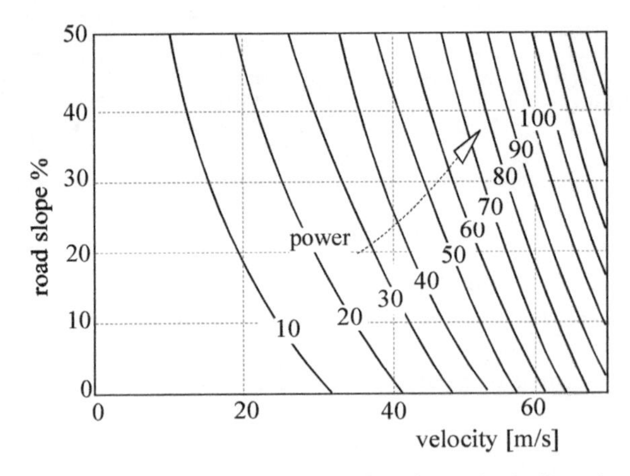

Fig. 3-5 Power at the wheel (kW) as a function of the forward velocity and road slope.

Example 2

Consider a motorcycle with the following characteristics:

- mass: $m = 200$ kg;
- frontal area: $A = 0.6$ m^2;
- drag coefficient: $C_D = 0.7$.

What is the driving force and power necessary to sustain the velocity under the conditions given?

Case 1: Flat road traveling at 200km/h.

The driving force necessary to maintain the motorcycle at a constant velocity of 200 km/h or 55.6 m/s along a horizontal road ($\alpha = 0$) must be equal to the sum of the aerodynamic resistance force F_D and the rolling resistance force F_w:

- force of aerodynamic resistance: $F_D = 0.5 \cdot 1.167 \cdot 0.6 \cdot 0.7 \cdot 55.6^2 = 756.4$ N
- force of rolling resistance : $F_w = 0.02 \cdot 200 \cdot 9.8 = 39.2$ N

Therefore the required driving force is : $S = F_D + F_w = 795.6$ N

For this value of the driving force, the power at the wheel (P) is equal to:

$$P = 796.8 \text{ N} \cdot 55.6 \text{ m/s} = 44.20 \text{ kW}$$

Case 2: Uphill road at 200km/h.

If the motorcycle travels at the same velocity, but along a road with a constant slope of 12% (angle $\alpha = 6.84°$) the resistant force caused by the slope of the road plane must also be taken into account.

- resistant force caused by the road slope: $F_P = \sin(6.84) \cdot 200 \cdot 9.8 = 233.5$ N

In this case the required driving force is: $S = F_D + F_w + F_P = 1029.1$ N

For this value of the driving force, the power at the wheel (P) is equal to:

$$P = 1029.1 \text{ N} \cdot 55.6 \text{ m/s} = 57.2 \text{ kW}$$

3.2 The center of gravity and the moments of inertia

3.2.1 Motorcycle center of gravity

The position of a motorcycle's center of gravity has a significant influence on the motorcycle's dynamic behavior. Its position depends on the distribution and quantity of the masses of the individual components of the motorcycle (engine, tank, battery, exhaust pipes, radiators, wheels, fork, frame, etc.). Since the engine is the heaviest component (about 25% of the total mass), its location greatly influences the location of the motorcycle's center of gravity.

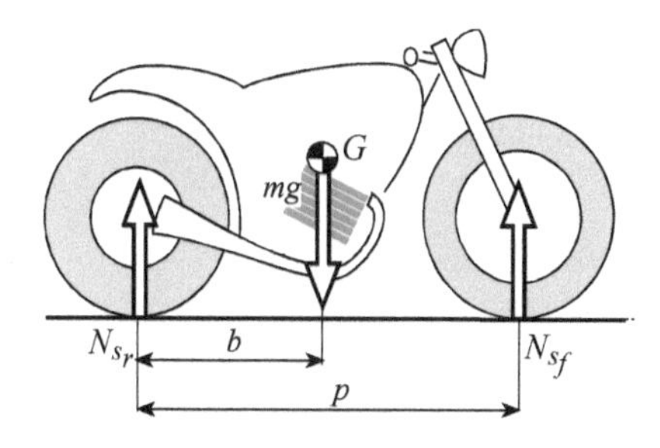

Fig. 3-6 The longitudinal position of the center of gravity.

The longitudinal distance b between the contact point of the rear wheel and the center of gravity can easily be determined by measuring the total mass of the motorcycle and the loads on the wheels under static conditions (front load N_{S_f} , rear load N_{S_r}):

$$b=\frac{N_{s_f}p}{mg}=p-\frac{N_{s_r}p}{mg}$$

In general, a motorcycle is characterized by the static loads that act on the wheels, expressed in a percentage formula:

$$\frac{\%\text{ front load}}{\%\text{ rear load}}=\frac{N_{s_f}/mg}{N_{s_r}/mg}=\frac{b/p}{(p-b)/p}$$

The distribution of the load on the two wheels under static conditions is generally greater on the front wheel for racing motorcycles (50-57% front, 43-50% rear); and conversely, it is greater on the rear wheel in the case of touring or sport motorcycles (43-50% front, 50-57% rear).

When the center of gravity is more forward (front load >50%), wheeling the motorcycle becomes more difficult, or in other words, there is an easier transfer of the power to the ground. This is one reason racing motorcycles are more heavily loaded in front. In addition, the greater load in the front partially compensates for the aerodynamic effects that unload the front wheel; this fact becomes important at high velocities. When the position of the center of gravity is more towards the rear of the motorcycle, braking capacity is increased reducing the danger of a "stoppie" or even forward flip over during a sudden stop with the front brake.

Modern sport motorcycles tend to have a 50 ÷ 50% distribution so as to perform well in both acceleration and braking phases. It is important to keep in mind that it is preferable, as a question of safety, to have longitudinal slip of the rear wheel in an acceleration phase, rather than longitudinal slip of the front wheel in a braking phase. The ratio b/p without rider varies from 0.35 to 0.51: the smallest values for the scooter and the highest for racing motorcycles.

In general, the position of the rider moves the overall center of gravity towards the rear (Fig. 3-7), and therefore, his or her presence increases the load on the rear wheel thereby diminishing the percentage of load on the front wheel (for example the ratio b/p goes from 0.53 to 0.50).

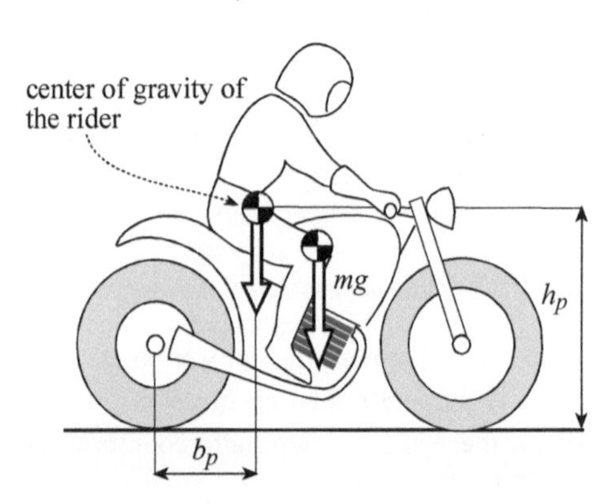

Fig. 3-7 The position of the center of gravity of the motorcycle and the rider.

Once the longitudinal position of the center of gravity has been found, its height can be determined by measuring the load on only one wheel, for example, the rear one with the front wheel raised by a known amount as in Fig. 3-8.

$$h = \left(\frac{N_{S_r}\, p}{mg} - (p-b) \right) \cot\left[\arcsin\left(\frac{H}{p} \right) \right] + \frac{R_r + R_f}{2}$$

The height of the center of gravity has a significant influence on the dynamic behavior of a motorcycle, especially in the acceleration and braking phases. A high center of gravity, during the acceleration phase, leads to a larger load transfer from the front to the rear wheel. The greater load on the rear wheel increases the driving force that can be applied on the ground, but the lesser load on the front wheel makes wheeling more probable.

In braking, a higher center of gravity causes a greater load on the front wheel and a resulting lower load on the rear. The greater load on the front wheel improves braking but it also makes the forward flip-over more likely, which occurs when the rear wheel is completely unloaded.

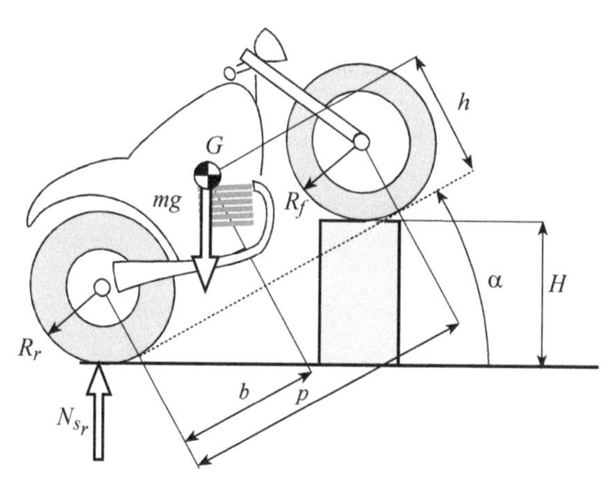

Fig. 3-8 Measure of the height of the center of gravity.

The optimal height of the center of gravity also depends on the driving/braking traction coefficient between the tires and road plane. With low values of the driving/braking traction coefficient (when the road is wet and/or dirty) it is good to have a high center of gravity to improve both the acceleration and braking capacities. With high values of the driving/braking traction coefficient it is good to have a lower center of gravity in order to avoid the limit conditions of wheeling and forward flip over.

It is clear that the choice of the height of the center of gravity and its longitudinal position is a compromise that must take into account the intended use and power of the motorcycle. All-terrain motorcycles are characterized by rather high centers of gravity, while very powerful motorcycles typically have a lower center of gravity. The main effects of the location of the center of gravity may be summarized in the following diagram:

Forward center of gravity	The motorcycle tends to over-steer (in curves the rear wheel slips laterally to a greater extent).
Rear center of gravity	The motorcycle tends to under-steer (in curves the front wheel slips laterally to a greater extent).
High center of gravity	The front wheel tends to lift in acceleration. The rear wheel may lift in braking.
Low center of gravity	The rear wheel tends to slip in acceleration. The front wheel tends to slip in braking.

The height of the center of gravity of the motorcycle alone has values varying from 0.4 to 0.55 m, but the presence of the rider raises the center of gravity to values ranging from 0.5 to 0.7 m. Obviously, the displacement of the center of gravity due the presence of the rider depends on the relation between the mass of the rider and that of the motorcycle.
The ratio h/p without rider and with fully extended suspension varies in the range 0.3-0.4; the smallest values for the cruiser and scooter and the highest for dual sport and enduro type motorcycles.

Example 3

A motorcycle with a mass of 196 kg has a static weight distribution (*50%* to *50%*) and a 1390 mm wheelbase. A 77 kg rider mass has his own center of gravity at 600 mm from the center of the rear wheel. How does the rider change the overall percentage weight distribution?

The percentage load on the rear wheel with rider increases to 52%, while that on the front wheel is reduced to 48%.

3.2.2 The moments of inertia

The dynamic behavior of a motorcycle also depends on the inertia of the motorcycle and the rider. The measurement of the moments of inertia is based on complex identification methodologies, which are outside the purpose of this book. The most important moments of inertia are the roll, pitch and yaw moments of the main frame, the moment of inertia of the front frame with respect to the steering axis, the moments of the wheels and the inertia moment of the engine. In the following table, the values of the gyration radii of the motorcycle and rider, with respect to the center of gravity, are presented (the moment of inertia is given by the product of the mass times the square of the radius of gyration).

The yaw moment of inertia influences the maneuverability of the motorcycle. In particular, high values of the yaw moment (obtained, for example by heavy baggage placed on the luggage rack) reduce handling. The roll moment of inertia influences the speed of the motorcycle in roll motion. High values of the roll inertia, maintaining the same height of the center of gravity, slow down the roll motion in both entry and exit of a curve.

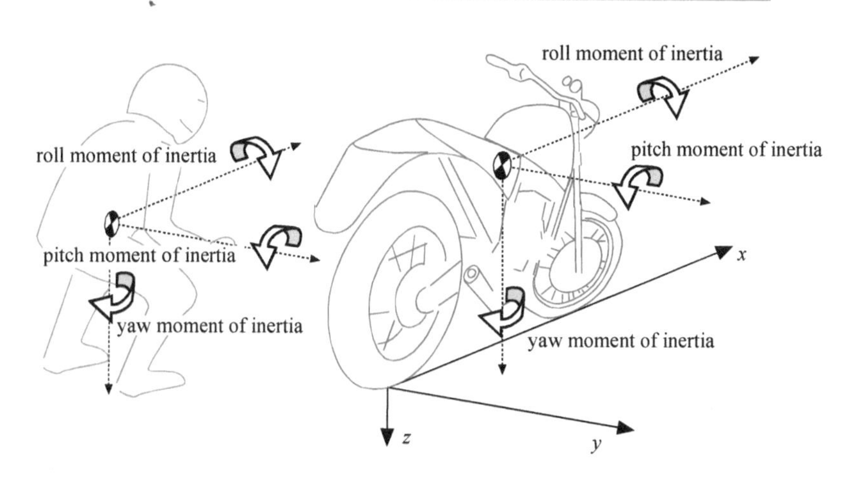

Fig. 3-9 Moments of inertia.

	Roll gyration radius [m]	Pitch gyration radius [m]	Yaw gyration radius [m]
Motorcycle	0.18 – 0.28	0.45 to 0.55	0.41 to 0.52
Rider	0.23 to 0.28	0.23 to 0.28	0.15 to 0.19

Table 3-1.

3.3 Motorcycle equilibrium in steady state rectilinear motion

We will introduce the following three hypotheses regarding the model of the motorcycle-rider system depicted in Fig. 3-10.

- the rolling resistance force is zero ($F_W = 0$);
- the aerodynamic lift force F_L is also considered zero;
- since the road surface is flat, the force resisting the forward motion of the motorcycle is reduced to just the aerodynamic drag force F_D.

The pressure center of the motorcycle (in which the drag force is applied) coincides with its center of gravity.

In addition to the drag force, the following forces act on a motorcycle:

- the weight mg that acts at its center of gravity;
- the driving force S, which the ground applies to the motorcycle at the contact point of the rear wheel;
- the vertical reaction forces N_f and N_r exchanged between the tires and the road plane.

The equations of equilibrium of a motorcycle enable us to determine the unknown values of the reaction forces N_f and N_r, once the weight force mg, driving force S, and drag force F_D are known.

($\Rightarrow$) Equilibrium of horizontal forces: $S - F_D = 0$

($\Uparrow$) Equilibrium of vertical forces: $mg - N_r - N_f = 0$

($\frown$) Equilibrium of moments with respect to the center of gravity:

$$S\,h - N_r b + N_f (p-b) = 0$$

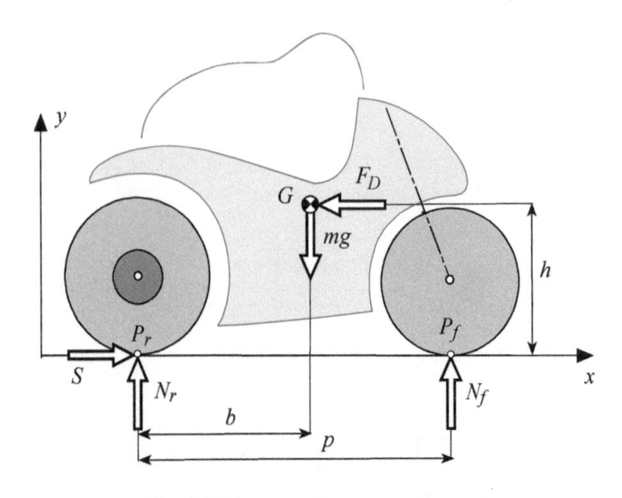

Fig. 3-10 Forces acting on a motorcycle.

The vertical forces exchanged between the tires and the road plane are therefore:

- dynamic load on the front wheel: $$N_f = mg\frac{b}{p} - S\frac{h}{p}$$

- dynamic load on the rear wheel: $$N_r = mg\frac{(p-b)}{p} + S\frac{h}{p}$$

These reaction forces are composed of two elements.

The first term (static load on the wheel), depends on the distribution of the weight force.

$$N_{s_f} = mg\frac{b}{p} \qquad N_{s_r} = mg\frac{(p-b)}{p}$$

The second term (load transfer), is directly proportional to the driving force S and the height h of the center of gravity, and inversely proportional to the motorcycle's wheelbase p.

$$N_{tr} = S\frac{h}{p}$$

We will now focus on the second term. "Load transfer" refers to the fact that there is a decrease in the load on the front wheel and a corresponding increase in the load on the rear wheel; "load is transferred from the front to the rear wheel," hence the designation.

The ratio between the height of the center of gravity and the wheelbase is much higher in motorcycles than in cars, where h/p is usually in the interval 0.3 to 0.45.

The loads on the wheels can be represented in non-dimensional form with respect to the weight:

- Normalized load on the front wheel: $$N_{a_f} = \frac{N_f}{mg} = \frac{b}{p} - S_a \frac{h}{p}$$
- Normalized load on the rear wheel: $$N_{a_r} = \frac{N_r}{mg} = (1 - \frac{b}{p}) + S_a \frac{h}{p}$$

where S_a indicates the ratio between the driving force S and the total weight mg (non-dimensional driving force).

Figure 3-11 illustrates the phenomenon of load transfer. The variations in the normalized loads are indicated as a function of the normalized driving force for two motorcycles with the following characteristics.

- The 1st motorcycle $b/p = 0.45$ and $h/p = 0.3$.
- The 2nd motorcycle $b/p = 0.45$ and $h/p = 0.43$.

It can be observed that the variations in the loads on the wheels are greater for the motorcycle having the higher h/p value.

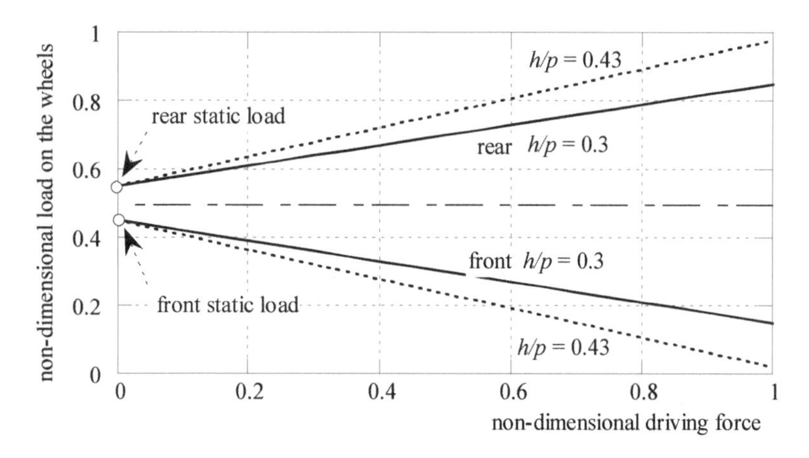

Fig. 3-11 Normalized loads on the wheels as a function of normalized driving force (b/p = 0.45).

Now let's consider the forces acting on the motorcycle, illustrated in Fig. 3-12. The weight mg is equal to the sum of the static loads acting on the wheels N_{s_f} and N_{s_r}. The driving force S and the force caused by the transfer of the load N_{tr}, turned upward because it has a positive sign, are applied at the rear wheel contact point.

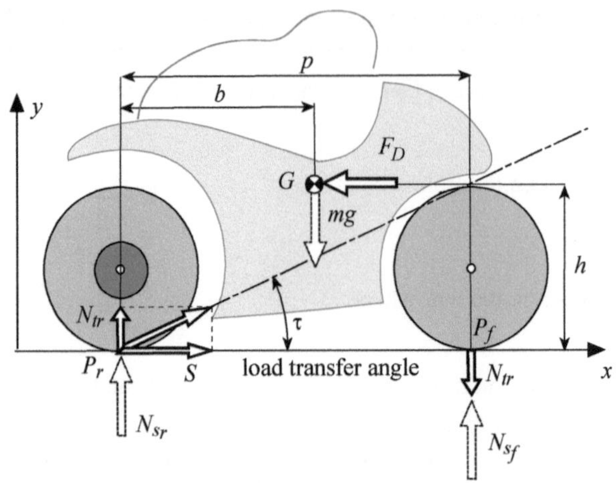

Fig. 3-12 Load transfer angle τ.

The direction of the resultant of these two forces is inclined with respect to the road by the angle:

$$\tau = \arctan \frac{h}{p}$$

which is therefore called the load transfer angle.

In order for a motorcycle to maintain equilibrium, this resultant force must be equal to and opposite in sign to the resultant of the drag force F_D and the load transfer N_{tr}, which acts on the front wheel (directed downward because it has negative sign).

Example 4

What maximum velocity can be reached by a 200 kg motorcycle ($C_D A = 0.3$) with a driving traction coefficient equal to 1.0, ignoring the lift and rolling resistance forces?

The maximum hypothetical forward velocity of the motorcycle depends on the load transfer from the front to the rear wheel. When the front wheel is completely unloaded, and thus the whole load moves to the rear wheel, the limiting condition is reached when maximum velocity is attained. Under these conditions, the maximum driving force that can be applied with a driving force coefficient equal to 1.0 is equal to 1962 N. This force is equal to the drag force and generates a velocity of 381 km/h. The power on the rear wheel is 208 kW.

Example 5

Calculate the maximum velocity at a motorcycle's limit condition represented by the wheeling phenomena, supposing that the center of the pressures coincides with the center of gravity.

- mass: $m = 200$ kg;
- drag area: $C_D A = 0.3$ m^2;
- lift area: $C_L A = 0.1$ m^2;
- aerodynamic pitching moment = 0;
- longitudinal distance of the center of gravity: $b = 0.7$ m;
- height of the center of gravity: $h = 0.65$ m;
- wheelbase: $p = 1.40$ m.

In conditions approaching the limit of wheeling phenomena, the front vertical load becomes zero. The sum of the load transfer generated by the drag force and the front component of the lift force are equal to the front static load. If we consider that the center of pressure coincides with the center of gravity, the lift force is distributed equally between the two wheels in the motorcycle under consideration. We therefore have:

$$F_D \frac{h}{p} + F_L \frac{b}{p} = mg \frac{b}{p}$$

The velocity, corresponding to this equation, is:

$$V = \sqrt{\frac{mg}{\left(\frac{1}{2}\rho \cdot C_D \cdot A \cdot \frac{h}{p} + \frac{1}{2}\rho \cdot C_L \cdot A \frac{b}{p}\right)} \frac{b}{p}}$$

The maximum velocity is therefore V = 339 km/h. The power at the rear wheel, ignoring the rolling resistance force should be at least 147 kW. At this velocity the front wheel is totally unloaded, so that it becomes impossible to control the motorcycle. This value should therefore be considered a maximum limit never to be reached.

3.4 Motorcycles in transitory rectilinear motion

We would like to consider a motorcycle in transitory rectilinear motion assuming the hypotheses presented in the preceding paragraph to be valid. The motorcycle's equilibrium equations which were written for steady state motion can still be considered valid for vertical translation and rotation.

(⇑) Equilibrium of the vertical forces: $mg - N_r - N_f = 0$

(∩) Equilibrium of the moments with respect to the center of gravity:

$$Sh - N_r b + N_f (p - b) = 0$$

where S indicates the driving force during acceleration (+) or the braking force during deceleration (-).

The equation of equilibrium for a motorcycle in horizontal motion takes on certain characteristics according to whether the motorcycle is in an acceleration or braking phase.

3.4.1 Acceleration

In this case, the driving force is equal to the sum of the inertial and resistance forces.

($\Rightarrow$) Equilibrium of the horizontal forces: $S^* = F_D + m^* \ddot{x}$

where $S^* = T(\omega_m / V)$ is the equivalent driving force obtained by multiplying T, the engine torque, by ω_m / V, the ratio between the engine speed and the forward velocity, and m^* indicates the equivalent mass of a motorcycle, which also takes into account the elements of rotational inertia. The latter is calculated by equating the total kinetic energy of the motorcycle (the sum of rotational kinetic energy of the rotating parts and the kinetic energy of translation) to the kinetic energy of an equivalent system constituted by one mass m^* (equivalent mass). From the viewpoint of dynamics, the motion law of the equivalent mass is equal to that of a real motorcycle (Fig. 3-13).

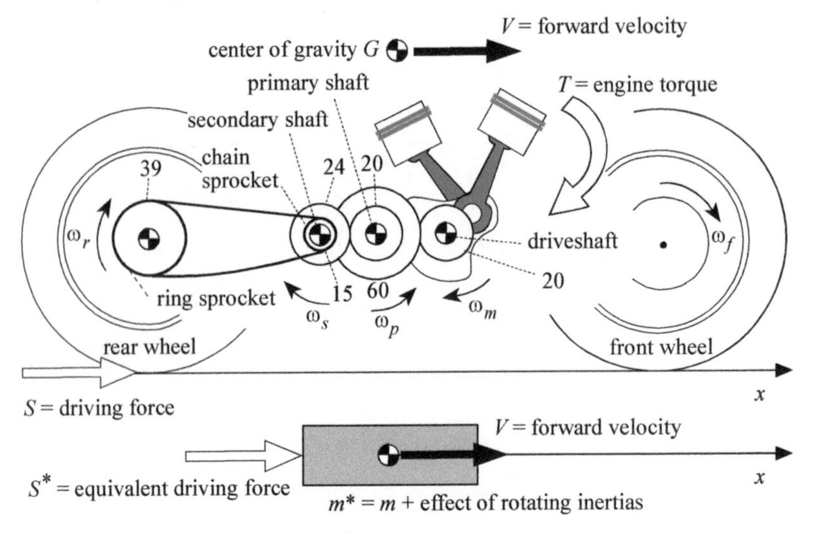

Fig. 3-13 Rotating parts of a motorcycle.

Equating the kinetic energies, we have the following expression.

$$m^* = m + I_{w_r} \cdot \tau_r^2 + I_{w_f} \cdot \tau_f^2 + I_{w_m} \cdot \tau_m^2 + I_{w_p} \cdot \tau_p^2 + I_{w_s} \cdot \tau_s^2$$

Where:

- m is the mass of the motorcycle;
- I_{w_r} is the inertia of the rear wheel;
- I_{w_f} is the inertia of the front wheel;
- I_{w_p} is the inertia of the primary shaft (including clutch);

- I_{w_s} is the inertia of the secondary shaft;
- τ is the velocity ratio.
- I_{w_m} is the inertia of the engine (crankshaft, counter-rotation shafts) , reduced to the crankshaft. This inertia can be considered constant in an initial approximation if we ignore the fluctuating terms of the masses having reciprocating motion.

The ratio of the angular velocity of the rear wheel and a motorcycle's forward velocity is:

$$\tau_r = \frac{\omega_r}{V} = \frac{1}{R_r}$$

The ratio between the angular velocity of the front wheel and a motorcycle's forward velocity is:

$$\tau_f = \frac{\omega_f}{V} = \frac{1}{R_f}$$

The ratio between the angular velocity of the secondary shaft and a motorcycle's forward velocity is:

$$\tau_s = \frac{\omega_s}{V} = \frac{\omega_s}{\omega_r}\frac{\omega_r}{V} = \tau_{s,r}\frac{1}{R_r}$$

where $\tau_{s,r}$ indicates the transmission ratio between the pinion and rear sprockets.

The ratio between the velocity of the primary gear shaft and the velocity of a motorcycle is:

$$\tau_p = \frac{\omega_p}{V} = \frac{\omega_p}{\omega_s}\frac{\omega_s}{\omega_r}\frac{\omega_r}{V} = \tau_{p,s}\tau_{s,r}\frac{1}{R_r}$$

where $\tau_{p,s}$ is the transmission ratio between the primary and secondary gear shafts. The drive sprocket is keyed on the secondary shaft.

The ratio between the velocity of the engine shaft and a motorcycle's forward velocity is:

$$\tau_m = \frac{\omega_m}{V} = \frac{\omega_m}{\omega_p}\frac{\omega_p}{\omega_s}\frac{\omega_s}{\omega_r}\frac{\omega_r}{V} = \tau_{m,p}\tau_{p,s}\tau_{s,r}\frac{1}{R_r}$$

where $\tau_{m,p}$ indicates the transmission ratio between the engine crank shaft and primary shaft.

As is the case of steady state motion, the dynamic loads on the wheels are given by the equations:

- dynamic load on the front wheel: $N_f = mg\frac{b}{p} - S\frac{h}{p}$
- dynamic load on the rear wheel: $N_r = mg\frac{(p-b)}{p} + S\frac{h}{p}$

Example 6

Consider a racing motorcycle with the following characteristics. What is the equivalent mass?

- total mass (motorcycle + rider): $m = 205$ kg;
- front wheel radius: $R_f = 0.30$ m;
- rear wheel radius: $R_r = 0.32$ m;
- front wheel moment of inertia: $I_{w_f} = 0.6$ kgm^2;
- rear wheel moment of inertia: $I_{w_r} = 0.8$ kgm^2;
- engine moment of inertia: $I_{w_m} = 0.05$ kgm^2;
- primary shaft moment of inertia: $I_{w_p} = 0.005$ kgm^2;
- secondary shaft moment of inertia: $I_{w_s} = 0.007$ kgm^2;
- transmission ratio for the driving-wheel sprockets: $\tau_{s,r} = 2.6$;
- transmission ratio for the primary-secondary gear shafts: $\tau_{p,s} = 0.9$ (in IV gear);
- transmission ratio for the engine - primary shafts: $\tau_{m,p} = 2$.

The velocity ratios thus become:

- $\tau_r = 3.125$;
- $\tau_f = 3.33$;
- $\tau_m = 2 \cdot 0.9 \cdot 2.6 \frac{1}{0.32} = 14.63$;
- $\tau_p = 0.9 \cdot 2.6 \frac{1}{0.32} = 7.313$;
- $\tau_s = 2.6 \frac{1}{0.32} = 8.125$.

Once the velocity ratios are known, it is possible to calculate the equivalent mass m^* as follows:

$$m^* = m + Iw_r \cdot \tau_r^2 + Iw_f \cdot \tau_f^2 + Iw_m \cdot \tau_m^2 + Iw_p \cdot \tau_p^2 + Iw_s \cdot \tau_s^2$$

$$205 + 7.81 \quad + 6.67 \quad + \quad 10.69 \quad + \quad 0.27 \quad + \quad 0.46 \quad = 230.9 \text{ kg}$$

It is worth pointing out that the engine plays a very important role since the velocity ratio is high, even if the value in the moment of inertia is low.

The equivalent mass obviously depends on the gear engaged. The transmission ratio of the gear shift varies from values equal to about 3 for the first gear (the gear-shift functions as a reduction gear), to values that can approach or be slightly lower than unity (down to about 0.7) in the tallest gear. The maximum value of reduced inertia is reached when first gear is engaged.

Figure 3-14 shows the variation of the driving force on the wheel versus velocity, in the various gears, for a racing motorcycle. The curve traced also shows the variation in the resistance force (the sum of aerodynamic and rolling resistance forces) in terms of the motorcycle's velocity.

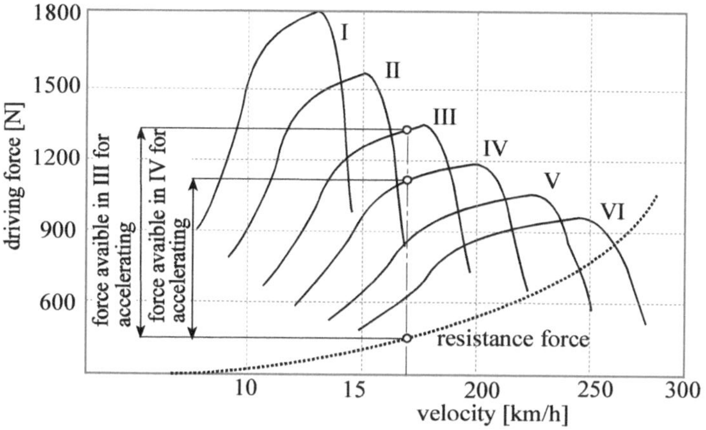

Fig. 3-14 Driving force as a function of velocity.

Let's suppose that the motorcycle proceeds in third gear at a determined velocity, as given in the figure. The driving force, though lower than the maximum available in that gear, is greater than the resistance force, so that the remaining driving force can be used to accelerate or go up a slope at the same velocity. If we consider the same set velocity with higher ratios, we see that there is less driving force available for accelerating. As the velocity gradually increases, the passage to the higher ratio makes a lower quota available. Maximum velocity is obviously reached when the resistant force is the same as the driving force in the highest gear.

The comparison between the driving force curves and the resistance curve can also be made in terms of power. A diagram of useful power to the wheel can be obtained for each ratio by multiplying each curve by its corresponding forward velocity. Fig. 3-15 shows the curves of useful power to the wheel and the resistance power in terms of forward velocity. In this case as well, the intersection point of the resistance power curve with the useful power curve, in the highest gear, determines the maximum velocity that can be reached.

Let's now suppose that the engine power remains constant under an increase in velocity. This is an ideal case in which the efficiencies are independent of the velocities, represented in Figure 3-15 by a horizontal line. Maximum motorcycle acceleration can be determined by integrating the following differential equation:

$$m^* \ddot{x} = \frac{P_{max}}{\dot{x}} - \frac{1}{2} C_D A \rho \cdot \dot{x}^2$$

where P_{max} indicates the maximum power of the engine. By carrying out the numerical integration of the preceding differential equation, we can calculate the maximum acceleration the motorcycle is capable of reaching. This is clearly an ideal value, since the maximum acceleration of the motorcycle can actually be limited both by the rear tire's adherence, the possible wheeling of the motorcycle and the finite number of gears etc.

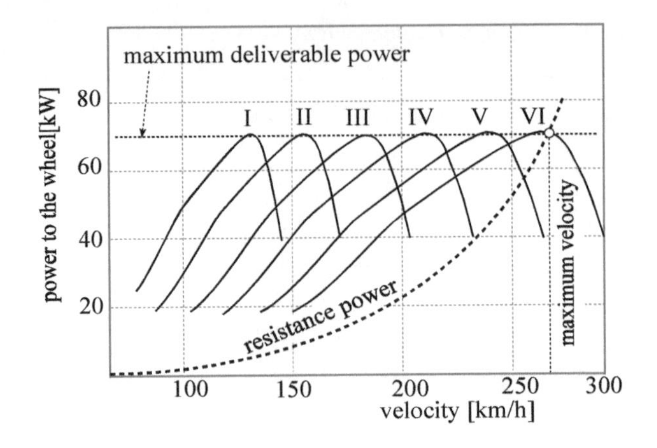

Fig. 3-15 Power at the wheel as a function of velocity.

Example 7

Consider a motorcycle with the following characteristics. Let use examine how changing the mass affects the velocity and acceleration.

- Equivalent mass: $m^* = 230.9$ kg;
- frontal area: $A = 0.6$ m^2;
- drag coefficient: $C_D = 0.7$.

The maximum power of the engine is equal to $P_{max} = 70$ kW while the maximum transmissible driving force to the road is equal to 4000 N.

The curve of the velocity versus time, obtained by carrying out a numerical integration of the differential equation of motion, is given in Fig. 3-16. A 30% lighter motorcycle presents a greater acceleration but the maximum velocity remains the same.

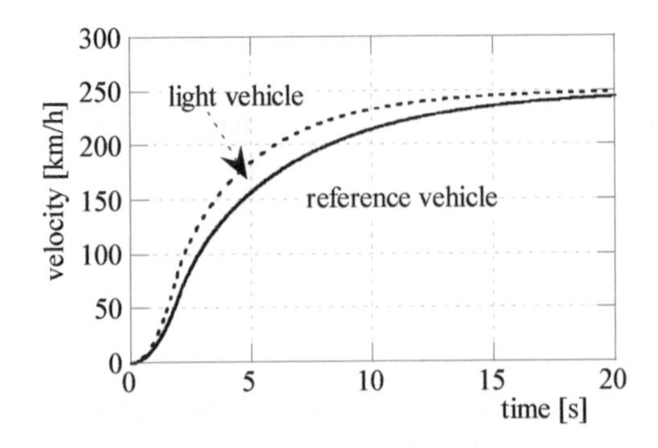

Fig. 3-16 An example of acceleration.

In reality the accelerations (the gradient of the curve) are actually lower because of the time intervals needed to change gears, during which the useful driving force is zero. Furthermore, in the initial phase, as previously anticipated, it is not always possible to apply the entire driving force because of rear tire slippage and/or possible wheeling.

3.4.2 Traction-limited acceleration

If we take into consideration a motorcycle accelerating as in Fig. 3-17 and assume it is possible to ignore the rolling resistance force F_w, then the motion equation can be written as follows:

$$S = m\ddot{x} + F_D$$

where S indicates the driving force on the wheel and F_D the drag force. Presuming that the engine has adequate power, the driving force must be lower, or at most, equal to the maximum force given by the product of the driving traction coefficient μ_p with the vertical load N_r.

$$S \leq \mu_p \cdot N_r$$

If we remember that

$$N_r = mg\frac{(p-b)}{p} + S\frac{h}{p}$$

we now have:

$$\ddot{x} \leq \frac{\mu_p \cdot g \cdot \frac{(p-b)}{p}}{\left(1-\mu_p \frac{h}{p}\right)} - \frac{F_D}{m}$$

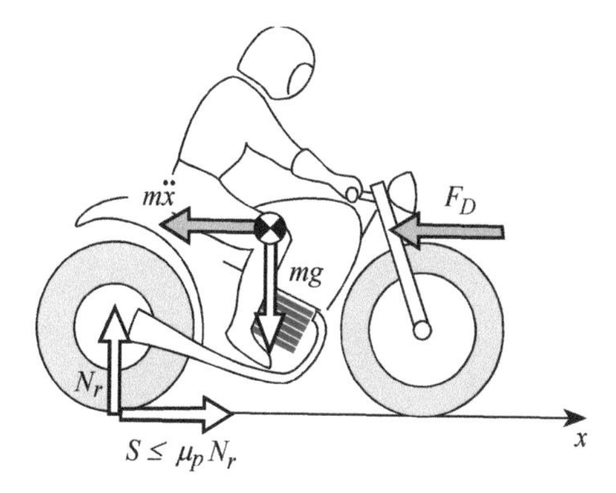

Fig. 3-17 Acceleration limited by the driving traction coefficient.

Maximum acceleration is reached when the resistance force F_D is zero, i.e. starting from low speed. As the velocity increases, the acceleration under limiting friction conditions diminishes. This happens because part of the driving force is equated to the resistance force and therefore cannot be used to accelerate.

3.4.3 Wheeling-limited acceleration

The limiting condition at the onset of wheeling is achieved when the load on the front wheel is reduced to zero, as seen in Fig. 3-18. This situation is expressed by the following relation:

$$N_f = g \cdot m \; \frac{b}{p} - S\frac{h}{p} = 0$$

from which we have

$$\ddot{x} = g\frac{b}{h} - \frac{F_D}{m}$$

Acceleration which corresponds to impending wheelie therefore depends on the ratio b/h.

As the forward velocity gradually increases, the acceleration at which the wheeling phenomenon begins, decreases. This is the case since the motion of wheeling is also favored by the drag force F_D, the value of which increases with velocity.

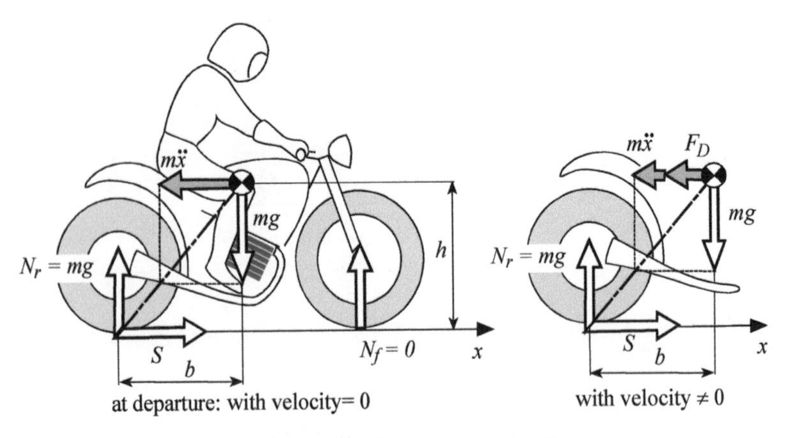

Fig. 3-18 Acceleration limited by wheeling.

Example 8

Consider a motorcycle with the following properties. Determine the wheeling limited acceleration at initial speed 0 km/h and 100 km/h.

- total mass: $m = 200$ kg;
- frontal area: $A = 0.7$ m^2;
- drag coefficient : $C_D = 0.6$;
- lift coefficient : $C_L = 0$;
- longitudinal distance of the center of gravity: $b = 0.58$ m;
- height of the center of gravity: $h = 0.62$ m;
- wheelbase: $p = 1.35$ m.

The maximum acceleration of the motorcycle, at the rear wheel friction limit, is represented in the graph in Fig. 3-19 in terms of the driving force coefficient between the tire and the road at a velocity of zero and 100 km/h. As the velocity increases, the maximum acceleration decreases, since part of the driving force is equated to the resistance force and cannot be used to accelerate.

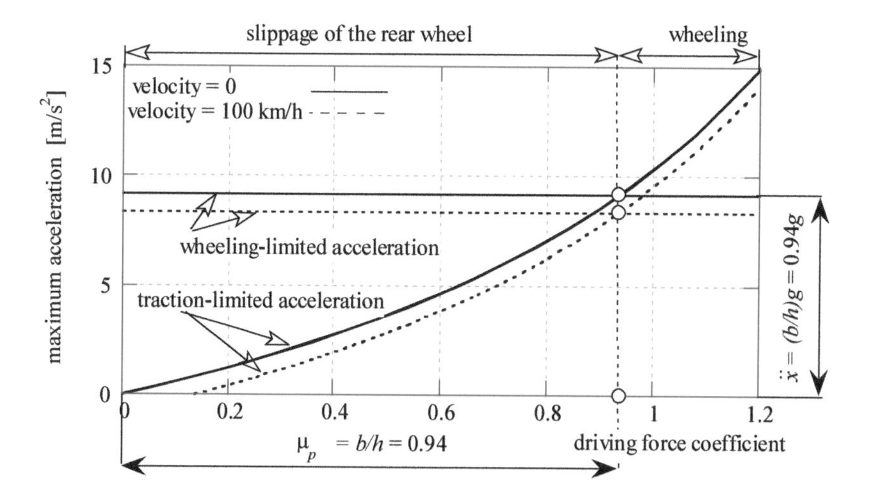

Fig. 3-19 Maximum acceleration at limit conditions.

The acceleration at the motorcycle's wheeling limit at zero km/h and 100 km/h is also shown in the graph. The horizontal line representing wheeling-limited acceleration is explained by the fact that acceleration does not depend on the driving force coefficient.

The wheeling-limited acceleration is equal to the traction-limited acceleration when the driving traction coefficient is:

$$\mu_p = \frac{b}{h}$$

Acceleration is:

$$\frac{\ddot{x}}{g} = \frac{b}{h}$$

For values of the coefficient below the ratio b/h, the motorcycle, in its acceleration maneuver, will not lift the front wheel, since rear wheel slippage prevents it from reaching wheeling acceleration. Analogously, for values of the coefficient above the ratio b/h, the motorcycle cannot reach maximum acceleration at the driving traction coefficient since the front wheel rises before reaching that limiting value. These considerations suggest that it is appropriate to limit the maximum torque the engine can deliver, if the intention is to avoid motorcycle wheeling and rear wheel slippage.

In the case under consideration, the acceleration as shown in Fig. 3-19 is equal to 9.18 m/s^2. This value is obtained by applying a driving force equal to 1835 N to the motorcycle. Whenever the motorcycle's engine is unable to deliver a useful force to the wheel of that magnitude, wheeling cannot occur naturally (nonetheless, the rider could cause the front wheel to rise by moving and making use of the pitching motion).

3.4.4 Braking

Driving safety requires, in addition to an efficient braking system, that the rider be able to judge the stopping distance required under various conditions and brake in the best way, using all of the braking system's possibilities and in particular those of the rear brake. In fact, many motorcycle riders tend to forget the rear brake, which in certain circumstances provides a useful contribution. Its correct use is important both in braking when entering a curve and in braking during rectilinear motion when an unforeseen obstacle appears in front of the motorcycle (especially when road adherence is precarious).

Role of the rear brake in sudden stops

During curve entry the use of the rear brake can be quite useful. Expert riders use the rear brake not only to decelerate the motorcycle but also to control the yaw motion. Rear braking in entering the curve increases the sideslip angle and therefore the yaw motion of the motorcycle.

In sudden deceleration a dangerous condition could arise especially when the load on the rear wheel diminishes toward zero due to load transfer.

If the motorcycle is not in perfectly straight running the force of the front brake and the inertial force of deceleration generate a moment that tends to cause the motorcycle to yaw. This is illustrated in Fig. 3-20.

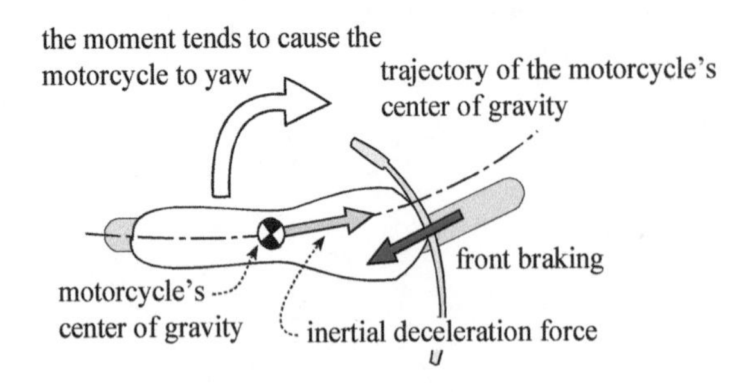

Fig. 3-20 Motorcycle in a curve with a braking force applied only in the front.

As shown in the in Fig. 3-20, the torque generated by the front braking force and the inertial force tends to yaw the vehicle. On the contrary, the presence of a rear braking force generates a torque which tends to align and stabilize the vehicle as can be seen intuitively in Fig. 3-21.

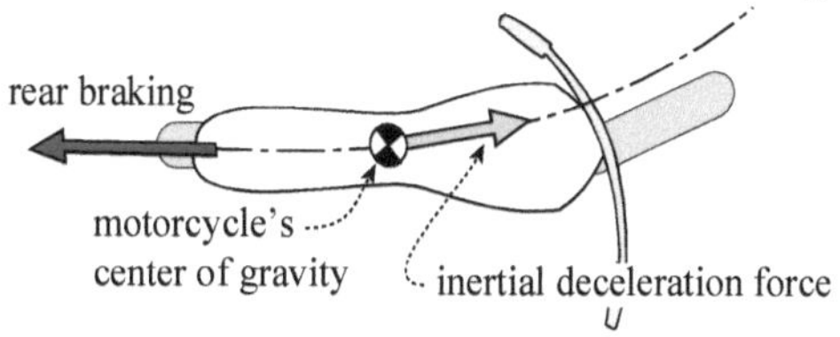

trajectory of the motorcycle's center of gravity
rear braking
motorcycle's center of gravity
inertial deceleration force

Fig. 3-21 Motorcycle in a curve with a braking force applied only in the rear.

These simple considerations suggest that proper utilization of front and the rear brakes has a positive effect on vehicle stability.

Load transfer during braking

In order to evaluate the role of the rear brake during a braking event at the limit of slippage, we need to bring up some points regarding the forces acting on a motorcycle. During deceleration, the load on the front wheel increases, while that on the rear wheel decreases and thus there is a load transfer from the rear to the front wheel. If we consider a motorcycle in a braking phase (Fig. 3-22) and apply Newton's law to the motorcycle, we can calculate the load transfer from the rear to the front wheel.

($\Rightarrow$) Equilibrium of the horizontal forces: $m\ddot{x} = -F_f - F_r$

($\Uparrow$) Equilibrium of the vertical forces: $mg - N_r - N_f = 0$

($\curvearrowleft$) Equilibrium of the moments around the center of gravity:

$$-Fh - N_r b + N_f (p-b) = 0$$

where F (overall braking force) indicates the sum of the front braking force F_f and the rear braking force F_r. The dynamic load on the front wheel is equal to the sum of the static load and the load transfer:

$$N_f = mg\frac{b}{p} + F\frac{h}{p}$$

while the dynamic load on the rear wheel is equal to the difference between the static load and the load transfer:

$$N_r = mg\frac{(p-b)}{p} - F\frac{h}{p}$$

It can be seen that the load transfer Fh/p is proportional to the overall braking force, and to the height of the center of gravity, and is inversely proportional to the wheelbase. To prevent a tire from slipping during braking, the value of the braking force applied to it must not exceed the product of the dynamic load acting on that tire times the local braking traction coefficient. This latter product represents the maximum braking force applicable to the tire, that is, the braking force at the limit of slippage.

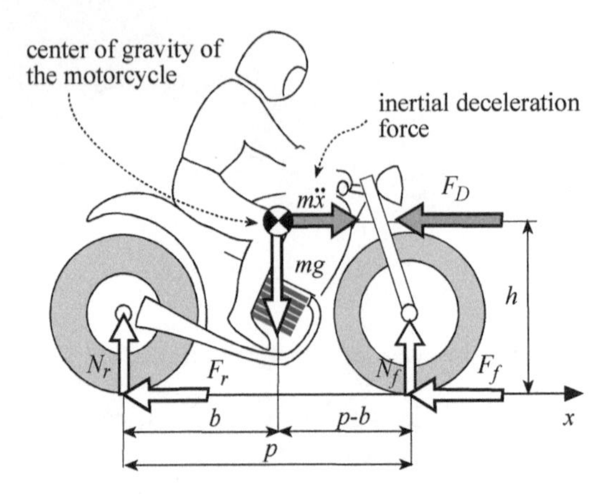

Fig. 3-22 A motorcycle under braking.

Given μ_{p_f} and μ_{p_r}, the braking traction coefficients relative, respectively, to the front wheel and the rear wheel, the overall braking force at the limit of slippage is given by the following expression:

$$F_{\max} = F_{f_{\max}} + F_{r_{\max}} = \mu_{p_f} N_f + \mu_{p_r} N_r$$

The limits of slippage are not usually attained during braking and therefore the braking force depends on the braking force coefficients used (indicated by μ and defined as the ratios of the longitudinal force and the corresponding vertical load) of the front and rear wheels.

$$F = F_f + F_r = \mu_f N_f + \mu_r N_r$$

Figure 3-23 shows the variation of dynamic loads on the wheels in terms of the braking force. Both the loads on the wheels and the braking force have been reduced to non-dimensional status with respect to the weight. The motorcycle under consideration has a 50% to 50% static load distribution on its two wheels, i.e. the center of gravity falls in the centerline of the wheelbase.

Let's suppose that the braking force coefficient used is very low, $\mu = 0.2$ for both wheels. We can note from the graph that the dynamic loads on the wheels are approximately equal to 0.4 on the rear wheel and 0.6 on the front one. Under these conditions, if the rear brake is not used, 40% of the maximum attainable braking force is not used. However, if the braking force coefficient used is very high, for example $\mu = 0.9$, as shown in Fig. 3-23, the load on the front wheel is 0.95, while the load on the rear wheel is only 0.05. In this case, the possible contribution of the rear braking force is nearly negligible.

In conclusion, the following general principles can be stated.

- The optimal distribution of the braking force varies according to the braking traction coefficient.
- The rear brake is of little use on optimal roads and with high grip tires (high coefficient of friction), but becomes indispensable on slippery surfaces (reduced

coefficient of friction).

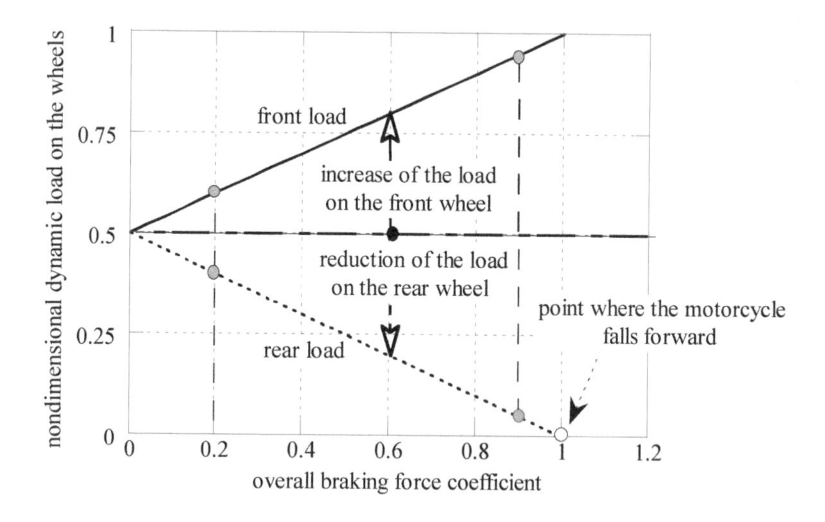

Fig. 3-23 Non-dimensional loads on the wheels versus the overall braking force coefficient.

3.4.5 Forward flip over of the motorcycle

Fig. 3.23 shows that, with an increase in the overall braking force, the load on the rear wheel becomes zero. This limiting condition represents the forward flip over of the motorcycle when the dynamic load on the rear wheel goes to zero.

In this situation, the dynamic load on the front wheel is equal to the weight of the motorcycle and the direction of the resultant of the dynamic load and braking force passes through the motorcycle's center of gravity. The equation of equilibrium of the moments with respect to the center of gravity provides the expression for the braking force at the point of turn over:

$$F = N_f \frac{(p-b)}{h} = mg \frac{(p-b)}{h}$$

A low value of this limit braking force indicates an increased propensity for a forward flip over. It can therefore be concluded that forward fall is favored when a motorcycle is light and when it has a high and forward position of the center of gravity.

The motion equation, in conditions where a fall is imminent, ignoring the aerodynamic resistance, is:

$$m\ddot{x} = -F = -mg \frac{(p-b)}{h}$$

The maximum deceleration, expressed in g's, is proportional to the longitudinal distance from the center of gravity to the contact point of the front wheel, and is inversely proportional to the height of the center of gravity.

$$\frac{\ddot{x}_{\max}}{g} \leq \frac{p-b}{h}$$

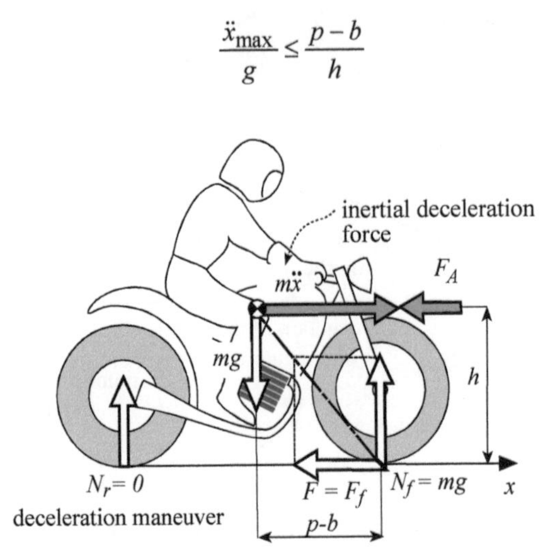

Fig. 3-24 Motorcycle at the point of falling forward.

It is important to note that the deceleration at the flip over limit depends only on the position of the center of gravity, and not on the weight of a motorcycle. To increase the value this limit, it is necessary to lower a motorcycle's center of gravity and place it as far back as possible. Taking into account the aerodynamic resistant force:

$$m\ddot{x} = -F - F_D$$

the maximum deceleration depends on both the mass and the velocity:

$$\frac{\ddot{x}_{\max}}{g} \leq -\frac{(p-b)}{h} - \frac{\frac{1}{2}\rho A C_D V^2}{mg}$$

Example 9

What is the maximum deceleration in braking to the limit of flip over, with a 50% to 50% distribution of the loads on the wheels, a wheelbase of 1400 mm and a height of the center of gravity of 700 mm?

It is easy to verify that the maximum deceleration is equal to gravity. If the velocity is also taken into account, deceleration increases as the velocity increases due to the effect of the aerodynamic resistance force. With a velocity of 100 km/h, a drag area equal to 0.4 m^2 and a mass of 200 kg, the maximum deceleration is equal to 1.26 g while at a velocity of 200 km/h the maximum deceleration increases to 1.54 g. Obviously it is very difficult, if not impossible, to brake at the flip over limit with a zero load on the rear wheel. In this condition nearing the limit, the best riders are able to attain decelerations equal to 1.1 to 1.2 g's.

3.4.6 Optimal braking

The equilibrium equation of the horizontal forces: $m\ddot{x} = -F_f - F_r$ and of the vertical forces. $mg - N_r - N_f = 0$: allows us to express the braking deceleration as a function of the front and rear braking force coefficients used during the event:

$$\frac{\ddot{x}}{g} = \frac{p\mu_r + b(\mu_f - \mu_r)}{p + h(\mu_r - \mu_f)}$$

It can be observed that deceleration depends on the geometric characteristics (wheelbase p, height of the center of gravity h, longitudinal distance of the center of gravity b) and the braking force coefficients used, and does not depend on the motorcycle's mass.

The braking force of the front and rear wheels, with respect to the total braking force, also depends only on the geometric magnitudes, and the braking force coefficients of the two wheels:

$$\frac{F_f}{F} = \frac{\mu_f(b + h\mu_r)}{p\mu_r + b(\mu_f - \mu_r)} \qquad \frac{F_r}{F} = \frac{\mu_r((p-b) - h\mu_f)}{p\mu_r + b(\mu_f - \mu_r)}$$

In the limiting conditions of friction, with equal braking traction coefficients for the two wheels $\mu = \mu_{p_f} = \mu_{p_r}$, the value of the maximum possible deceleration becomes:

$$\ddot{x}_{\max} = \mu g$$

The relation between the braking forces, to attain the limit condition at both wheels (equal braking traction coefficients), simultaneously must be equal to:

$$\frac{F_f}{F} = \frac{b + \mu h}{p} \qquad \frac{F_r}{F} = \frac{(p-b) - \mu h}{p}$$

This relation indicates how to distribute the braking forces in order to have optimal braking, given the value of the braking force coefficient μ.

The distribution curves for braking and deceleration are shown in Fig. 3-25 (considering the acceleration of gravity to be $g = 9.81$ m/s^2) in terms of the braking force coefficients used on each wheel. In Fig. 3-25 we can see that deceleration increases as the braking force coefficients increase, especially with regard to the front wheel. This behavior is understandable since, as has already been explained, during braking there is a load transfer from the rear to the front wheel. The solid lines represent the distribution of braking between the front and rear wheels.

The horizontal axis corresponds to braking with the rear wheel alone (0/100) while the vertical axis represents the case of braking with the front wheel alone (100/0). The figures show the utility of using the rear brake, especially when the braking traction coefficient is low. Its usefulness diminishes until it becomes almost negligible in the presence of very high braking traction coefficients.

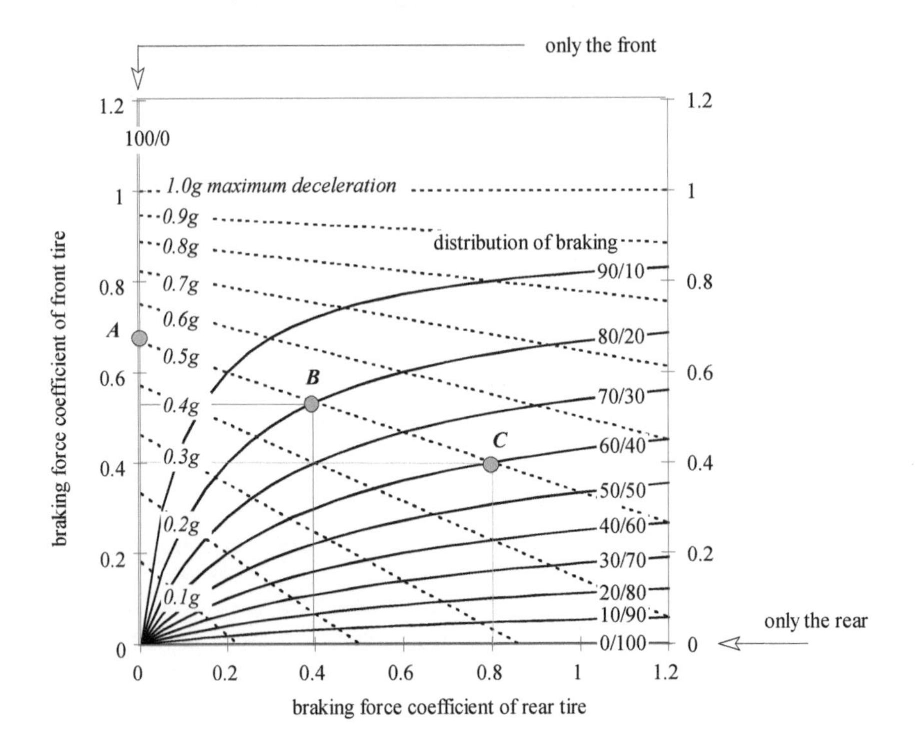

Fig. 3-25 Curves of deceleration and distribution of braking.
[p = 1.4 m; h = 0.7m; b = 0.7m]

Considering this data, it can be seen that the limiting condition of a flip over occurs when deceleration is close to 1.0 g . In this case, the curve (1.0 g) is, therefore the maximum attainable deceleration.

Suppose we wanted to brake the motorcycle with a deceleration equal to 0.5 g . The possible combinations of use of the front and rear brakes that could provide the desired deceleration are infinite. For example, braking only with the front brake, the deceleration of 0.5 g is obtained by using a braking force coefficient in front equal to 0.68 (point A). On the other hand, with a distribution of the braking forces of 80% front and 20% rear, a braking force coefficient of 0.55 in front and 0.4 in back must be used (point B). Another possibility is given by point C which shows a distribution of the braking force of 60% front and 40% rear, in which there is a greater use of the rear tire and a corresponding lesser use of the front one.

Let's now suppose that the braking traction coefficients of the front and rear tires are the same. Figure 3-26 shows that by using the same braking coefficient for the two tires, we obtain the maximum possible deceleration. For example, if the braking force coefficient is equal to 0.8 for both the front and rear wheels, the maximum deceleration (equal to $0.8\,g$) is obtained with a 90/10 braking distribution. The maximum use of the two tires is attained with this distribution. The figure further shows that using only the front brake gives a deceleration that is lower at $0.67\,g$ and that using only the rear brake yields only $0.29\,g$. If the road is more slippery and the braking traction coefficient of both the wheels is 0.4, the optimal braking occurs with a different distribution (70/30) and gives a deceleration of $0.4\,g$.

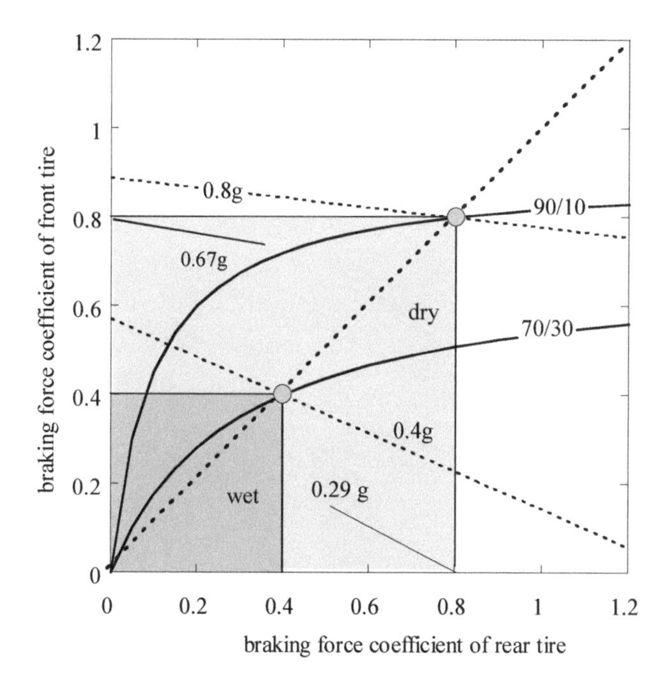

Fig. 3-26 Example of braking on dry and wet surfaces.

This example shows that optimal braking requires a different distribution of braking between the two wheels when varying the desired deceleration. In fact, the 45° line corresponding to $\mu_f = \mu_r$, which represents the condition for optimal braking, intersects different curves of braking distribution when varying the desired deceleration. This means that the devices for automatic distribution of braking that are present on some motorcycles, should adapt the distribution to the conditions of the road.

Furthermore, it is worth pointing out that in the example considered it is not a good idea to use a rear braking force greater than the front one. Figure 3-27 shows that the optimal distribution of braking (dotted line) is tangent at the point of origin to the curve of 50% to 50% distribution; but it does not intersect the curves of distribution characterized by greater force to the rear.

This is also valid for motorcycles with a different distribution of the static load on the two wheels, for example 45% on the front and 55% on the rear wheel. The optimal line of braking is always tangent at the origin to the braking distribution line having the same distribution between the static loads on the two wheels. For example, with a load distributed 45% to the front and 55% to the rear, the optimal braking line is tangent to the distribution curve of the braking force 45% to the front and 55% to the rear.

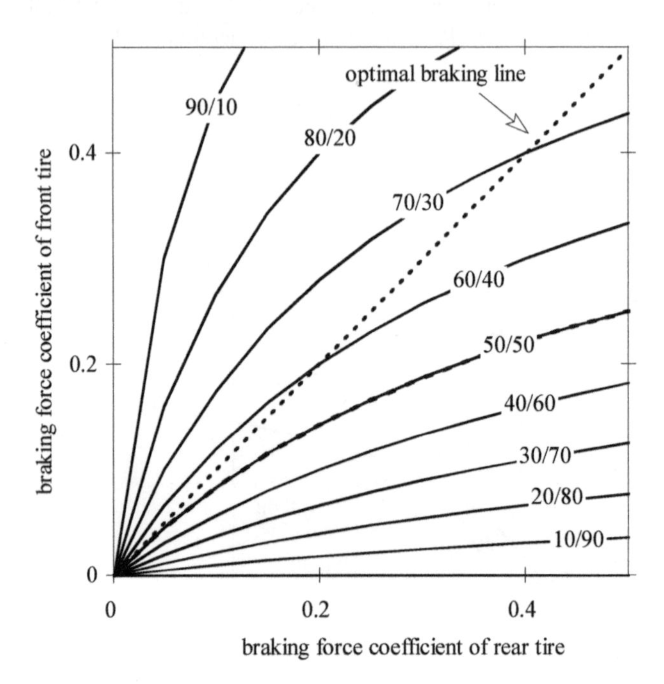

Fig. 3-27 Optimal braking.
[p = 1.4 m; h = 0.7m; b = 0.7m]

Parilla 250 cc of 1946

4 Steady Turning

During steady turning motion the motorcycle can have neutral, under or over-steering behavior. To maintain equilibrium the rider applies a torque to the handlebars that can be zero, positive, in the same direction of the handlebar rotation, or negative, i.e., applied in the direction opposite to the rotation of the handlebar. These characteristics are important and concur to define the sensation of the motorcycle's handling.

4.1 The motorcycle roll in steady turning

4.1.1 Ideal roll angle

The motorcycle, in steady turning, is subject to both a restoring moment, generated by the centrifugal force that tends to return the motorcycle to a vertical position, and to a tilting moment, generated by the weight force, that tends to increase the motorcycle's inclination, or roll angle (Fig. 4-1).

We introduce the following simplifying hypotheses:

- the motorcycle runs along a turn of constant radius at constant velocity (steady state conditions);
- the gyroscopic effect is negligible.

Considering the cross section thickness of the tires to be zero, the equilibrium of the moments allows us to derive the roll angle in terms of the forward velocity V and the radius of the turn R_c (the radius of the turn in this case is measured from the center of gravity to the turning axis):

$$\varphi_i = \arctan\frac{R_c \Omega^2}{g} = \arctan\frac{V^2}{g R_c}$$

where Ω indicates the angular yaw rate, while $V = \Omega R_c$ indicates the vehicle's forward velocity.

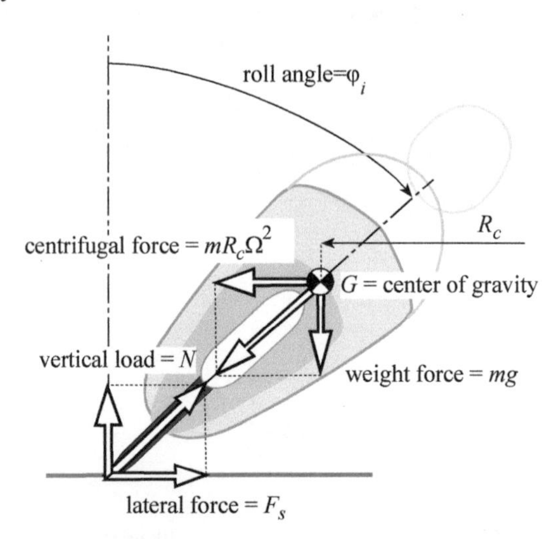

Fig. 4-1 Steady turning: roll angle of the motorcycle equipped with zero thickness tires.

In conditions of equilibrium the resultant of the centrifugal force and the weight force passes through the line joining the contact points of the tires on the road plane. This line lies in the motorcycle plane if the wheels have zero thickness and the steering angle is very small.

In reality, if a non-zero steering angle is assigned, the front contact point is displaced laterally with respect to the x-axis of the rear frame and the line joining the contact points of the tires is not contained in the plane of the rear frame.

4.1.2 Effective roll angle

Now consider a motorcycle with tires of thickness $2t$ which describes the same turn radius R_c at the same yaw velocity Ω. Since the thickness of the tires is not zero, the roll angle φ that is necessary for the equilibrium of the moments exerted by the weight force and the centrifugal force, is greater than the ideal one φ_i (Fig. 4-2):

$$\varphi = \varphi_i + \Delta\varphi$$

The increase $\Delta\varphi$ of the roll angle is given by the equation:

$$\frac{\sin\Delta\varphi}{t} = \frac{\sin\varphi_i}{h-t}$$

The effective roll angle is:

$$\varphi = \varphi_i + \Delta\varphi = \arctan\frac{V^2}{gR_c} + \arcsin\frac{t \cdot \sin\left(\arctan\frac{V^2}{gR_c}\right)}{h-t}$$

The preceding equation shows that $\Delta\varphi$ increases both as the roll angle and the cross section radius increase and as the height of the center of gravity h decreases. Therefore, the use of wide tires forces the rider to use greater roll angles with respect to the angle necessary with a motorcycle equipped with tires that have smaller cross sections. Furthermore, with equal cross sections of the tires, to describe the same turn with the same forward velocity, a motorcycle with a low center of gravity needs to be tilted more than a motorcycle with a higher center of gravity.

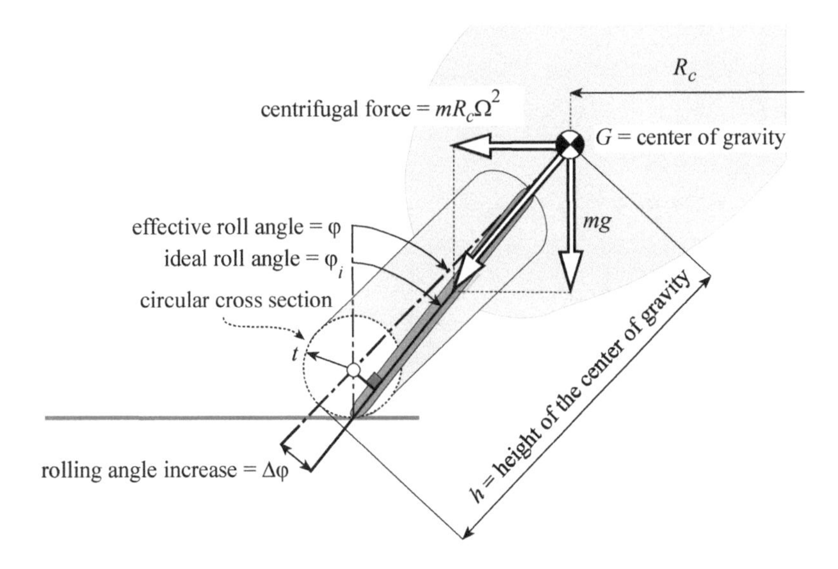

Fig. 4-2 Steady turning: roll angle of the motorcycle equipped with real tires.

The motorcycle roll angle on a turn is influenced, in a significant way, by the rider's driving style. By leaning with respect to the vehicle, the rider changes the position of the his center of gravity with respect to the motorcycle. Figure 4-3 illustrates the possible situations.

If the rider remains immobile with respect to the chassis (Fig. 4-3a), the center of gravity of the motorcycle-rider system remains in the motorcycle plane. Undoubtedly, this is an elegant way of handling the turns, but not the best. In fact, in this case (and only in this case), the actual roll angle corresponds exactly to the theoretical roll angle φ that was previously calculated.

If the rider leans towards the exterior of the turn (Fig. 4-3b), the center of gravity is also moved to the exterior of the turn with respect to the motorcycle. As a result, he needs to incline the motorcycle further so that the tires, being more inclined than necessary, operate under less favorable conditions. Certainly this rider is not an expert.

If the rider leans his torso towards the interior of the turn and at the same time rotates his leg so as to nearly touch the ground with his knee, he manages to reduce the roll angle of the motorcycle plane (Fig. 4-3c).

When racing, the riders move their entire bodies to the interior of the turn, both to reduce the roll angle of the motorcycle and to better control the vehicle on the turn. The displacement of the motorcycle-rider system's center of gravity towards the interior is carried out both by moving the leg and by the movement of the body in the saddle (Fig. 4-3d). The displacement of the body towards the interior and in particular, the rotation of the leg cause an aerodynamic yawing moment that facilitates entering and rounding the turn.

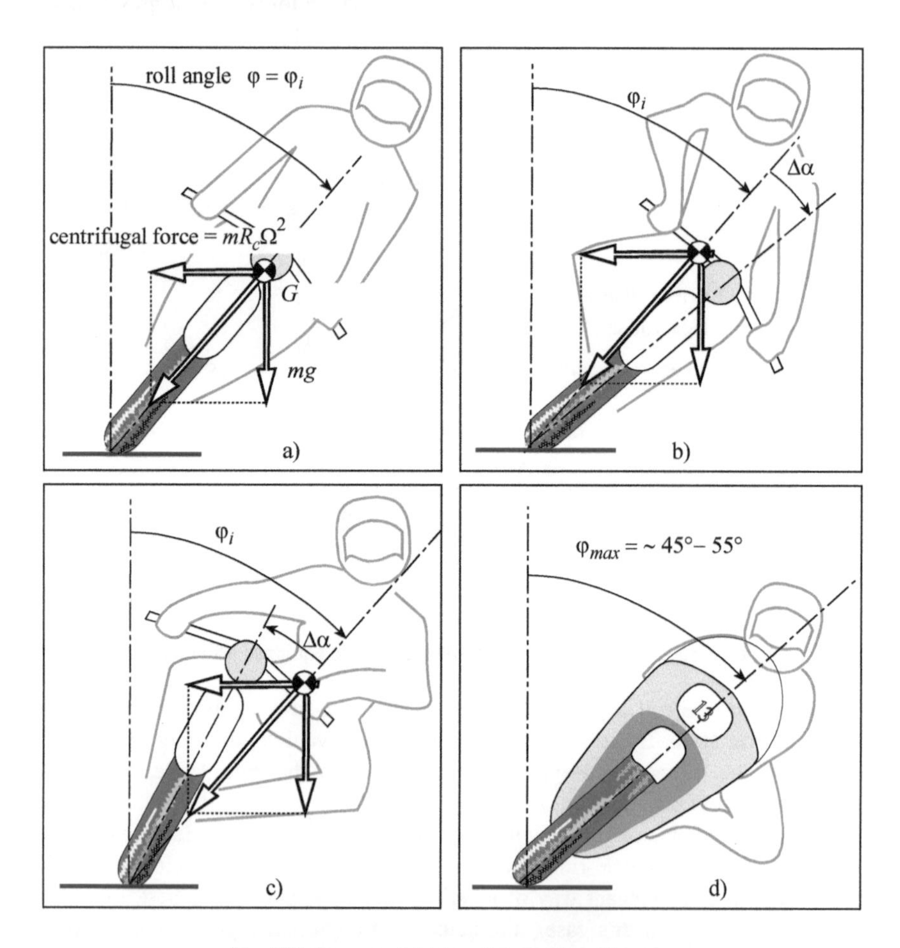

Fig. 4-3 Influence of driving style on the roll angle.

4.1.3 Wheel velocity in a turn

The velocity of the vehicle is represented by the forward velocity of the contact point of the rear wheel. Therefore, the yaw velocity Ω is:

$$\Omega = \frac{V}{R_{C_r}}$$

If we suppose no longitudinal slippage between the tires and the road surface (in the forward direction of the wheels), the spin velocity of the wheels, in terms of the vehicle forward velocity, roll angle and kinematic steering angle, is then:

$$\omega_r = -\frac{V}{\rho_r + t_r \cos\varphi} \qquad \omega_f = -\frac{V}{(\rho_f + t_f \cos\beta)\cos\Delta}$$

In reality, it must be observed that during the thrust and braking phases there is always a longitudinal slippage between the rear wheel and the road plane. In the front wheel there is longitudinal slippage in the braking phase, while under steady state conditions the slippage is negligible because it is only due to rolling resistance.

It is important to note that, with the same forward velocity, the angular velocity of the wheels increases during turning with respect to the angular velocity of the wheels in straight running, because contact does not occur on the largest circumference of the wheels.

4.2 Directional behavior of the motorcycle in a turn

Let us now consider a motorcycle in a steady turning condition. If each wheel advances ideally with a pure rolling motion, the velocity vector of the wheel's center would be contained in the plane of the wheel.

The lateral slip is expressed by the sideslip angle λ, defined as the angle formed by the direction of forward motion and the plane of the wheel. When sideslip angles approach zero, steering is called kinematic steering.

The lateral reaction forces depend on the sideslip angles of the tires, roll angle and vertical loads. The forces can be expressed by the following linear expressions, when slip and roll angles are small:

$$F_{s_f} = (k_{\lambda_f}\lambda_f + k_{\varphi_f}\varphi_f)N_f$$
$$F_{s_r} = (k_{\lambda_r}\lambda_r + k_{\varphi_r}\varphi_r)N_r$$

The constant k (expressed in radians^{-1}) represent the stiffness coefficients of the tires:

k_φ = camber stiffness coefficient;

k_λ = cornering stiffness coefficient.

The larger the sideslip and camber stiffnesses are, the smaller the sideslip angle necessary to generate the lateral force on the tire is.

4.2.1 Effective steering angle and path radius

The effective steering angle of a motorcycle (Fig. 4-4) also depends on sideslip angles; its value is given by the equation:

$$\Delta^* = \Delta + \lambda_r - \lambda_f \cong \frac{\cos\varepsilon}{\cos\varphi}\delta + \lambda_r - \lambda_f$$

where Δ indicates the kinematic steering angle that depends on the steering angle δ, caster angle ε and roll angle φ.

The turning radius of the trajectory described by the rear wheel is also a function of the sideslip angles and of the kinematic steering angle:

$$R_{C_r} = \frac{p}{\tan(\Delta - \lambda_f)\cos\lambda_r + \sin\lambda_r}$$

If the sideslip angles and the kinematic steering angle are small, the radius can be calculated with the following approximate formula:

$$R_{C_r} \cong \frac{p}{\Delta + (\lambda_r - \lambda_f)} = \frac{p}{\Delta^*}$$

where p indicates the motorcycle wheelbase (Fig.4-4).

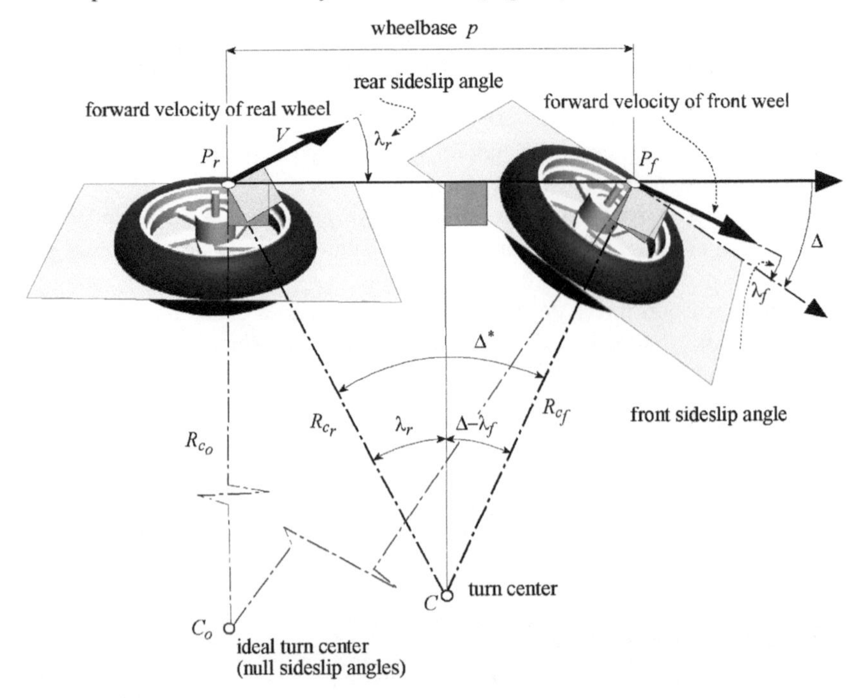

Fig. 4-4 Path radius, steering angle and sideslip angles.

4.2.2 Steering ratio

The motorcycle steering behavior depends on various geometric parameters (wheelbase, offset, caster angle, wheel radii and cross section radii), on the mass distribution and tire properties. Tire properties, in particular, are very important because the effective steering angle depends on the difference between the side slip angles.

The effective steering angle Δ^* is only equal to the kinematic steering angle Δ chosen by the rider if the slip angles of both wheels are equal. In this case the steering system has "neutral" behavior. Otherwise, the effective steering angle is smaller or larger than what is expected by the rider and the vehicle has under or over-steering behavior. It is worth pointing out that wheel's steering angle Δ may be smaller than the sideslip angles when the steady turning radius is large and the speed high.

The steering behavior can be expressed by means of the steering ratio ξ:

$$\xi = \frac{R_{C_o}}{R_{C_r}} = \frac{\Delta^*}{\Delta} \approx 1 + \frac{\lambda_r - \lambda_f}{\Delta}$$

where R_{C_o} is the kinematic radius of curvature.

The vehicle's behavior is:

- neutral if $\xi = 1$: the sideslip angles are equal ($\lambda_f = \lambda_r$);
- over-steering if $\xi > 1$: the difference of the sideslip angles is positive ($\lambda_r > \lambda_f$);
- under-steering if $\xi < 1$: the difference is negative ($\lambda_r < \lambda_f$).

Neutral behavior

Figure 4-5 shows a vehicle in a turn, in the special case where the sideslip angles of the two wheels are equal ($\lambda_f = \lambda_r$). If the sideslip angles were zero, the turn center (point C_o) would be determined by the intersection of the lines perpendicular to the planes of the wheels and passing through the contact points. Since the sideslip angles of the two tires are equal, the effective steering angle Δ remains constant as the values of the sideslip angles vary, while the center of rotation C moves towards the front and along a path passing through the point C_o. The radius of curvature remains approximately constant and equal to that relating to kinematic steering. This behavior is defined as neutral, since the curvature radius depends only on the steering angle selected by the rider and not on the value of the sideslip angles.

Under-steering

If the sideslip angle of the rear tire is less than the front tire ($\lambda_r < \lambda_f$) the center of curvature C is displaced to the exterior of the path of the neutral center (Fig. 4-6a). Here the radius of curvature R_{C_r} is greater than the ideal one R_{C_o} associated with the kinematic steering. The vehicle's behavior is therefore defined as under-steering.

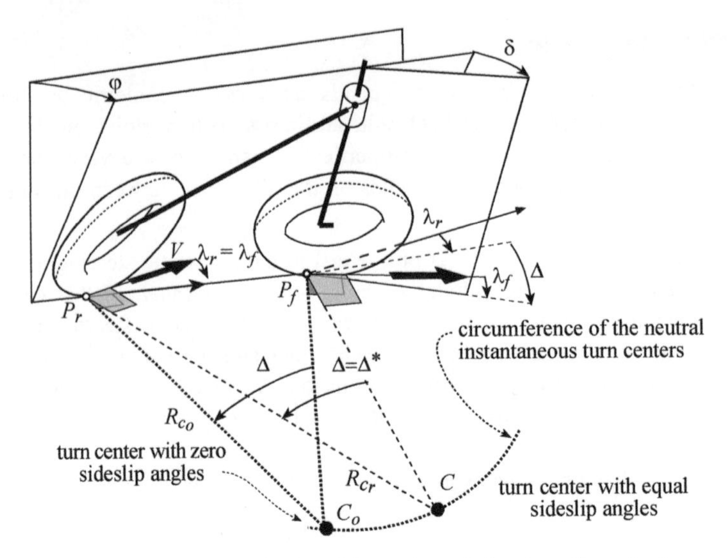

Fig. 4-5 Neutral behavior of the motorcycle in a turn ($\lambda_r = \lambda_f$).

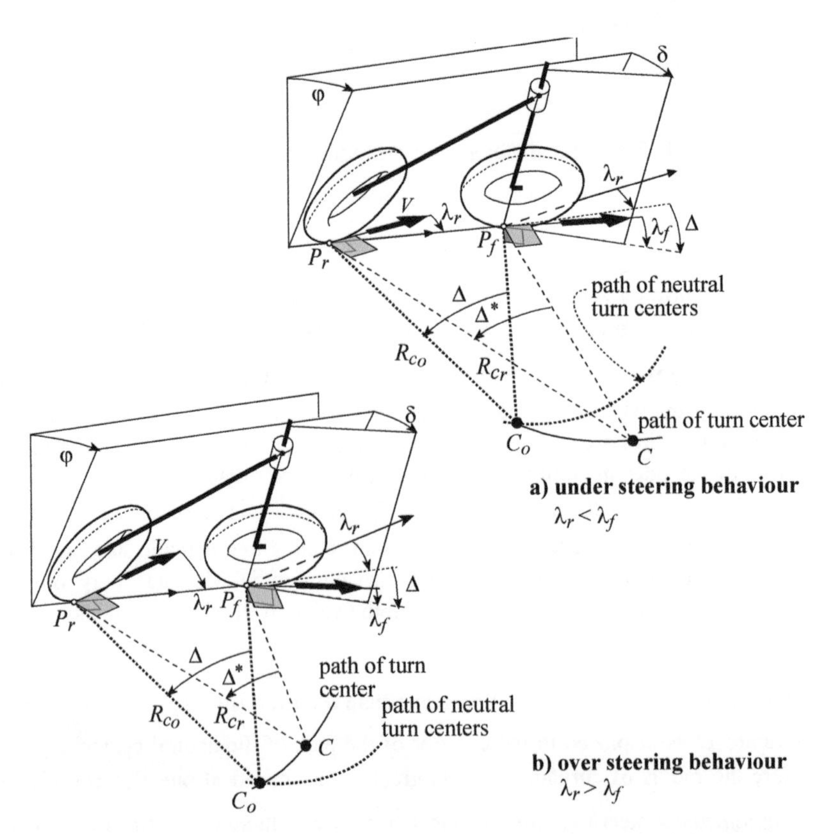

Fig. 4-6 Under-steering and over-steering behavior of the motorcycle in a turn.

Over-steering

If the sideslip angle of the rear tire is greater than that of the front tire ($\lambda_r > \lambda_f$), the center of curvature C is inside the path of the neutral center, so that the curvature radius R_{C_r} is less than the ideal radius R_{C_o} associated with the kinematic steering (Fig. 4-6 b). The vehicle in this case has an over-steering behavior.

Now consider a motorcycle that is under-steering while it is rounding a turn. Since the vehicle tends to expand the turn, in order to correct the trajectory the rider is obliged to increase the steering angle so that the lateral reaction force of the front wheel will be increased.

When the rotation of the handlebars becomes considerable, the force needed for equilibrium can overcome the maximum friction force between the front tire and the road plane, with the result that the wheel slips and the rider falls.

A motorcycle that is under-steering is therefore dangerous, since the rider cannot control the vehicle once the front wheel has lost adherence.

On the other hand, with an over-steering motorcycle, if the force needed for equilibrium overcomes the maximum friction force between the tire and the road plane, the rear wheel slips, but the expert rider, with a counter steering maneuver, has a better chance of controlling the vehicle equilibrium and avoiding a fall.

4.3 Cornering forces

Figure 4-7 shows the motorcycle in kinematic turning without driving force, rotating about the ideal turn center C_o. Rolling resistance and aerodynamic forces are neglected. The tire lateral forces are perpendicular to the wheels and their resultant is equal to the desired centripetal force directed towards the turn center point.

Figure 4-8 shows a motorcycle with sideslip angles not equal to zero and with the driving force required to give steady tangential velocity. Front rolling resistance is considered whereas aerodynamic forces are neglected. The turn center C is the intersection of the lines perpendicular to the directions of the forward velocity of the two wheels. For equilibrium the intersection of the total front and rear forces must intersect the line GC. The resultant of the two tire forces gives the centripetal component directed towards the turn center C.

If the aerodynamic force is included the driving force necessary for the equilibrium will increase.

The aerodynamic force also influences the vertical loads on the wheels:

$$N_f = mg\frac{b}{p} - F_A\frac{h}{p}\cos\varphi \qquad\qquad N_r = mg\frac{p-b}{p} + F_A\frac{h}{p}\cos\varphi$$

The front vertical load is equal to the static load less the load transfer due to the aerodynamic force. Alternatively, the rear load is equal to the sum of the static load and load transfer.

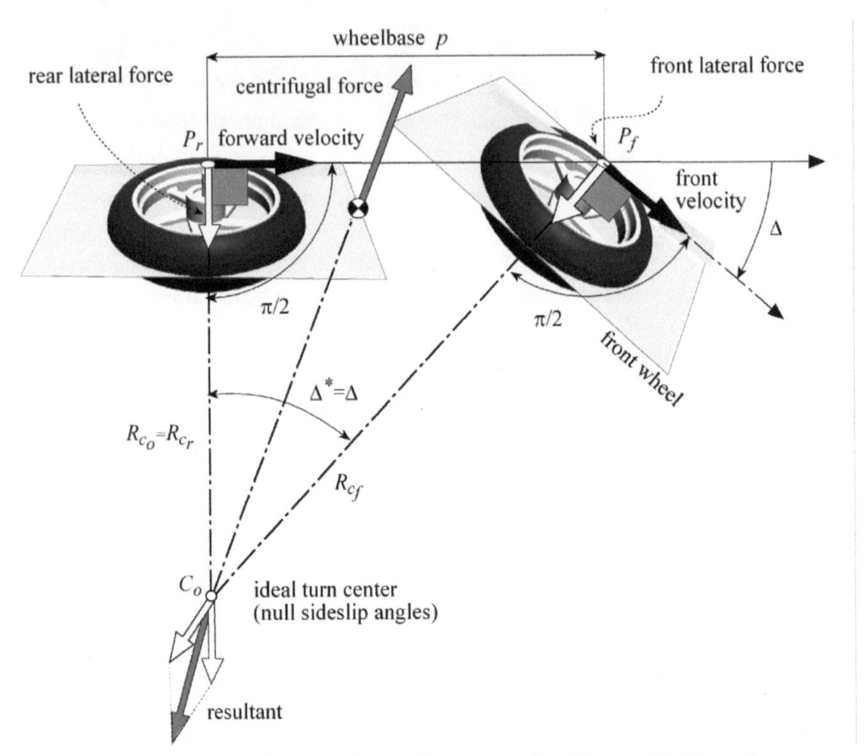

Fig. 4-7 Plan view: Forces acting on the motorcycle with zero sideslip angles.

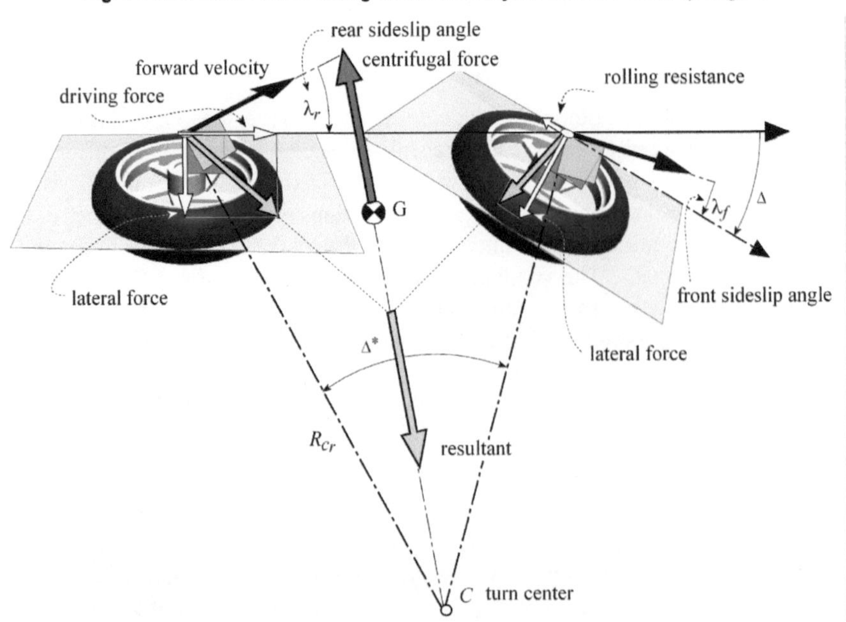

Fig. 4-8 Plan view: Forces acting on the motorcycle with non-zero sideslip angles.

The vertical load on the front wheel increases slightly in the passage from rectilinear motion to the turn, while it decreases on the rear wheel, due to the dependence of the load transfer on the roll angle.

It is worth highlighting that, even if rolling resistance is neglected, a driving force is necessary for equilibrium in steady turning when sideslip angles are present.

4.4 Linearized model of the motorcycle in a turn

Now consider a motorcycle rounding a turn with large radius, with respect to the motorcycle's wheelbase, and suppose that the aerodynamic force is negligible so that the load transfer is negligible with respect to the vertical loads on the wheels.

$$N_f \cong mg\frac{b}{p} \qquad N_r \cong mg\frac{p-b}{p}$$

If we consider small roll, steering and sideslip angles the lateral forces acting on the wheels are equal to (see Fig.4-7):

$$F_{s_f} = \frac{b}{p\cos\Delta} m \frac{V^2}{R_{c_r}} = \frac{N_f}{g\cos\Delta} \frac{V^2}{R_{c_r}} \qquad F_{s_r} = \frac{p-b}{p} m \frac{V^2}{R_{c_r}} = \frac{N_r}{g} \frac{V^2}{R_{c_r}}$$

The lateral forces depend on the distribution of the static loads on the two wheels, while the front lateral force also depends on the effective steering angle. The ratios between the lateral forces acting on the wheels, and the vertical loads are equal to:

$$\frac{F_{s_f}}{N_f} = \frac{V^2}{gR_{c_r}\cos\Delta} \cong \frac{V^2}{gR_{c_r}} \approx \varphi \qquad \frac{F_{s_r}}{N_r} = \frac{V^2}{gR_{c_r}} \approx \varphi$$

Keeping in mind the expressions for the lateral forces in terms of the sideslip and roll angles, the sideslip angles can be written in the following way:

$$\lambda_f = \frac{1}{k_{\lambda_f}}\left(\frac{F_{s_f}}{N_f} - k_{\varphi_f}\varphi_f\right) \qquad \lambda_r = \frac{1}{k_{\lambda_r}}\left(\frac{F_{s_r}}{N_r} - k_{\varphi_r}\varphi_r\right)$$

It can be noted that the sideslip angles depend on two terms of opposite sign. The negative term is proportional to the camber stiffness and its increase brings about a reduction in the sideslip angle that can also become negative, as we have seen in the chapter on tires. The first, remaining term depends on the lateral force (numerator) and the vertical load (denominator).

The sideslip angles can therefore be expressed in the form:

$$\lambda_f \approx \frac{1-k_{\varphi_f}}{k_{\lambda_f}}\varphi \qquad \lambda_r \approx \frac{1-k_{\varphi_r}}{k_{\lambda_r}}\varphi$$

It can be noted that the sideslip angles are inversely proportional to the cornering stiffness and that they depend on both the camber stiffness and the roll angle (and therefore on the forward velocity and the radius of curvature). Observe that if the camber stiffness coefficient is greater than one the sideslip angle is negative.

Now let us consider the steering ratio ξ that characterizes the directional behavior of the vehicle:

$$\xi = \frac{R_{c_o}}{R_{c_r}} = \frac{\Delta^*}{\Delta} \approx 1 + \frac{\lambda_r - \lambda_f}{\Delta}$$

Taking into account the sideslip expressions we obtain:

$$\xi = \frac{1}{1 - \left(\frac{1-k_{\varphi_r}}{k_{\lambda_r}} - \frac{1-k_{\varphi_f}}{k_{\lambda_f}}\right)\frac{V^2}{g\,p}}$$

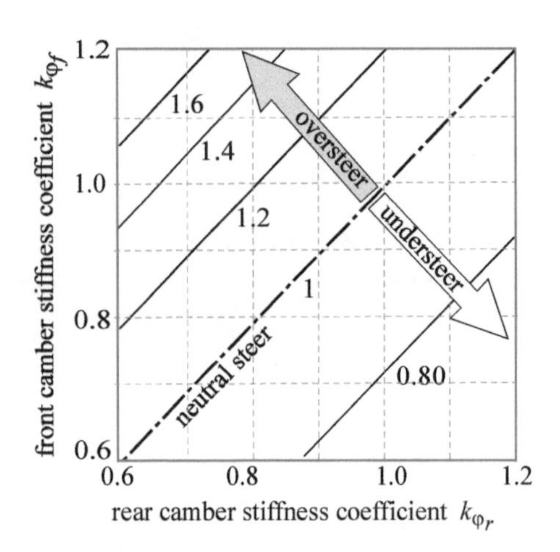

Fig. 4-9 Steering ratio ξ as the camber stiffness varies [V=12 m/s, $k_{\lambda_f} = k_{\lambda_r}$ =12 rad^{-1}].

The directional behavior is **neutral** if the following relation is satisfied:

$$\xi = 1 \quad \Rightarrow \quad R_{c_r} = R_{c_o} \quad \Rightarrow \quad \frac{1-k_{\varphi_f}}{k_{\lambda_f}} = \frac{1-k_{\varphi_r}}{k_{\lambda_r}}$$

In this case, the vehicle's response at any velocity coincides with that present under ideal conditions of kinematic steering.

The directional behavior is **over-steering** if the ratio ξ is greater than one:

$$\xi > 1 \quad \Rightarrow \quad R_{c_r} < R_{c_o} \quad \Rightarrow \quad \frac{1-k_{\varphi_r}}{k_{\lambda_r}} > \frac{1-k_{\varphi_f}}{k_{\lambda_f}}$$

In this case, the curvature radius diminishes with an increase in velocity.

The directional behavior is **under-steering** if the ratio ξ is less than one:

$$\xi < 1 \quad \Rightarrow \quad R_{c_r} > R_{c_o} \quad \Rightarrow \quad \frac{1-k_{\varphi_r}}{k_{\lambda_r}} < \frac{1-k_{\varphi_f}}{k_{\lambda_f}}$$

In this case, as the velocity gradually increases, the radius of curvature also increases and thereby, increasingly greater steering angles are required to go through the same trajectory.

Figure 4-9 shows the steering ratio under variation in the camber stiffnesses, in the case in which the cornering stiffnesses are equal. The graph shows that for neutral behavior the two tires must have equal camber stiffness coefficient values.

Example 1

Consider a motorcycle with a wheelbase $p = 1.4$ m, equipped with different tires. The tires have the following three combinations of the cornering stiffness coefficient while the camber stiffness coefficient is assumed to be constant $k_{\varphi_f} = k_{\varphi_r} = 0.8$.

Motorcycle 1:

$k_{\lambda_f} = 12$ rad $^{-1}$, $k_{\lambda_r} = 12$ rad $^{-1}$, $k_{\varphi_f} = 0.80$ rad $^{-1}$, $k_{\varphi_r} = 0.80$ rad $^{-1}$.

Motorcycle 2:

$k_{\lambda_f} = 14$ rad $^{-1}$, $k_{\lambda_r} = 10$ rad $^{-1}$, $k_{\varphi_f} = 0.80$ rad $^{-1}$, $k_{\varphi_r} = 0.80$ rad $^{-1}$.

Motorcycle 3:

$k_{\lambda_f} = 10$ rad $^{-1}$, $k_{\lambda_r} = 14$ rad $^{-1}$, $k_{\varphi_f} = 0.80$ rad $^{-1}$, $k_{\varphi_r} = 0.80$ rad $^{-1}$.

Figure 4-10 shows the steering ratio versus the forward velocity. It can be observed that the behavior of the vehicle is:

- over-steering when the front cornering stiffness coefficient is greater than that of the rear tire ($k_{\lambda_f} > k_{\lambda_r}$);
- neutral when the stiffnesses are equal ($k_{\lambda_f} = k_{\lambda_r}$);
- under-steering when the rear cornering stiffness coefficient is greater than that of the front tire ($k_{\lambda_f} < k_{\lambda_r}$).

This is due to the fact that the sideslip angle of a tire is greater to the extent that its stiffness is less; with high values of the cornering stiffness, the sideslip angles could be zero (kinematic steering).

It is interesting to observe how, on the basis of this linear model (expressed in terms of stiffness coefficients), neither the steering angle nor the longitudinal position of the center of gravity influence the vehicle's directional behavior.

Actually, as the sideslip angle increases, the lateral forces increase at an increasingly lower rate than that predicted by the linear law forming the basis of the tire model.

Furthermore, in reality the directional behavior of the motorcycle is also influenced by the longitudinal position of the center of gravity and by the value of the driving force.

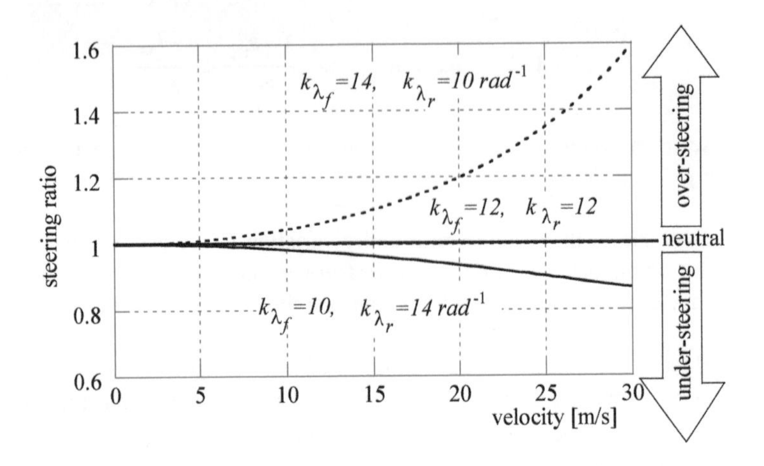

Fig. 4-10 Steering ratio ξ versus the velocity [$k_{\varphi_f} = k_{\varphi_r} = 0.8$].

4.4.1 Critical velocity

The previous example has shown that the under/over-steer behavior in a turn depends mainly on the camber and cornering stiffnesses. Their influence is important especially if the stiffness values of the front and rear tires are different.

Consider again a motorcycle that rounds a turn with a large radius with respect to its wheelbase.

When the behavior of the motorcycle is over-steering, the steering ratio ξ approaches ∞ at a certain value of the velocity, called critical velocity:

$$V_c = \sqrt{\frac{g\,p}{\dfrac{1-k_{\varphi_r}}{k_{\lambda_r}} - \dfrac{1-k_{\varphi_f}}{k_{\lambda_f}}}} \qquad \Rightarrow \quad \xi \rightarrow \infty$$

The control of the motorcycle over the critical velocity is possible by performing counter-steering maneuvers. This strategy is adopted by riders in speedway and motard racing.

Example 2

Consider a motorcycle with a wheelbase $p = 1.4$ m, characterized by the following tire properties: $k_{\lambda_f} = 12$ rad^{-1}, $k_{\lambda_r} = 14$ rad^{-1}, $k_{\varphi_f} = 0.9$ rad^{-1}, $k_{\varphi_r} = 0.85$ rad^{-1}.

The motorcycle is over-steering because the steering ratio is greater than one. The critical velocity is equal to 50.6 m/s.

Figure 4-11 gives the progress of the ratio ξ as a function of the velocity. It can be observed here that the value of ξ increases rapidly. This means that at high velocities even small values of the steering angle suffice to turn the vehicle.

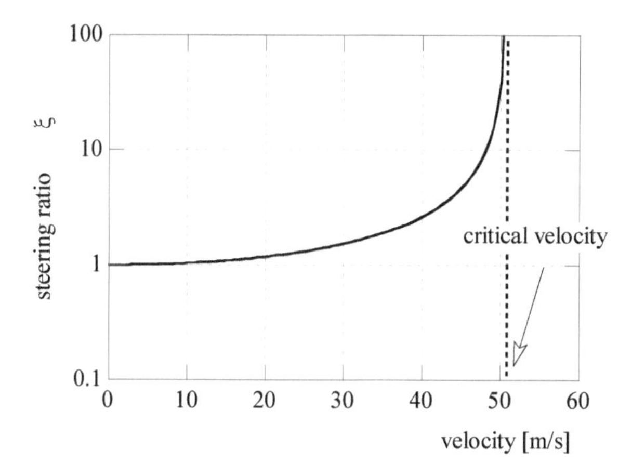

Fig. 4-11 Over-steering motorcycle and critical velocity.

Figure 4-12 illustrates the variation of critical velocity in terms of the cornering stiffness coefficient of the tires. It can be noted in the left plot that if $k_{\varphi_f} = k_{\varphi_r}$, in the field of values ($k_{\lambda_f} < k_{\lambda_r}$) critical velocity does not exist; the equation shows that critical velocity is imaginary.

The right plot refers to a motorcycle with camber stiffness coefficient in the front tire greater than that of the rear tire. In this case the vehicle is always in over-steering. Critical velocity increases by increasing the rear tire cornering stiffness and decreasing the front tire cornering stiffness.

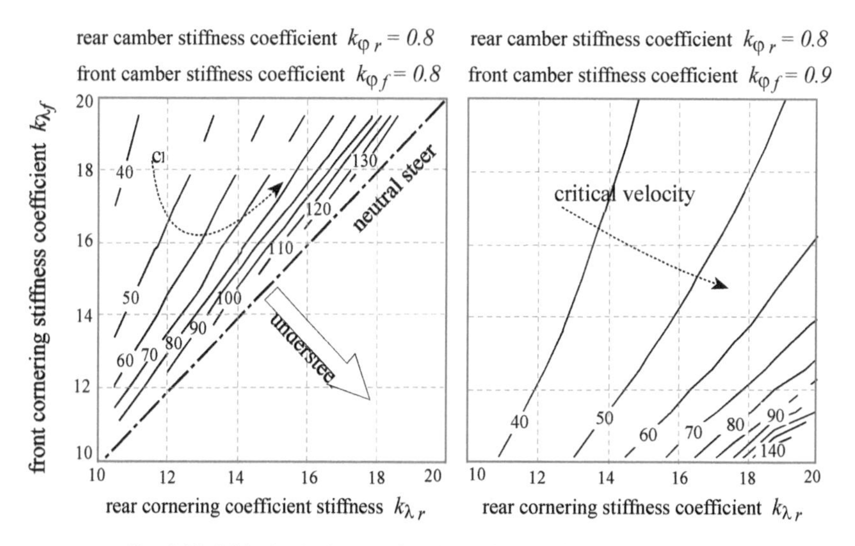

Fig. 4-12 Critical velocity as a function of the stiffnesses of the tires.

4.5 Multi-body model of motorcycles in steady turning

A motorcycle can be described as a system of six rigid bodies: sprung steering components, unsprung steering components, rear frame (including frame, engine, tank and driver), rear swinging arm and the two wheels.

The driver is considered to be a rigid body firmly attached to the frame. The following Fig. 4.13 shows a sketch of a motorcycle in steady turning motion.

The speed of travel V is the speed of the contact point of the rear wheel. When no slip is present, V is directly proportional to the angular velocity of the rear wheel and is directed along the wheel symmetry plane.

The distributed aerodynamic forces which air exerts on the motorcycle are taken into account by considering drag, lift and lateral forces acting at the center of mass of the rear frame $\left(F_D, F_L, F_S\right)$ and three aerodynamic torques $(M_{A_x}, M_{A_y}, M_{A_z})$.

The interaction between each tire and the road is represented by three forces (vertical load, longitudinal and lateral forces) acting at the geometric contact point and by three torques (overturning, rolling resistance and yaw torque), acting along the three independent axes. The tire forces and torques are non-linearly dependent on the roll angle and slip quantities.

4.5.1 Mathematical model of motorcycle

The equations of motion of the motorcycle in steady turning motion are described in the paper [Cossalter et al., 1999].

The equilibrium conditions give six equations:

- three force equilibrium equations;
- three moment equilibrium equations: equilibrium around the X-axis (roll), the Y-axis (pitch) and the Z-axis (yaw).

In addition we have two equations that give the lateral forces as functions of sideslip and camber angles.

Once the roll angle φ and the steering angle δ are assigned, the eight equations allow us to obtain the eight unknowns:

- forward velocity V;
- vertical forces N_f and N_r applied respectively to the front and rear wheels;
- lateral forces F_{s_f} and F_{s_r} applied respectively to the front and rear wheels;
- sideslip angles λ_f, λ_r,
- the driving force S.

Finally, the equilibrium of the front and/or the rear frame around the steering axis gives the torque exerted by the rider and applied to the handlebars, which provides and equal and opposite reaction on the rear frame.

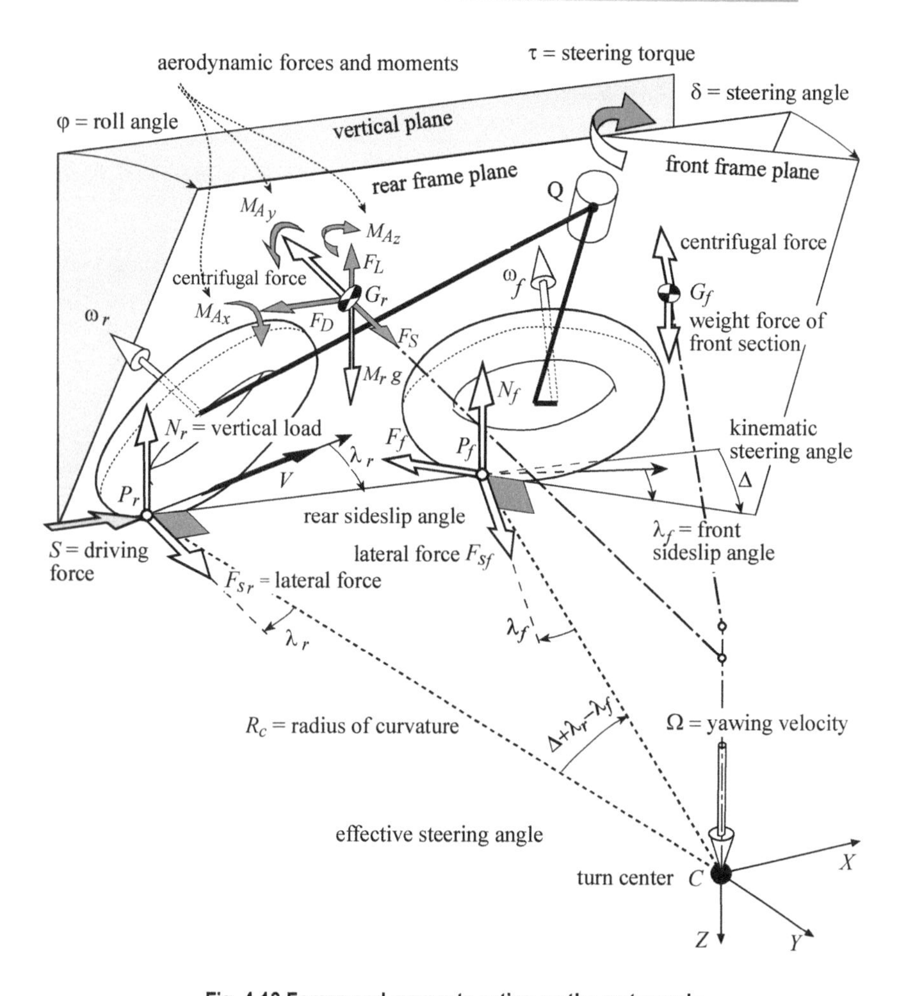

Fig. 4-13 Forces and moments acting on the motorcycle.

The inertial and geometrical properties are defined with respect to the coordinate systems represented in Fig. 4-14.

Let us examine the rear frame:

- it has mass M_r;
- it is characterized by the center of gravity G_r having coordinates $(b_r, 0, -h_r)$ with respect to the rear coordinate system (A_r, X_r, Y_r, Z_r);
- it is considered symmetrical with respect to the $x-z$ plane, hence its inertial characteristics are represented by the following four terms:

– I_{x_r} = mass center moment of inertia about x_r axis (roll moment of inertia);

– I_{y_r} = mass center moment of inertia about y_r axis (pitch moment of inertia);

– I_{z_r} = mass center moment of inertia about z_r axis (yaw moment of inertia);

- I_{xz_r} = mass center inertia product about x_r - z_r -axes.

Let us examine the front frame:

- it has mass M_f;
- it is characterized by the center of gravity G_f having coordinates $(b_f, 0, -h_f)$ with respect to the front coordinate system (A_f, X_f, Y_f, Z_f);

The coordinate system axes (A_f, X_f, Y_f, Z_f) are assumed to be principal axes of inertia so that the inertia tensor is diagonal.

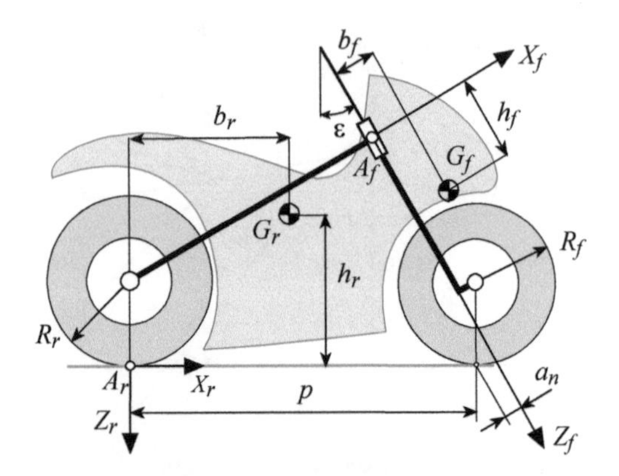

Fig. 4-14 Sketch of the motorcycle.

4.5.2 Simplified model of motorcycles

Ignoring the small displacements of the wheels' contact points (with respect to the radius of curvature) which, as we have seen in the chapter on kinematics, depend on the angles of pitch, roll and steering as well as the geometry of the wheels, the six equations of equilibrium for the motorcycle in a turn can be easily derived (Fig- 4-15):

(⇒) equilibrium of the forces along the X axis:

$$S - F_{s_f} \sin\Delta + m \cdot X_G \Omega^2 - F_A = 0$$

(⇒) equilibrium of the forces along the Y axis:

$$F_{s_f} \cos\Delta + F_{s_r} + m \cdot Y_G \, \Omega^2 = 0$$

(⇑) equilibrium of the forces along the Z axis:

$$-N_f - N_r + mg = 0$$

equilibrium of the moments:

(∩) around the X axis:

$$-I_{YZ} \Omega^2 - (N_f + N_r) Y_r + mg \cdot Y_G + (I_{w_f} \omega_f + I_{w_r} \omega_r) \Omega \cos\varphi = 0$$

(∩) around the Y axis:

$$-I_{XZ}\Omega^2 - N_f(p+X_r) + mg\cdot X_G + F_A\cdot Z_G - N_r X_r = 0$$

($\cap$) around the Z axis:

$$(p+X_r)F_{s_f}\cos\Delta + Y_r F_{s_f}\sin\Delta + F_{s_r}X_r - S\,Y_r + F_A Y_G = 0$$

The meaning of the symbols as the follows:

- S the thrust which is necessary for holding the motorcycle stationary in a turn;
- F_A the aerodynamic resistant force assumed to be applied to the center of gravity;
- F_{s_f}, F_{s_r} the lateral forces applied to the tires by the road;
- N_f, N_r the vertical loads;
- I_{w_f}, I_{w_r} spin moments of inertia of the wheels;
- ω_f, ω_r angular velocities of the wheels;
- Ω yaw velocity;
- Δ kinematic steering angle measured on the road plane.
- X_G, Y_G, Z_G coordinates of the motorcycle center of gravity with respect to the reference system (C,X,Y,Z):

$$X_G = b - R_{c_r}\sin\lambda_r$$
$$Y_G = h\sin\varphi - R_{c_r}\cos\lambda_r$$
$$Z_G = -h\cos\varphi$$

X_r, Y_r coordinates of the contact point of the rear wheel with respect to the reference system (C,X,Y,Z);

$$X_r = -R_{c_r}\sin\lambda_r$$
$$Y_r = -R_{c_r}\cos\lambda_r$$

I_{XZ}, I_{YZ} products of inertia of the motorcycle with respect to the axes $X-Z$ and $Y-Z$. These products of inertia depend on the mass center moments of the motorcycle, $I_{z_G}, I_{y_G}, I_{xz_G}$, mass m, roll angle φ, and on the coordinates X_G, Y_G, Z_G of the motorcycle mass center:

$$I_{YZ} = mY_G Z_G + (I_{z_G} - I_{y_G})\cos\varphi\sin\varphi$$
$$I_{XZ} = mX_G Z_G + I_{xz_G}\cos\varphi$$

The six equations constitute a non-linear system. Expressing the roll angle φ as a function of the yaw velocity Ω and of the radius R_{C_r}, and expressing the lateral forces of the tires as linear functions of the sideslip angles λ_f, λ_r and roll angle φ, we can calculate the six unknowns.

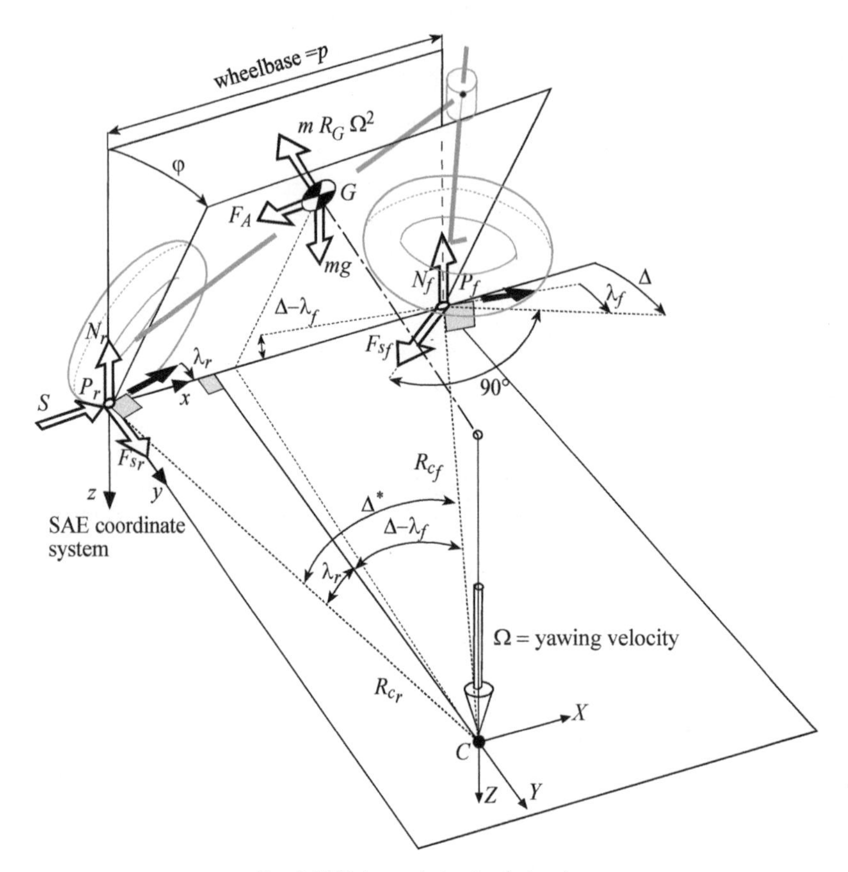

Fig. 4-15 Motorcycle in steady turning.

For example setting the steering angle δ and the yaw velocity Ω the six unknowns are:

- the sideslip angles λ_f, λ_r;
- the radius R_{C_r}:
- the vertical loads N_f, N_r;
- the thrust S necessary for assuring motion at a constant velocity.

If the sideslip angles λ_f, λ_r, the roll angle φ and the effective steering angle Δ are known, it is possible to calculate the radius of the circular trajectory covered by the rear wheel R_{C_r}.

4.6 Roll, steering and sideslip angles

We will show how a motorcycle in a turn, with assigned forward velocity, roll and steering angles, describes a trajectory whose path radius depends on the sideslip angles of the tires. If the sideslip angles are zero, the trajectory coincides with the kinematic one.

The conditions of stationary equilibrium of a motorcycle in a turn, in the curvature-forward velocity diagram, are represented by means of the contour lines of the steering angle δ and the roll angle φ. These curves provide the necessary values for equilibrium on a turn, once the radius of the trajectory and the velocity of the motorcycle have been set (Fig 4-16).

The roll angle contour lines show how the velocity and the turning radius have to vary to assure vehicle equilibrium, maintaining the roll angle constant. The steering angle needed under various stationary equilibrium conditions is represented by the intersection of the roll curve with the steering curve.

In the same way we can consider a motorcycle that rounds a turn holding the steering angle constant. The steering contour lines show how the forward speed and the radius of the turn need to vary in order to assure the vehicle's equilibrium. The roll angle necessary under various equilibrium conditions is represented by the intersection of the steering contour line with the roll line.

Three motorcycles having the same geometry and inertial properties but equipped with different tires are considered. The different behavior of the three motorcycles depends on the various sideslip angles of the tires required to generate the lateral forces that are necessary for equilibrium. It should be noted that the radius of the turn depends on the difference in the sideslip angles as well as on the steering angle.

4.6.1 Case 1: reference motorcycle

Roll and steering angles

Figure 4-16 represents the case of a motorcycle with tires having equal camber and cornering stiffness.

The horizontal straight lines represent the equilibrium conditions of a motorcycle rounding turns of increasing radius, at constant velocity. The graph shows how the roll and steering angles vary in terms of the curvature.

The vertical lines represent motorcycles rounding turns of constant radius with variable velocity. The graph shows how the roll and steering angles need to vary in terms of velocity. To round a turn with constant radius, the steering angle must diminish as velocity increases. This phenomenon derives from the fact that the effective steering angle also depends on the roll angle (see chapter 1).

It must be recalled that the range of steering angles used is normally much more restricted than that indicated in the figure, especially at high velocities.

We can see in the figure that even if $\varphi=0$ a steady state turning motion, at low velocity, is possible. If we consider a perfectly vertical motorcycle (roll angle φ = 0) with the handlebars turned to the right, we are led to suppose that it cannot attain an equilibrium velocity since we imagine that the centrifugal force generated as the turn is rounded to the right, tends to make the motorcycle fall outside the turn, i.e., to the left.

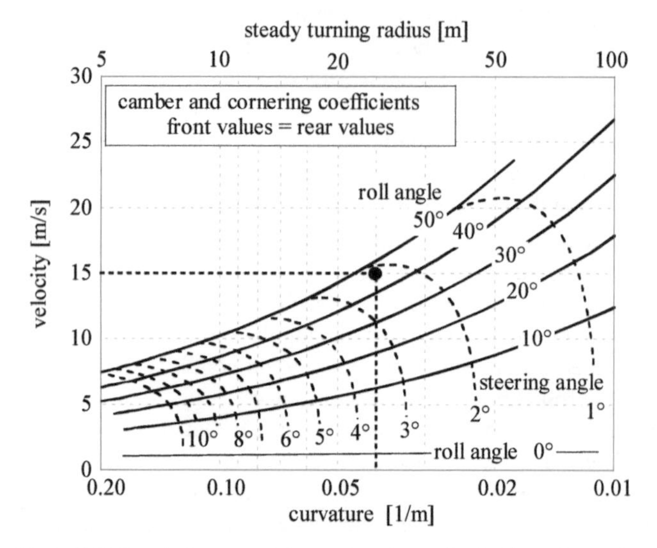

Fig. 4-16 Roll and steering angles as functions of velocity and curvature.

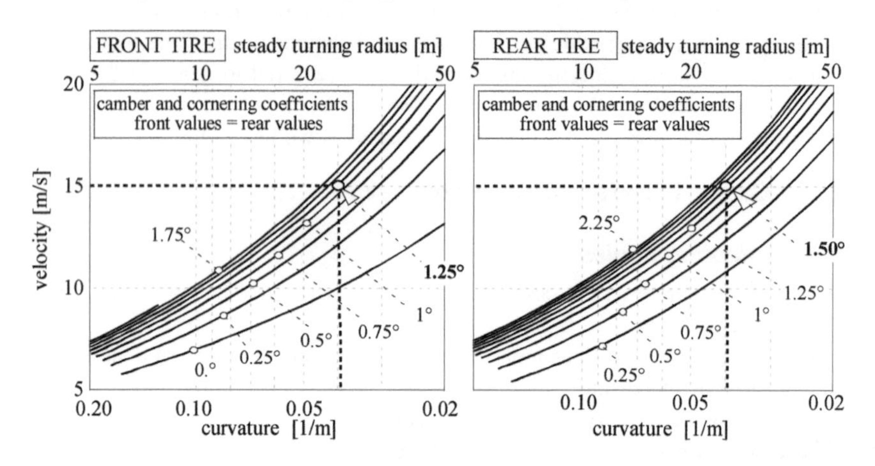

Fig. 4-17 Sideslip angles in terms of velocity and curvature. Reference motorcycle.

Actually, it must be recalled that the presence of the motorcycle trail causes a displacement to the left of the front wheel's contact point and that, furthermore, the roll angle of the front section is not zero, but increases with the steering angle even if the roll of the rear section is zero. It follows intuitively that both the center of gravity of the front section and that of the rear section are displaced to the right of the straight line joining the contact points. Non-zero equilibrium velocities are therefore possible, since the overturning moment due to the centrifugal force is balanced out by the moments generated by the weight forces.

Sideslip angles.

The sideslip angles required for a motorcycle with tires of equal stiffness are represented in Fig. 4-17. Note that the vehicle has nearly neutral behavior, with nearly equal front and rear sideslip angles ($\lambda_r - \lambda_f = 0.25°$ for a velocity of 15 m/s and turn radius of 30 m).

4.6.2 Case 2: front tire stiffness (+10%), rear tire stiffness (-10%)

Roll and steering angles

If the values of both the camber and cornering stiffnesses of the two tires are changed, the vehicle behaves differently in the turn. If the front tire has larger (+10%) stiffness values and the rear tire smaller (-10%) than the preceding case, the behavior will be different, as can be observed in Fig. 4-18.

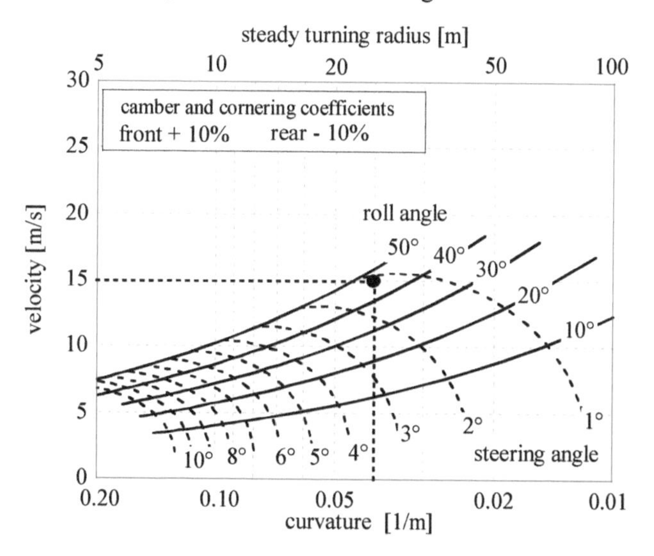

Fig. 4-18 Roll and steering angles as functions of velocity and curvature.

At the same velocity and curvature radius, the steering angle necessary for equilibrium on a turn is notably smaller (for example, at a velocity of about 15 m/s and with a curvature ratio of 30 m, the steering angle diminishes from approximately 2° to approximately 1°).

Furthermore, as in the previous case, to round a turn with constant radius, the steering angle must diminish as the velocity increases.

Sideslip angles.

With a front tire having higher camber and cornering stiffnesses, i.e., with a front tire performing more than the rear one, the sideslip angles change significantly. The front sideslip angle is very small or even negative, while the rear sideslip angle increases (Fig. 4-19). The difference between the sideslip angles has positive sign and becomes: $\lambda_r - \lambda_f = 1.9°$.

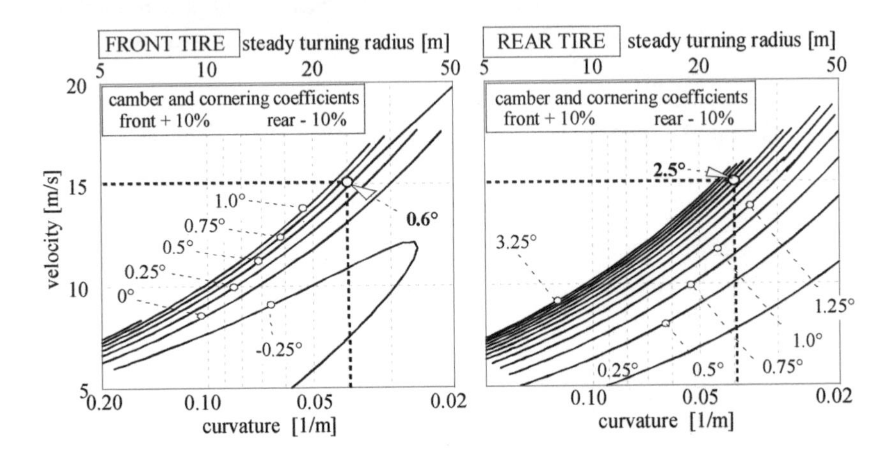

Fig. 4-19 Sideslip angles as functions of velocity and curvature. Over-steering motorcycle.

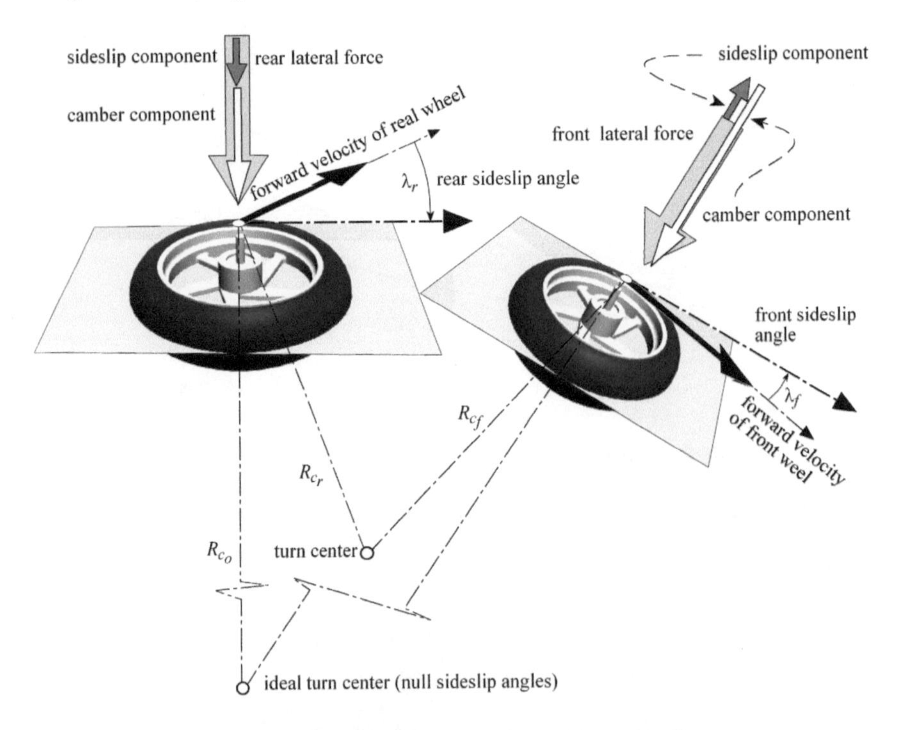

Fig 4-20 Motorcycle with negative front sideslip angle.

Fig. 4-20 shows the plan view of a motorcycle with a negative value of the front sideslip angle. In this case the front lateral force is the sum of a positive component due to the camber angle and a negative component due to the sideslip angle. Such a condition would exist below the $\lambda_f = 0$ curve for the front tire in Fig. 4-19.

4.6.3 Case 3: front tire stiffness (-10%), rear tire stiffness (+10%)

Roll and steering angles

Consider a third vehicle with larger stiffness values of the rear tire (+10%) and smaller stiffness values of the front tire (-10%). In this case (Fig. 4-21) the vehicle behaves very differently from the previous two cases: to round a turn of equal radius, greater steering angles are necessary. For example to round a turn of radius 30 m at a velocity of 15 m/s, requires a steering angle about 3° greater than that of the previous cases.

The greater difference between this case and the previous two is highlighted by the plot of the steering angle contour lines. Consider for example the vehicle while it rounds a turn of 50 m with increasing velocity. With an increase in velocity of up to approximately 15 m/s, the steering angle necessary for equilibrium must increase slightly, while, for greater velocities the steering angle has to diminish.

Sideslip angles.

In this case the rear tire has better characteristics than the front one, the front sideslip angle increases while the rear diminishes significantly (Fig. 4-22). The difference between the sideslip angles has a negative sign and becomes: $\lambda_r - \lambda_f = -1.8°$.

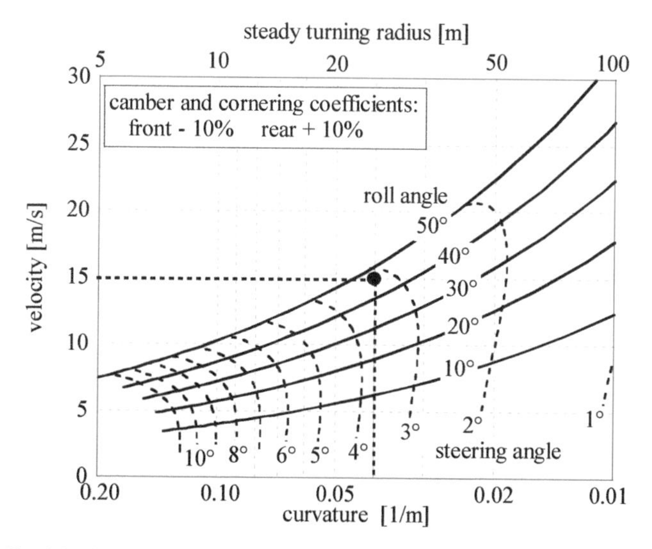

Fig. 4-21 Case 3: roll and steering angle in terms of velocity and curvature.

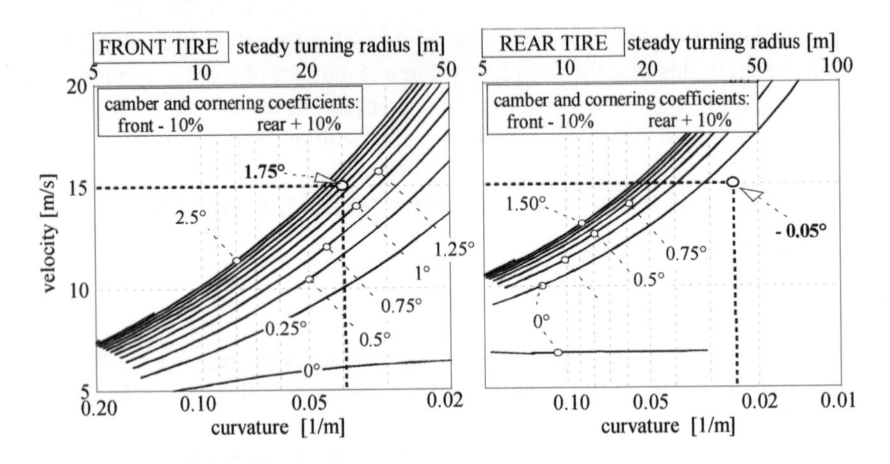

Fig. 4-22 Sideslip angles in terms of velocity and curvature. Under-steering motorcycle.

4.7 Steering ratio

Case 1.

Let us examine a motorcycle in steady turning equipped with front and rear tires which have the same cornering stiffness coefficient ($k_{\lambda_r} = k_{\lambda_f} = 15\ \text{rad}^{-1}$) and camber stiffness coefficient ($k_{\varphi_r} = k_{\varphi_f} = 0.8\ \text{rad}^{-1}$).

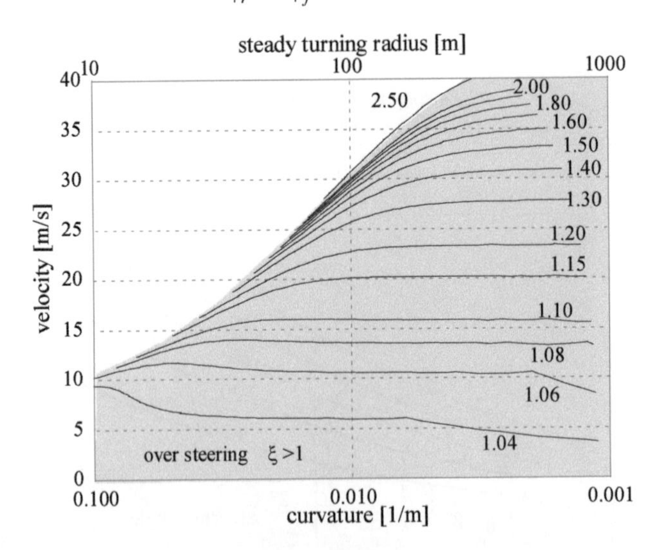

Fig. 4-23 Steering ratio: front and rear tires with cornering stiffness $k_{\lambda} = 15\ \text{rad}^{-1}$ and camber stiffness $k_{\varphi} = 0.8\ \text{rad}^{-1}$.

This case is not realistic because the front and rear tires usually have different properties, but it helps us understand the effect of tire properties on steering behavior. The contour plot of ξ, represented in Fig. 4-23, shows that the vehicle's behavior is almost "neutral" when the speed is very low and becomes over-steering when the speed increases.

The sideslip angle of the front wheel is smaller that of the rear wheel for at least two reasons. First of all, the driving force that acts on the rear tire necessitates a larger sideslip angle to generate the lateral force. This effect becomes more important when the speed increases because the aerodynamic force (and the driving force) increases. Secondly, front wheel camber angle β is larger than rear wheel camber angle φ, hence, the effect of camber angle on the lateral force is more significant on the front wheel.

Case 2.

In this case the motorcycle is equipped with the same front tire (k_{λ_f} = 15 rad^{-1} and k_{φ_f} = 0.8 rad^{-1}) but with the rear tire having different properties: the rear camber stiffness coefficient is increased (k_{φ_r} = 0.87 rad^{-1}) while the cornering stiffness coefficient is the same as the front tire.

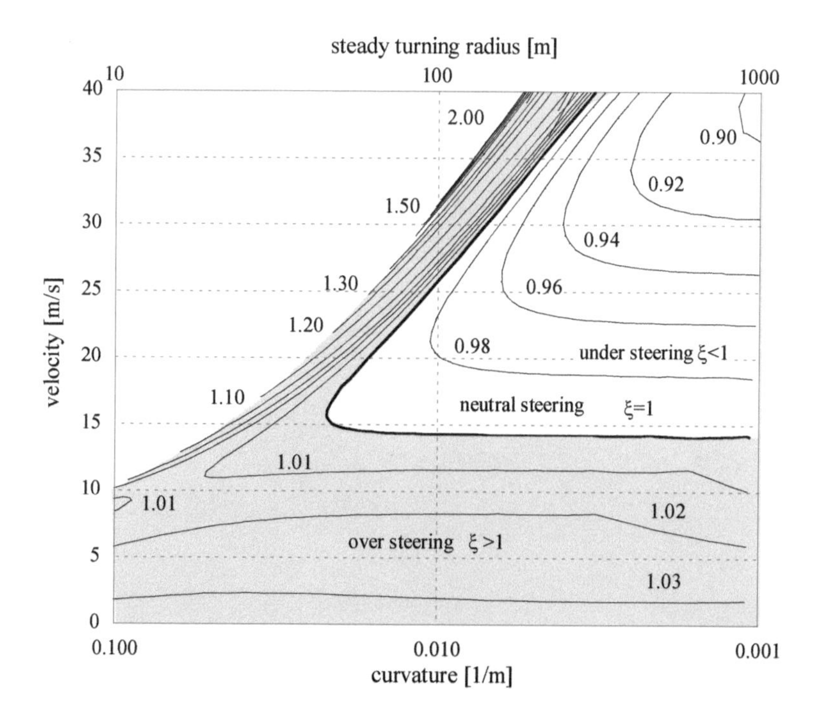

Fig. 4-24 Steering ratio: rear tire with increased camber stiffness coefficient (k_{φ_r} = 0.87 rad^{-1}).

Figure 4-24 shows that, when the speed is higher than 15 m/s, under-steering occurs for a wide range of values of the steady turning radius. Under-steering behavior does not take place when the speed is low for at least two reasons. First, if the steady turning radius is large, the roll angle is small (less than 5°) and the increased camber stiffness of the rear tire is unable to significantly influence the sideslip angle. Then, if the steady turning radius is small, both roll angle φ and steering angle δ are rather large, but the camber angle β of the front wheel is larger than φ and this effect compensates for the increased camber stiffness of the rear tire.

Case 3.

Finally, Fig. 4-25 deals with a rear tire which has increased both cornering stiffness coefficient and camber stiffness coefficient (k_{λ_r} =17.3 rad^{-1} and k_{φ_r} = 0.87 rad^{-1}). In this case under-steering behavior takes place even when the speed is lower than 15 m/s.

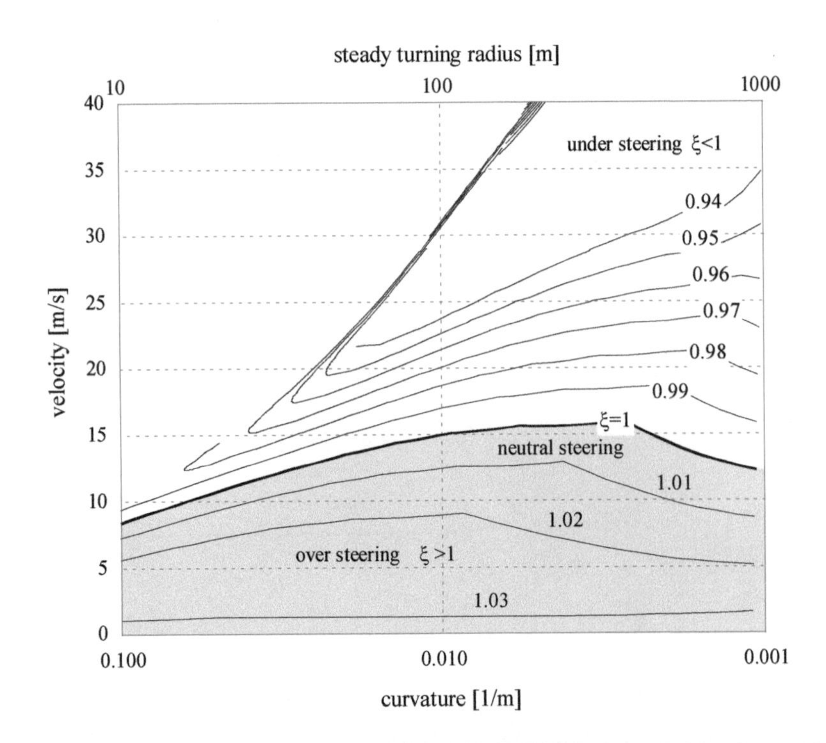

Fig. 4-25 Steering ratio: rear tire with increased cornering stiffness coefficient (k_{λ_r} = 17.3 rad^{-1} and camber stiffness coefficient k_{φ_r} = 0.87 rad^{-1}).

4.8 The torque applied to steering

The equilibrium of moments around the steering axis enables the evaluation of the torque τ that the rider must apply to the handlebars to assure the motorcycle's equilibrium in a turn (Fig. 4-26). It must be specified that this refers to steady turning, i.e., to a motorcycle at a constant velocity and turn radius. In transitory movement, in a turn with variable velocity and curvature radius, the torque the rider must exercise will be substantially different from that calculated in a steady state, especially if the variations in velocity and trajectory occur suddenly.

The torque applied by the rider is equal, but of opposite sign, to the resultant of all the moments generated by the forces acting on the front section. The resultant torque is composed of six terms:

$$\tau = -\tau_{P_f} - \tau_{C_f} - \tau_{N_f} - \tau_{F_f} - \tau_{W_f} - \tau_M$$

- disaligning influence (sign +) due to the weight force of the front section,

$$\tau_{P_f} = \left[\left(Y_{G_f} - Y_A\right)\sin\varepsilon + \left(X_{G_f} - X_A\right)\cos\varepsilon \, \sin\varphi\right] g m_f$$

aligning influence (sign -) due to the centrifugal force of the front section,

$$\tau_{C_f} = -[m_f\left(X_A Y_{G_f} - X_{G_f} Y_A\right)\cos\varepsilon\cos\varphi + \\ +\left(I_{yz_f} - m_f Y_{G_f} Z_A\right)\sin\varepsilon + \left(I_{xz_f} - m_f Y_{G_f} Z_A\right)\cos\varepsilon\sin\varphi]\Omega^2$$

disaligning influence (sign +) due to the normal load on the front wheel,

$$\tau_{N_f} = \left[\left(Y_{P_f} - Y_A\right)\sin\varepsilon + \left(X_{P_f} - X_A\right)\cos\varepsilon\sin\varphi\right] N_f$$

aligning influence (sign -) due to the lateral force on the front wheel,

$$\tau_{F_f} = -\left\{\cos\varepsilon\cos\varphi[\left(X_A - X_{P_f}\right)\cos\Delta + \left(Y_A - Y_{P_f}\right)\sin\Delta] + \\ +\left(\cos(\varepsilon)\sin\varphi\sin\Delta - \sin\varepsilon\cos\Delta\right) Z_A\right\} F_f$$

- aligning influence (sign -) due to the gyroscopic effect of the front wheel (Fig. 4-27),

$$\tau_{W_f} = \left[\sin\varepsilon\cos\varphi\cos\delta - \sin\varphi\sin\delta\right] I_{W_f} \omega_f \Omega$$

- disaligning influence (sign +) due to the twisting torque of the front tire,

$$\tau_M = M_{zf} \cos\varepsilon \cos\varphi$$

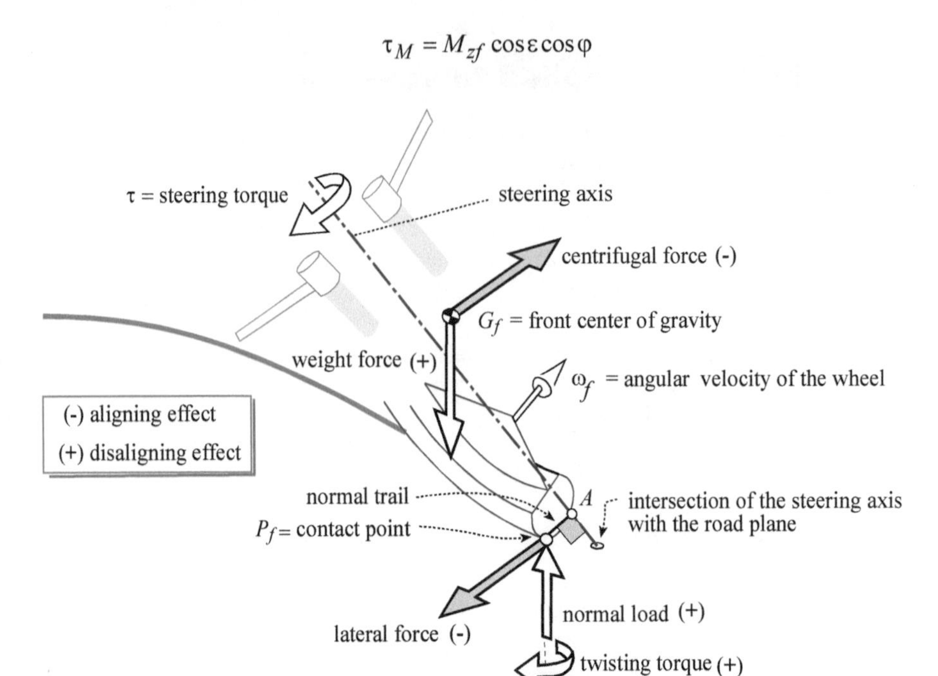

Fig. 4.26 Equilibrium of the front frame.

Point A indicates the intersection of the steering axis with the normal line passing through the front tire contact point. The distance between point A and contact point P_f represents the effective trail of the tire.

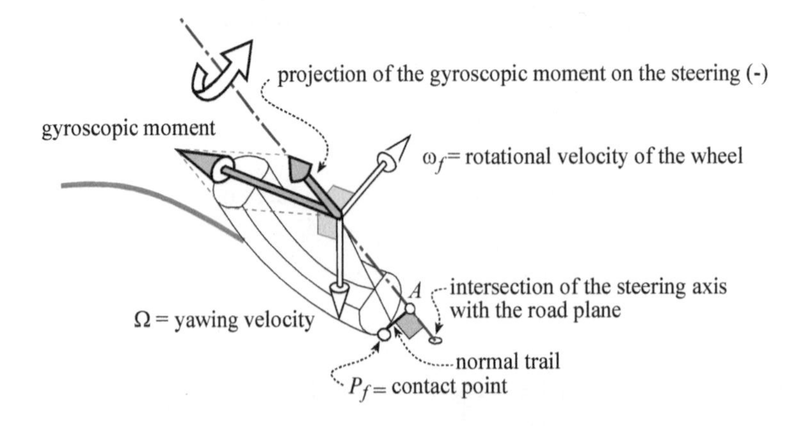

Fig. 4.27 Genesis of the gyroscopic moment on the front section.

4.8.1 Torque components

In Fig. 4-28 the variations of the torque applied by the rider to the handlebars of the sample motorcycle is shown. It is useful to recall that the torque exercised by the rider is, by definition positive if it tends to increase the steering angle into the turn.

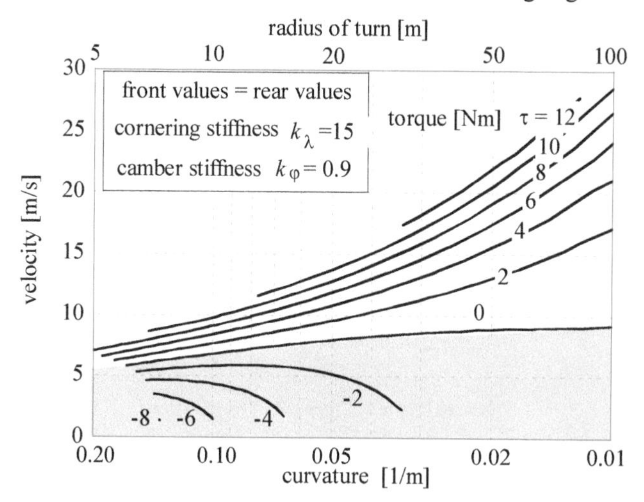

Fig. 4-28 Torques applied to the handlebars versus the turn radius and the velocity.

This means that there are essentially two possible situations:

- at low velocities the steering torque is negative. Therefore, in steering, the rider must block the handlebars, which otherwise tend to rotate further. When the values of the steering torque become strongly negative, the inclination and the entry into the turn become easier;
- with an increase in velocity, the torque to be applied to the handlebars becomes positive. This circumstance, if the value of the torque remains high, generates in the rider the unpleasant sensation of driving a motorcycle that is hard to incline and to insert into tight turns.

Figure 4-29 shows the various components combining to define the resulting couple that acts on the steering axis, in terms of the roll angle. The various contributions have the following effect:

- vertical load: the vertical reactive force generates a positive moment of high value;
- lateral force: the lateral reactive force generates a high value negative moment of the same order of magnitude generated by the vertical load;
- front weight force: the moment is positive;
- centrifugal force: the moment is negative, of the same order of magnitude as that generated by the weight force;
- gyroscopic moment: it generates an aligning effect;
- twisting moment: it generates a disaligning effect that increases with the roll angle.

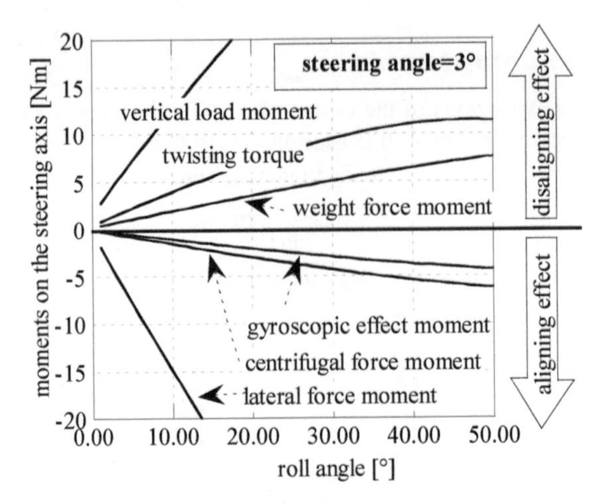

Fig. 4-29 Moments exercised around the steering axis.

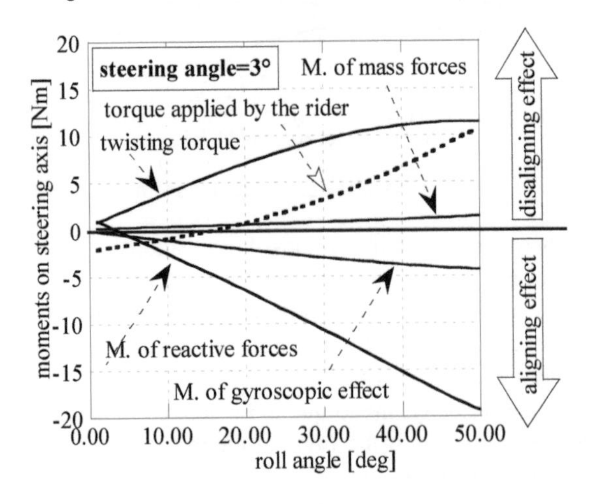

Fig. 4-30 Torque applied by the rider and moments exercised around the steering axis.

It should be observed that both the two forces applied to the center of gravity of the front section (weight force and centrifugal force) and the two reaction forces applied at the contact point (lateral force and vertical load) each contain influences of opposite sign.

Figure 4-30 shows the variation in the torque applied by the rider as the roll angle varies. The torque is equal to the sum, with signs changed, of the following influences:

- mass forces (vector sum of the moments of weight force and centrifugal force);
- reaction forces (vector sum of the moments of vertical load and lateral force);
- gyroscopic moment;
- twisting torque of the tire.

Figure 4-30 also shows an example in which for small roll angles the rider needs to exercise a negative couple, while for large roll angles, he must apply a positive couple.

The maximum maneuverability is obtained when the couple necessary for assuring equilibrium is zero or nearly so. In fact under these conditions, if the rider lets go of the handlebars the motorcycle continues to round the set turn.

4.8.2 The influence of motorcycle geometry on the steering torque

The steady turning behavior of a motorcycle is a function of vehicle geometry, inertia and tire properties.

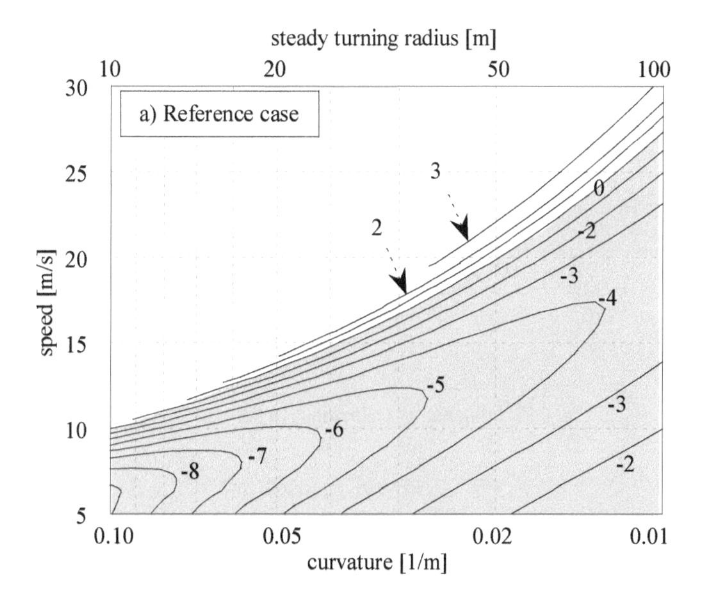

Fig. 4-31 Steering torque against turn radius and velocity.

Normal trail

Fig. 4-31 shows the reference case, while Fig. 4-32 shows the effect of a positive increment in the normal trail: the steering torque contour plot shifts towards lower values in the whole area under consideration, whereas there is no significant change in the shape of the curves.

This result can be explained by considering the fact that when the trail increases, the disaligning effect due to the front tire vertical load increases more than the aligning effect due to the lateral force. The result is a more stable steering behavior, in the area of interest.

Steering head angle

On the contrary, the increase in the caster angle (Fig. 4-33) has an aligning effect, since the steering torque increases (as it has negative values, its magnitude decreases). It is worth noting that the effect of the steering head angle is relevant. Considering that the real steering head angle is influenced by the motorcycle's attitude, depending on speed, mass distribution and suspension behavior, particular attention should be paid to this parameter.

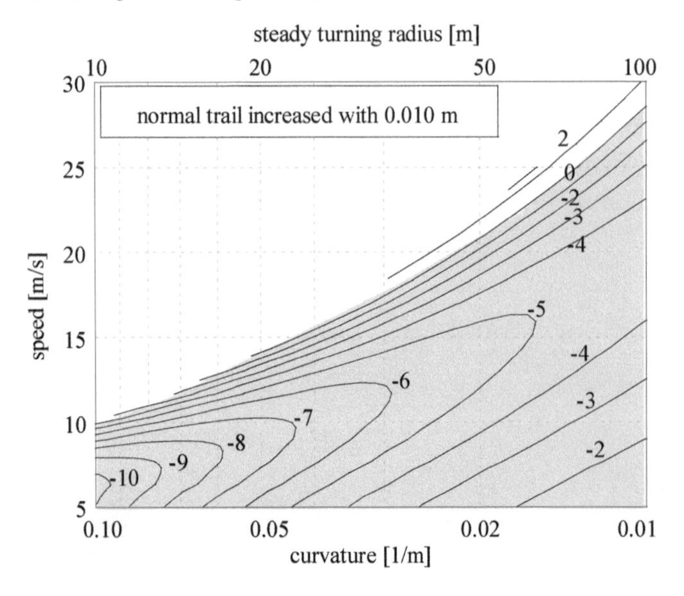

Fig. 4-32 Influence of an increase in the normal trail.

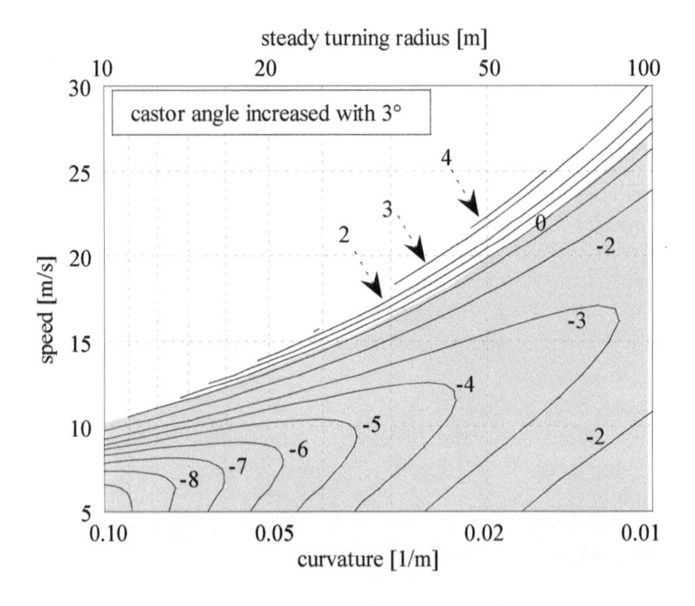

Fig. 4-33 Influence of an increase in the steering head angle.

Front tire cross section radius

The positive increment in front tire section radius has a strong aligning influence (Fig. 4-34): this effect is caused by the displacement of the front wheel contact point due to the roll angle. It can be seen that the zero steering torque curve shifts towards lower values of the forward speed, and the resulting behavior is quite different from the reference case.

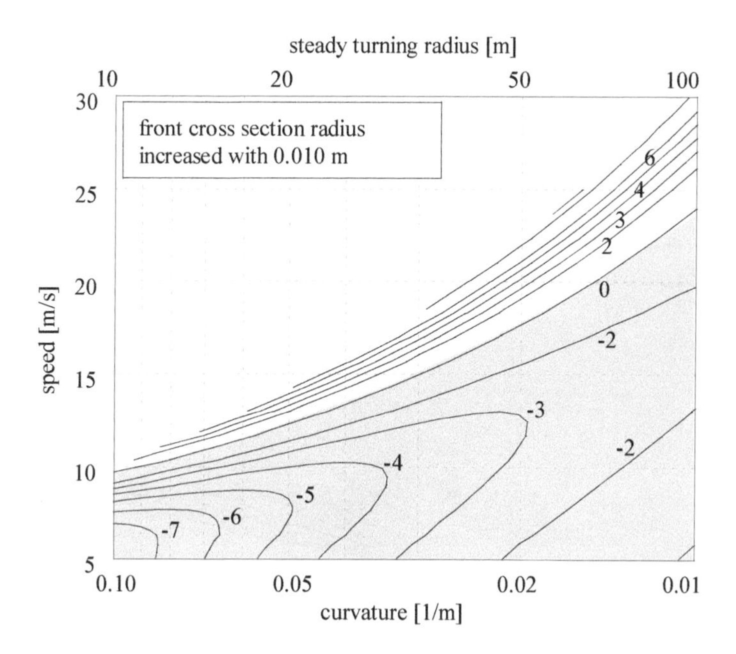

Fig. 4-34 Influence of an increase of the cross section radius.

Rider position

A forward displacement of the rider's center of mass has a slight self-steering effect. The vertical position of the rider's center of mass has a very small aligning effect. The result is that if the rider moves, always remaining in the plane of symmetry of the motorcycle, the steering behavior doesn't change significantly. On the contrary, a lateral displacement of the rider toward the inside of the curve has a strong aligning effect. Considering that sport riders usually move sideways considerably it is obvious that the steering characteristics of the motorcycle are strongly influenced by driving style. An expert rider can take advantage of this phenomenon and shift the zone of low steering torque to match the current steering conditions and thus gain better, easier steering control.

The presence of a passenger alters mass distribution of the motorcycle. The resulting effect is slightly aligning, but the steering turning steering torque is not substantially changed.

The influence of the rider's lateral position on the steering behavior of the motorcycle is represented in Fig. 4-35. A 0.05 m lateral displacement of the rider's center of mass in towards the curve was considered, this corresponds to a decrease in the roll angle of about 1°. The figure shows the acceleration index (ratio between the torque and the lateral acceleration) both in the presence of lateral displacement and under normal conditions (without lateral displacement).

In the presence of lateral displacement the acceleration index is positive for every value of forward speed if the steady turning radius is larger than 25 m. For each value of the steady turning radius the curve, which is calculated by taking into account the lateral displacement of the rider, lies above the curve calculated in nominal conditions. The difference between the two curves becomes very large when lateral acceleration (V^2/R) is low.

This behavior can be explained by taking into account the effect due to the decrease in roll angle caused by the rider's lateral displacement. The first effect is the decrease of the disaligning effect of the tire twisting torque. The second is the variation in the moment of tire forces about the steering axis. In particular the disaligning effect of the tire load, which tends to rotate the wheel towards the inside of the curve, decreases.

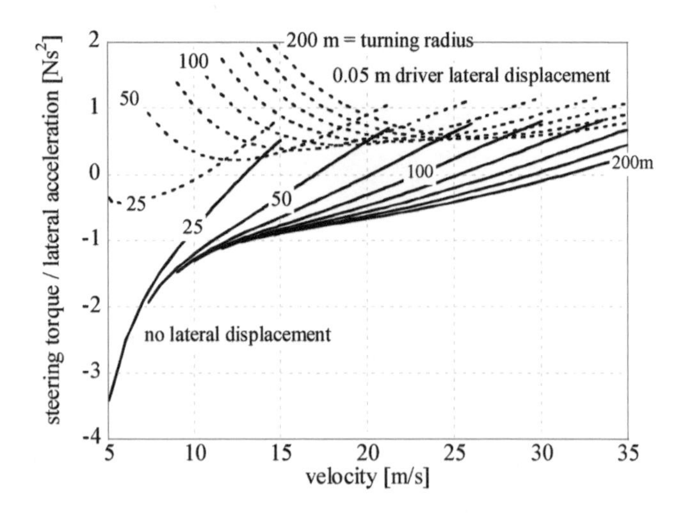

Fig. 4-35 Influence of rider's lateral displacement on the acceleration index.

Tire properties

All drivers know that tires have an important effect on the behavior of the motorcycle. The same motorcycle equipped with different tires sometimes behaves as a completely different motorcycle. An increase in the cornering stiffness or in the camber stiffness of the front tire causes only small variations in the acceleration index.

The more important tire parameter is the yaw torque of the front tire which, as we have seen in the second chapter, includes two terms:

- a term which tends to align the wheel with the forward speed and is due to the lateral force and pneumatic trail (which depends on sideslip angle λ);

- a term, which is named twisting torque, that tends to disalign.

Figure 4-36 shows the yaw moment of the front tire in the reference case (on the left) and with a tire having a decreased trail (-20%) and an increased twisting torque (+11%).

Figure 4-37 highlights that the decrease in the tire trail and the increase in the twisting torque make the steering torque negative in the whole range of velocities and steady turning radii. Nevertheless the two families of curves show the same general trends, like the increment of acceleration index when forward speed increases and steady turning radius decreases.

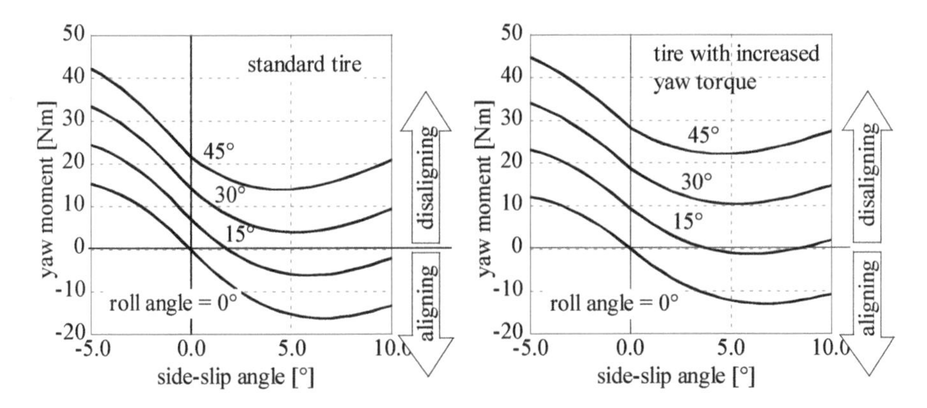

Fig. 4-36 Yaw torque characteristics.

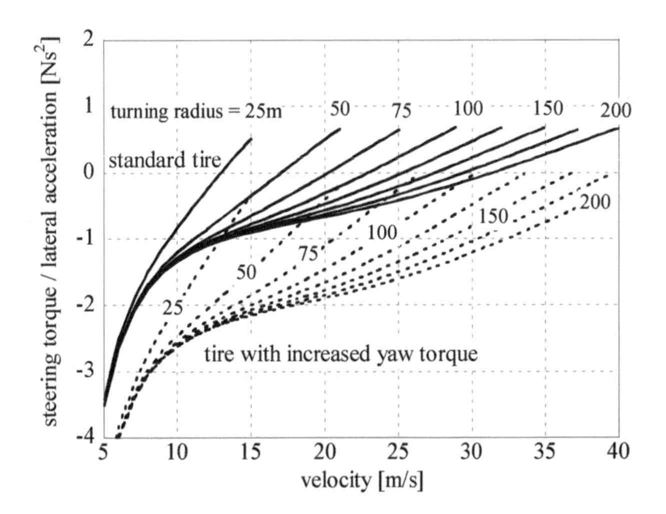

Fig. 4-37 Influence of the yaw torque on the acceleration index.

* * *

We have seen that the torque to be applied to the handlebars can have zero value to assure equilibrium, positive (a torque in accordance with the steering angle) or negative (a torque not in accordance with the steering angle).

Any modifications to the motorcycle will bring variations in the torque to be applied to the handlebars; the influence of the main geometric and inertial parameters on the steering torque is brought to light in Fig. 4-38.

It can be observed that the steering head angle, the front tire cross section radius, the height of the center of gravity and the normal trail are the parameters that most influence the value of torque.

	parameter increased
	rider displacement inside the curve of 0,05 m
ε	caster angle
t_f	radius of the cross-section of the front tire
h	height of the motorcycle mass center
a_{t_f}	front tire trail
b_f	distance front mass center-steering axis
I_{w_f}	front wheel spin inertia
	reference value
b	distance motorcycle mass center-rear wheel
t_r	radius of the cross-section of the rear tire
a	mechanical trail
M_{t_f}	twisting torque of the front tire

effect: disaligning ← | → aligning

Fig. 4-38 Influence of some parameters on the steering torque.

Aligning effect: the steering angle tends to decrease. The rider must steer into the turn (+) to counteract this effect. If the torque was negative it must becomes less negative.

Disaligning effect: the steering angle tends to increase. The rider must steer out of the turn (-) to counteract. If the torque was negative it becomes more negative.

Gilera Saturno "Sanremo" 500 cc version 1947

5 In-Plane Dynamics

A motorcycle without suspension moving over uneven ground presents difficulties in steering because of the loss of wheel grip on the road, and because of rider discomfort. Small bumps on the ground are easily absorbed by the tires, but for adequate absorption of larger bumps, the motorcycle needs appropriate suspension.

A motorcycle with suspension, from a dynamics point of view, can be considered as a rigid body connected to the wheels with elastic systems (*front and rear suspension*). The rigid body constitutes the sprung mass (*chassis, engine, steering head, rider),* while the masses attached to the wheels are called unsprung masses.

Suspension has to satisfy the following three purposes:

- allow the wheels to follow the profile of the road without transmitting excessive vibration to the rider. This purpose concerns rider comfort, that is the isolation of the sprung mass from the vibration generated by the interaction of the wheels with road irregularities;
- ensure wheel grip on the road plane in order to transmit the required driving, braking and lateral forces;
- ensure the desired trim of the vehicle under various operating conditions (acceleration, braking, entering and exiting turns).

The degree of required comfort varies according to the use of the vehicle. For example, with racing vehicles, comfort is less important than the motorcycle's capacity to keep the wheels in contact with the ground and to assume the desired trim.

However, in other vehicles the suspension is expected to serve other purposes. For example, in off-road vehicles the suspension serves to isolate the sprung mass from continuous impact generated by vehicle jumps. For this reason, suspension in off-road vehicles has greater wheel travel than in touring vehicles, and more so than in racing vehicles.

As for the trim, it should be highlighted that it depends on the stiffness of the suspension and on the loads. The load can be quite variable in motorcycles (one or two passengers, possibly with baggage); and furthermore, load transfer between the front and rear wheel occurs in both acceleration and braking.

5.1 Preliminary considerations

In the study of in-plane dynamics, the motorcycle is considered as an elastically suspended rigid body. It has three degrees of freedom: one degree of freedom is associated with the vehicle's forward motion, while the other two are associated with two vibrating modes and are, therefore, characterized by their respective natural frequencies.

The combination of distance between bumps on the road plane and forward velocity causes excitations of the vehicle in a range of frequency that can be evaluated from 0.25 Hz to 20 Hz. Since the tires have radial stiffness much greater than that of the suspension (6-12 times greater), their influence at low frequencies (below approximately 3 Hz) becomes negligible.

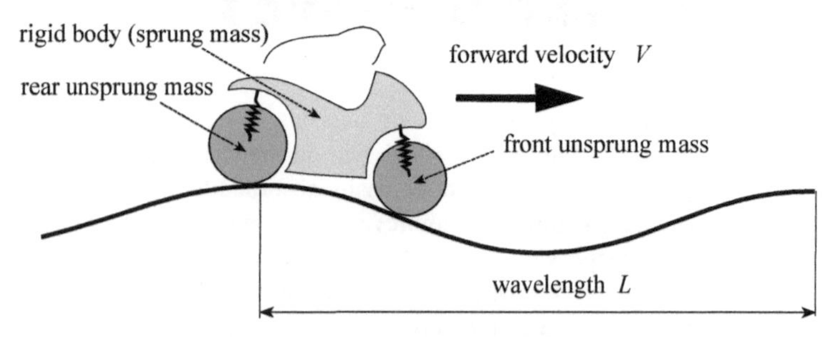

Fig. 5-1 Wavelength of a disturbance.

Now let us see when resonance conditions can be generated as a result of irregularities in the road surface. Suppose the motorcycle advances with constant velocity V on a road profile presenting equidistant irregularities - for example, the bays of a viaduct (Fig. 5-1). The time required for the motorcycle to cover the distance L_{wave} between the two irregularities (length of the bay) is equal to:

$$T = \frac{L_{wave}}{V}$$

T therefore represents the period of external excitation of the motorcycle.

The resonance condition occurs when the excitation frequency is equal to the natural frequency of one of the vibration modes of the vehicle in the plane.

Critical forward motorcycle velocity is defined as the forward velocity at which the motion imposed by the road disturbance has the same frequency as one of the vibration modes of the vehicle in the plane.

If L_{wave} is the wavelength of the disturbance and ν_n is the frequency of one of the motorcycle modes, (T_n is the natural period), the critical forward velocity is given by the following expression:

$$V = \frac{L_{wave}}{T_n} = L_{wave} \nu_n$$

For example, with a natural frequency ν_n equal to 2 Hz and a perturbation with wavelength L_{wave} equal to 6 m, the resonance condition occurs at a velocity of 12 m/s (critical velocity).

If the motorcycle proceeds at a velocity below critical velocity, the frequency ν of the motion imposed:

$$\nu = \frac{V}{L_{wave}}$$

is lower than the natural frequency ν_n. Alternatively, it is above the vehicle's critical velocity for velocities greater than the critical velocity.

It is also possible to follow a different approach. Assuming that the motorcycle forward velocity is 50 m/s. Given the natural frequency ν_n, at this velocity the resonance condition occurs when the perturbation has a (critical) wavelength equal to:

$$L_{wave} = \frac{V}{\nu_n}$$

which, under the assumption made ($\nu_n = 2$ Hz, $V = 50$ m/s), corresponds to a critical wavelength of 25 m. Critical wavelengths therefore diminish in proportion to the forward velocity.

5.2 Suspension overview

Suspension systems were introduced on motorcycles in the 1930s and 1940s and numerous architectures and kinematic models have been proposed. We will briefly analyze the kinematic schemes of the front and rear suspension that are now most common.

5.2.1 Front suspension

The most widespread front suspension is, undoubtedly, the telescopic fork. It is made up of two telescopic sliders which run along the inner tube of the fork and form a prismatic joint between the unsprung mass of the front wheel and the sprung mass of the chassis.

The constructive solution with the two telescopic sliders attached to the steering head is referred to as "conventional," and is currently the most common construction for street motorcycles (Fig. 5-2).

The solution with the two fork tubes fixed to the steering head and the two slider tubes on the lower end, called "upside down", is the one most commonly used in sport motorcycles, especially since it has more bending and torsional stiffness.

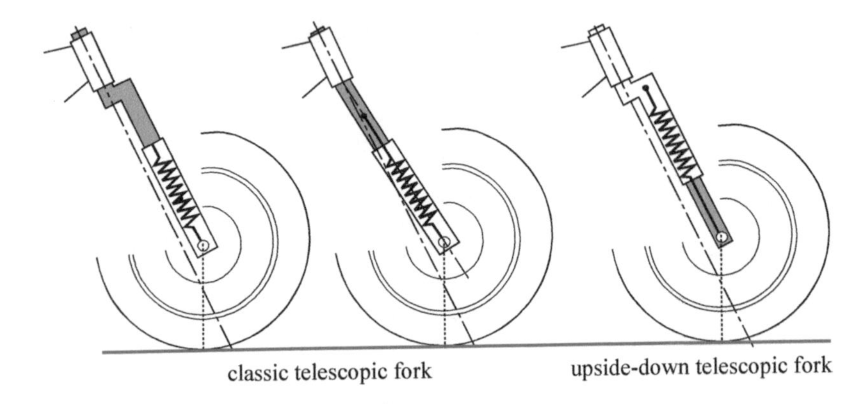

Fig.5-2 Schemes of classic and upside-down telescopic forks.

The telescopic fork is characterized by low inertia around the axis of the steering head. Its greatest disadvantage is represented by the high friction forces encountered when forces are applied orthogonal to the axis along which the sliders run - for example, in braking and on curves.

In braking, because of the load transfer, the telescopic fork compresses as the rear suspension is unloaded; thus, the vehicle pitches forward. The pitching changes the trim of the vehicle and further diminishes the steering head angle. A smaller angle of inclination of the fork causes a reduction of the value of the trail.

Two limitations of the telescopic fork are the impossibility of achieving progressive force/displacement and the rather high values of the unsprung mass that is an integral part of the wheel.

To overcome the typical defects of the telescopic fork, different suspension systems have been used. These can be classified from a kinematic point of view as:

- push arm;
- trailing arm;
- four-bar linkage.

In an arm front suspension, the arm can be "pushed" (Earles-type fork) or pulled back (a scheme used by the Piaggio Vespa), as illustrated in Fig. 5-3.

The four-bar linkage can also be used in a front suspension. In this case the axis of the steering head can be attached to the chassis, or to the connecting link, as shown in Fig. 5-4.

The front arm suspension and four-bar linkage suspension can be designed so as to provide total or partial anti-dive behavior in braking. In addition, not having prismatic joints the dry friction problems typical of telescopic forks are eliminated from the start.

The torsional stiffness (with respect to the steering axis and an axis normal to it) of these suspension systems depends on the design but in general it is easier to obtain greater values with the telescopic fork. Furthermore, these elaborate designs can also reduce the unsprung mass. An appropriate position of the spring or springs, especially in the case of the four-bar linkage suspension, makes progressive suspension possible.

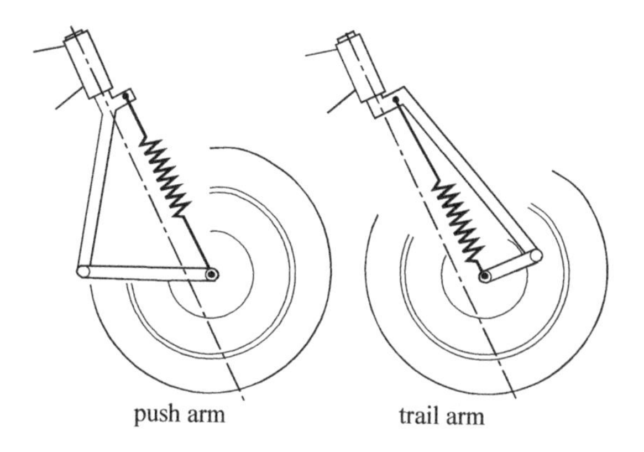

Fig. 5-3 Schemes of front suspension with pushed and pulled wishbones.

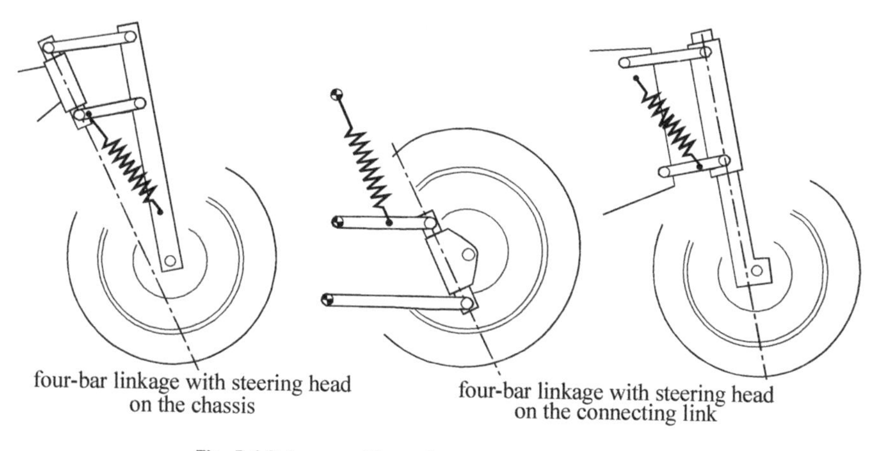

Fig. 5-4 Schemes of front four-bar linkage suspension.

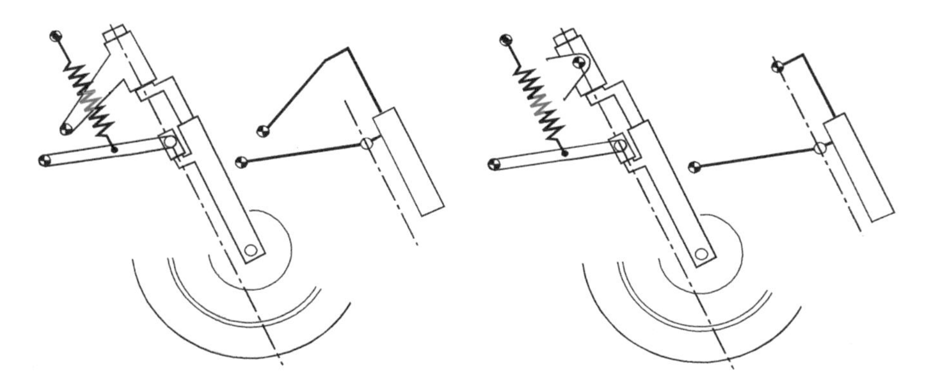

Fig. 5-5 Schemes of four-bar linkage front suspension with prismatic pairs.

A variant of the four-bar linkage front suspension is obtained by substituting a revolute joint with a prismatic joint, as illustrated in the diagram on the left in Fig. 5-5. This kinematic design has the disadvantage that the vertical movement of the wheel in relation to the chassis causes rotation in the handlebars around the upper revolute joint fastened to the chassis. This disadvantage is greatly reduced in the design on the right in Fig. 5-5, which is obtained by moving the revolute joint fastened to the chassis in relation to the steering head axis. This design solution is used in the BMW "Telelever" suspension.

5.2.2 Rear suspension

The classic rear suspension is composed of a large fork made up of two trailing-ing arms with two spring-damper units, one on each side, inclined at a certain angle with respect to the swinging arm. (Fig. 5-6).
The principal advantages of the traditional rear suspension are:

- simplicity of construction;
- ease of dissipation of the heat produced by the shock absorbers;
- large amplitude of the motion of spring-damper units which is nearly equal to the vertical amplitude of the wheel motion and which therefore causes high compression and extension velocities of the shock absorbers;
- the low reaction forces transmitted to the chassis.

The greatest disadvantages are:

- limitation of the vertical oscillation amplitude of the wheel;
- not very progressive force-displacement characteristic;
- possibility that the two spring-damper units generate different forces due to differences in the spring preloads or the characteristics of the shock absorbers, with consequent malfunctioning of the suspension, due to the generation of moments that torsionally stress the swinging arm.

One variant of the dual-strut suspension is the cantilever mono-shock system, characterized by only one spring-damper unit. It has the following advantages over the twin shock arm:

- ease of adjustment since there is only one shock absorber;
- smaller unsprung mass;
- high torsional and bending stiffnesses;
- high vertical wheel amplitude.

This suspension does not enable a progressive force-displacement characteristic and the positioning of the spring-shock absorber unit above or behind the engine can cause heat dissipation problems for the shock absorber.

In the classic and cantilever system the introduction of a linkage in the rear suspension makes it easier to obtain the desired stiffness curves. These designs are generally based on the four-bar linkage. They are distinguished only by the different attachment points of the spring-damper unit, which can be inserted between the chassis and the rocker (Unitrak design of Kawasaki) or between the connecting link and the chassis (Pro-Link design of Honda) or between the swinging arm and the rocker (Full Floater design of Suzuki) as shown in Fig. 5-7. Modest unsprung masses are obtained, as well as large wheel amplitude, but great reaction forces are exchanged among the various parts of the four-bar linkage.

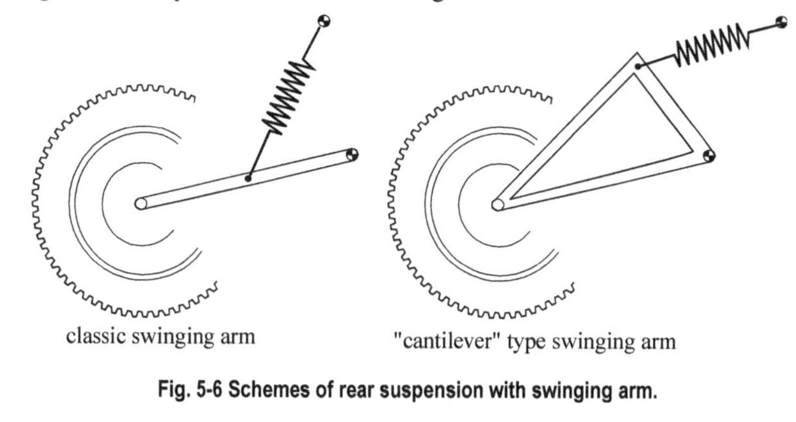

Fig. 5-6 Schemes of rear suspension with swinging arm.

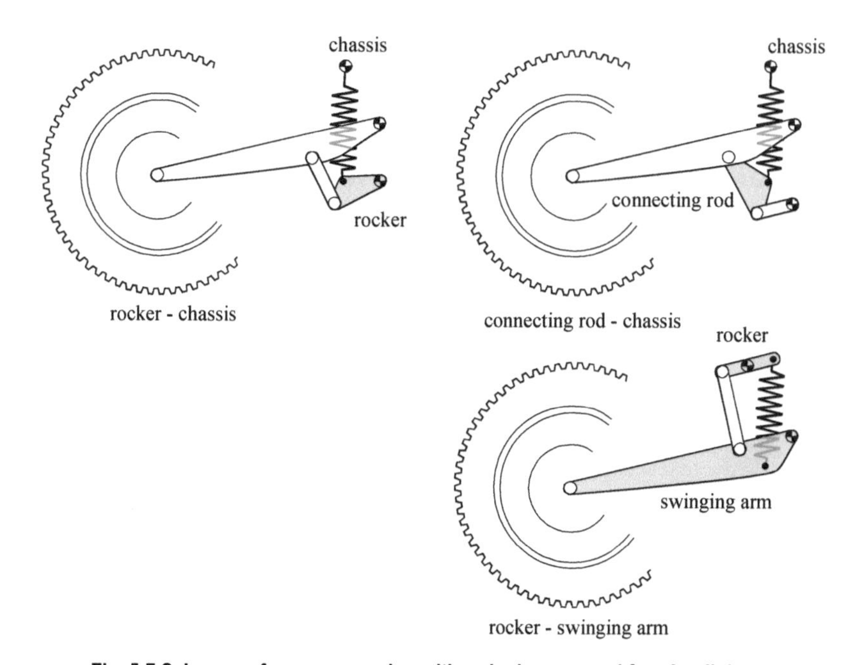

Fig. 5-7 Schemes of rear suspension with swinging arm and four-bar linkage.

The four-bar linkage is also commonly used to accommodate a shaft-drive with universal joints. The wheel is attached to the connecting rod of the four-bar linkage. Its center of rotation with respect to the chassis is therefore the point of intersection of the axes of the two rockers. The position of the rotation center depends on the angles of inclination of the two rockers. The suspension acts as if it were composed of a very long fork fastened to the chassis in the center of rotation. This kinematic design is used in the BMW "Paralever" suspension and in the Guzzi motorcycles modified by Magni (with parallel rockers).

A suspension based on a six-bar linkage has also been tried (Morbidelli 500 GP). This can potentially generate curves with more unique progression of suspension stiffness. This potential advantage, however, does not justify the highly complex construction.

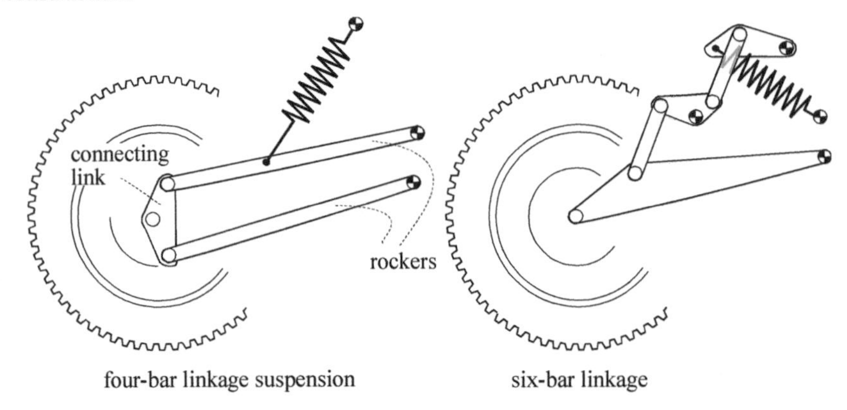

Fig. 5-8 Schemes of rear suspensions with four-bar and six-bar linkages.

5.3 Reduced suspension stiffness

The choice of front and rear suspension characteristics (stiffness, damping, pre-load) depends on many parameters: the weight of the rider and the motorcycle, the position of the center of gravity or the distribution of the loads on the wheels, the characteristics of stiffness and vertical damping of the tires, the geometry of the motorcycle, the conditions of use, the road surface, the braking performance, the motor power, the driving technique, etc..

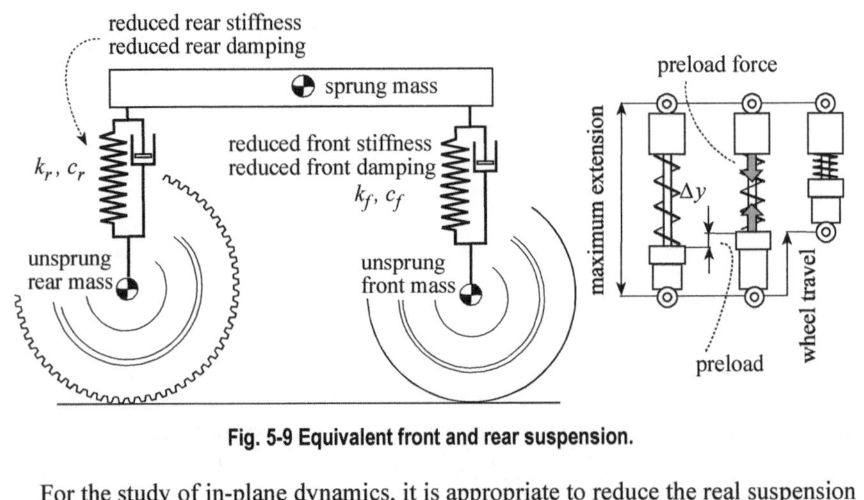

Fig. 5-9 Equivalent front and rear suspension.

For the study of in-plane dynamics, it is appropriate to reduce the real suspension to equivalent suspension, represented by two vertical spring-damper units that con-

nect the unsprung masses to the sprung mass.
The parameters defining equivalent suspension are:

- reduced stiffness,
- reduced damping,
- dependence of the reduced stiffness on the vertical displacement (progressive/degressive suspension),
- maximum travel and preload.

5.3.1 Reduced front suspension stiffness

Now consider the front fork suspension as depicted in Fig. 5-10.

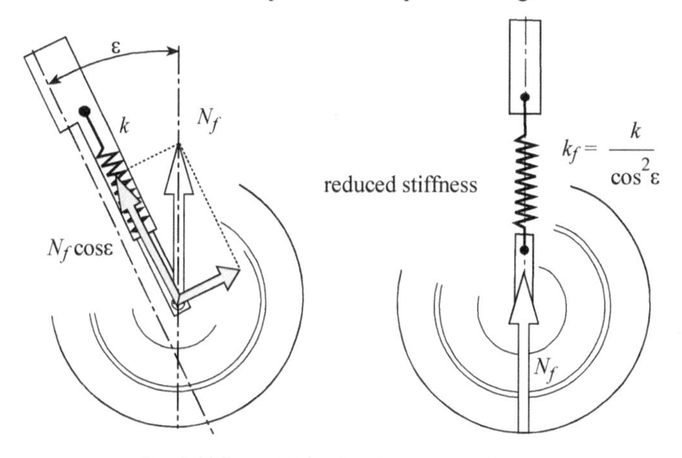

Fig. 5-10 Reduced front suspension stiffness.

If subjected to the same vertical load, the real front fork and the equivalent vertical suspension have equal vertical displacements. Hence, the reduced stiffness k_f and the real stiffness k have to satisfy the following equation:

$$k_f = \frac{k}{\cos^2 \varepsilon}$$

Consider now the fork and the equivalent suspension with just the damping devices. If the same vertical velocity on the wheel hub is imposed, the viscous vertical forces are equal. Then, the equivalent damping of the front suspension c_f must satisfy the equation:

$$c_f = \frac{c}{\cos^2 \varepsilon}$$

where c represents the damping constant of the fork. Since there are two groups of spring dampers arranged in parallel in the fork, the stiffness k is equal to the sum of the stiffnesses of the two springs, and the damping c is equal to the sum of the damping of the two dampers. It should be noted that the increase in the steering head angle of inclination causes a reduction in the stiffness and damping coefficients of the reduced suspension.

Example 1

A front suspension is required with reduced vertical stiffness $k_f = 14$ N/mm. The angle of inclination of the fork is equal to 30°. Determine the actual stiffness of each fork spring.

The overall stiffness of the fork is equal to:

$$k = k_f \cos^2 \varepsilon = 10.5\,\text{N/mm}$$

The stiffness of the single spring must therefore be equal to 5.25 N/mm. If the caster angle is less, and equal to 24°, the stiffness of the single spring must be greater or equal to 5.84 N/mm.

5.3.2 Reduced rear suspension stiffness

Now consider the classic rear suspension represented in Fig. 5-11. The elastic force F_e is proportional to the deformation of the spring:

$$F_e = k\,(L_m - L_{m_o})$$

where k indicates the stiffness of the spring, L_{m_o} its initial length and L_m the length of the deformed spring (it is a function of the swinging arm angle of inclination ϑ).

The elastic moment M_e exerted on the swinging arm is given by the product of the force and the velocity ratio $\tau_{m,\vartheta}$.

$$M_e = F_e \tau_{m,\vartheta}$$

$\tau_{m,\vartheta}$ is the ratio between the spring's deformation velocity and the swinging arm angular velocity:

$$\tau_{m,\vartheta} = \frac{\dot{L}_m}{\dot{\vartheta}}$$

We can consider that the swinging arm has, in place of the effective spring, a torsional spring that generates a moment equal to the one generated by the effective spring. The derivative of the elastic moment, with respect to the angle of rotation of the swinging arm, represents the reduced stiffness of the torsional spring:

$$k_\vartheta = \frac{\partial}{\partial \vartheta} M_e$$

from which, through substitution:

$$k_\vartheta = k\,\tau_{m,\vartheta}^2 + k\,(L_m - L_{m_0})\frac{\partial \tau_{m,\vartheta}}{\partial \vartheta}$$

The second term is less important than the first, and, in a first approximation, it can be ignored.

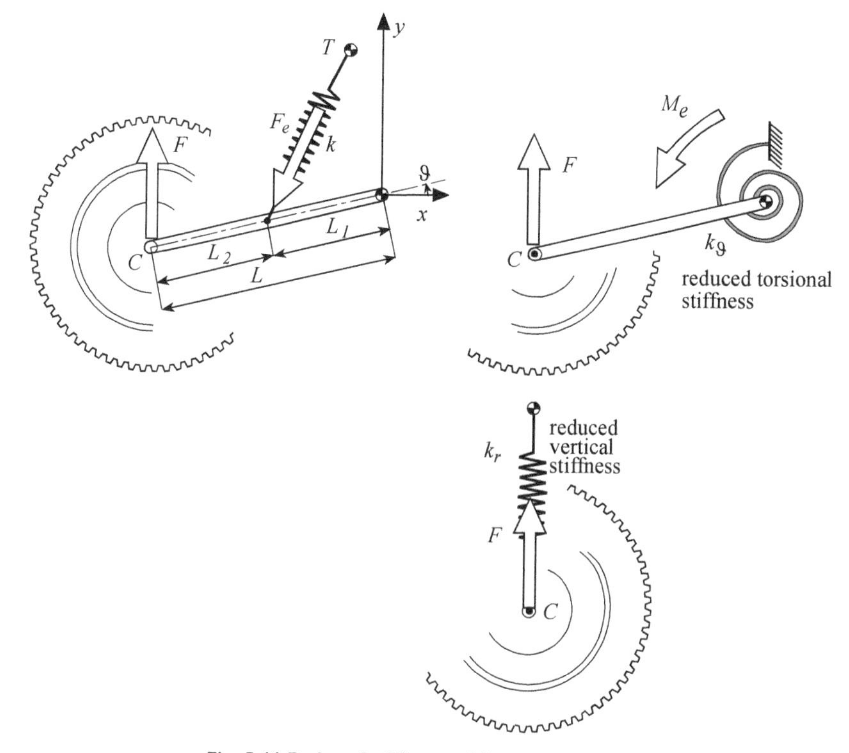

Fig. 5-11 Reduced stiffness of the rear suspension.

The rear suspension can also be substituted by a vertical spring attached to the wheel hub, rather than a torsional spring.

The reduced elastic force F is equal to the product of the elastic force exerted by the spring and the velocity ratio τ_{m,y_C} :

$$F = F_e \, \tau_{m,y_C}$$

where τ_{m,y_C} represents the ratio between the deformation velocity of the spring (which is obviously equal to the velocity of the damper) and the vertical velocity of the wheel.

$$\tau_{m,y_C} = \frac{\dot{L}_m}{\dot{y}_C} = \frac{\tau_{m,\vartheta}}{-L\cos\vartheta}$$

In this case, the reduced vertical stiffness is equal to the derivative of the vertical force applied to the wheel pin with respect to the vertical displacement of the wheel:

$$k_r = \frac{\partial}{\partial y_C} F \cong k\,\tau^2_{m,y_C}$$

The velocity ratio depends on the geometric characteristics of the rear suspension mechanism and varies with the vertical wheel travel.

In the case of the classic swinging arm, the velocity ratio is:

$$\tau_{m,y_C} = \frac{\dot{L}_m}{\dot{y}_C} = \frac{L_1(x_T \sin\vartheta - y_T \cos\vartheta)}{\sqrt{L_1^2 + 2L_1(x_T \cos\vartheta + y_T \sin\vartheta) + x_T^2 + y_T^2}} \frac{1}{L\cos\vartheta}$$

where x_T, y_T indicate the coordinates of the spring-damper pin attached to the chassis.

Now consider the damping force:

$$F_s = c\,\dot{L}_m$$

c indicates the damping constant of the damper.

The damping moment acting on the swinging arm is given by the product of the force and the velocity ratio $\tau_{m,\vartheta}$:

$$M_s = F_s \tau_{m,\vartheta}$$

The reduced damping of the torsional damper is obtained by the ratio between the damping moment and the angular velocity:

$$c_\vartheta = \frac{M_s}{\dot{\vartheta}}$$

Substituting, we obtain:

$$c_\vartheta = c\,\tau^2_{m,\vartheta}$$

The reduced vertical force is:

$$F = F_s \tau_{m,y_C}$$

The reduced vertical damping, on the other hand, is:

$$c_r = \frac{F_s}{\dot{y}_C} = c\,\tau^2_{m,y_C}$$

It should be noted that the vertical force depends on the velocity ratio, while the reduced stiffness and damping depend on the square of the velocity ratio. The velocity ratio τ_{m,y_C} in suspension systems based on the four-bar linkage varies from 0.25 to 0.5. This means that the stiffness and the damping of the spring-shock absorber group must be between 4 and 16 times greater than the values of the reduced spring-shock absorber group.

5.3.3 Stiffness curve

The curve representing the elastic force against the vertical displacement of the wheel can have a linear trace, or a progressively increasing or decreasing one, to which a constant, increasing or decreasing reduced stiffness corresponds, as shown in Fig. 5-12. These cases are referred to, respectively, as linear, progressive or degressive suspension.

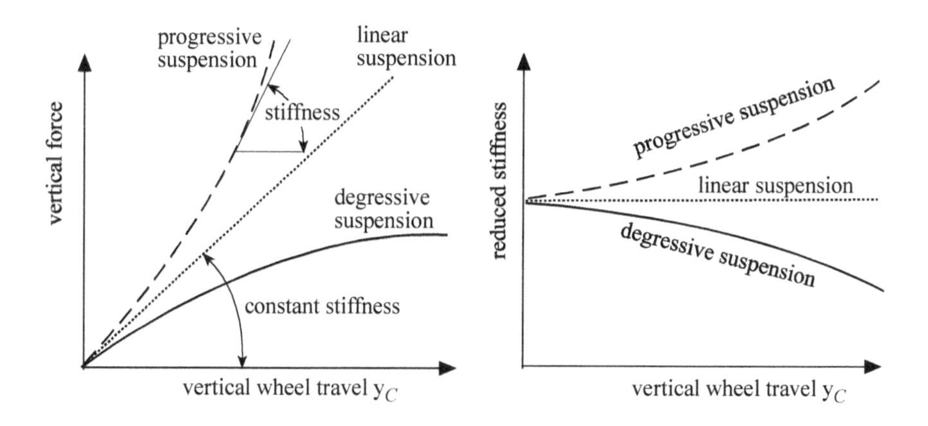

Fig. 5-12 Elastic force and stiffness of the suspension versus the vertical wheel travel.

For the sake of comfort in motion, it would be appropriate for the stiffness to be as low as possible, so as to minimize natural frequencies of the motorcycle vibration modes, in relation to the excitation frequencies of the motion imposed on the wheels by irregularities in the road plane. Very soft springs, however, cause wide variations in vehicle height as the load varies, as well as significant variations in trim, in the passage from rectilinear to curved motion, and during the acceleration and braking phases.

On the other hand, with irregularities in the road surface, very hard springs can cause, besides drastically reduced comfort, tire adherence problems in the rear section during acceleration and in the front section in braking.

To avoid these difficulties, more or less progressive suspension systems are employed in accordance with the type of use of the vehicle. Substantially, progressive suspension provides two important advantages:

- an increase in stiffness together with an increase in deformation, which enables the maintenance of more or less constant frequency of the modes of vibration in the plane as the vehicle mass increases (an increase caused, for example, by the passenger or the luggage);
- the suspension is soft in the case of small disturbances and thus in the case of small wheel travel, while it is rigid in the case of high wheel travel due to more severe disturbances. Riding comfort is thereby increased.

5.3.4 Preload

To regulate the trim of the motorcycle, for example, under variation of the load, preloading of the springs can be used. Preloading consists of a pre-compression of the spring. If the spring is stressed with forces that are lower than or equal to that of preloading, it is not deformed.

The force exerted by the spring, with preload, is:

$$F = k\,\Delta y + k\,y$$

where Δy indicates the deformation due to preload.

Preloading also makes it possible to limit deformation in compression of the spring-shock absorber group. The graph of Fig. 5-13 shows that with preloading, in order to obtain maximum amplitude greater forces must be applied or conversely, that with the same force applied the amplitude will be less.

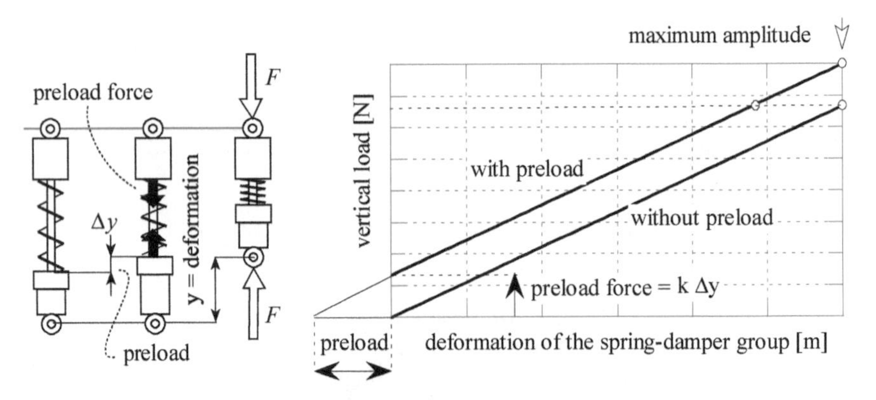

Fig. 5-13 Characteristics of the suspension as the preload varies.

Figure 5-14a shows a suspension with a spring that is not preloaded. The static load of the sprung mass compresses the spring-shock absorber group by an amount that depends on the stiffness of the spring, assuming that during forward motion the sprung mass is not displaced in the vertical direction, i.e. it ideally does not encounter irregularities in the road surface.

In order for the wheel to follow the profile when passing over a hole, the spring-shock absorber group needs to be able to extend by a quantity equal to the depth of the hole. In the case of suspension without preload, the extension, or better the hole's depth, can at most be equal to the ratio between the weight force of the sprung mass and the stiffness of the suspension.

However, in the case of suspension with a preloaded spring, illustrated in Fig. 5-14b, the maximum extension of the spring-shock absorber group is less, in this example, by a quantity equal to the preload.

The preload therefore governs the maximum value of the wheel travel in extension. The capacity of the suspension to follow the irregularities below the road plane depends on this value. These irregularities are called “negative”. For example, if a preload is applied that is equal to the static load, the wheel will not be able to follow the negative irregularities. In fact, in Fig. 5-15 it can be observed that, with an increase in the force of the preload, the field of the amplitudes of the suspension for the negative irregularities diminishes.

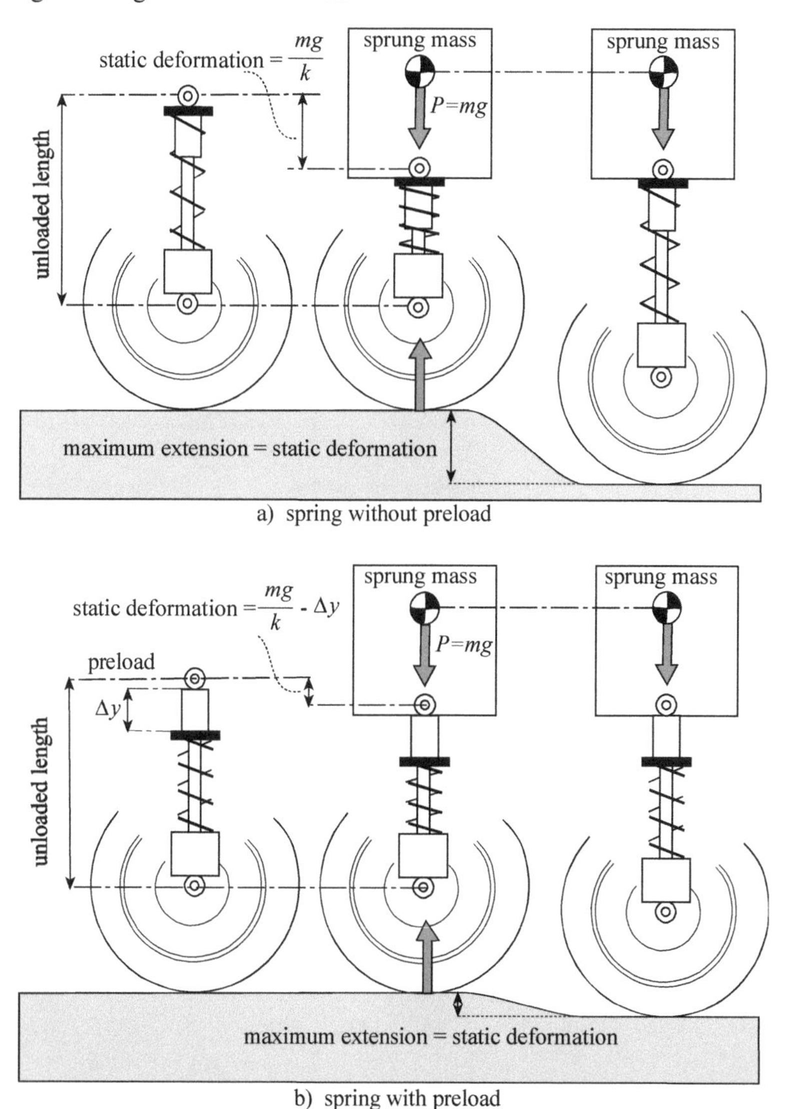

Fig. 5-14 Suspension with preloaded spring.

An approximate value of the reduced stiffness of a suspension can be determined on the basis of simple static considerations. The maximum load on either the front or rear wheels can be equal to the total weight of the motorcycle plus the rider. Such circumstances can occur under the limiting condition of a wheeling or forward fall of the motorcycle, respectively. Stiffness depends on the value of the wheel travel required in this condition:

$$k = \frac{mg}{y_{max}}$$

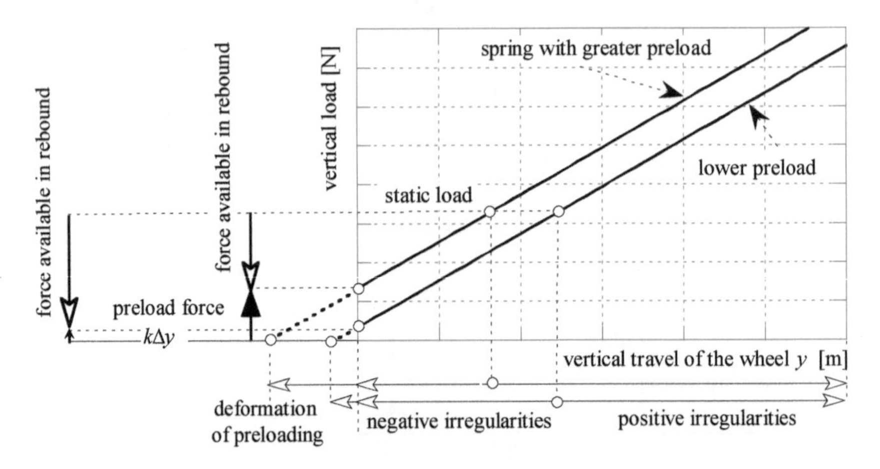

Fig. 5-15 Characteristics of the suspension as the preload varies.

If it is required that for a set load (for example, the static load acting on a wheel due to the weight of the motorcycle plus the rider) the same preload amplitude applies, then different preloads must be adopted by varying the values of the suspension stiffness, as illustrated in Fig. 5-16.

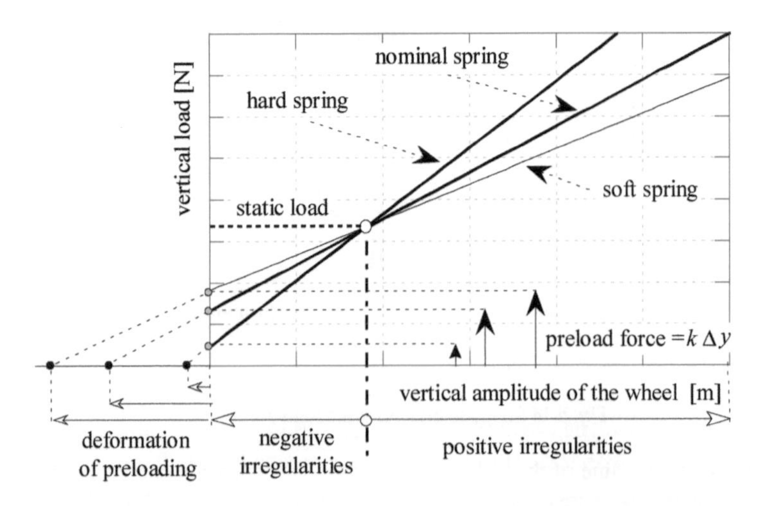

Fig. 5-16 Characteristics of the suspension with varying stiffness.

5.3.5 Front suspension stiffness

The front forks have a slightly progressive behavior because of the influence of air contained in the sleeves, which acts like a pneumatic spring positioned in parallel with the helicoidal spring.

The elastic force, with the influence of the pneumatic spring, is given by the equation:

$$F = k(\Delta y + y) + A\, p_1 \left(\frac{V_1}{V_1 - Ay} \right)^{1.4}$$

where:

$k\Delta y$ represents the force generated by the preload;

ky represents the linear force of the metal spring.

The third term represents the influence of the pneumatic spring; p_1 indicates the initial air pressure contained in the sleeve of the fork; V_1 the initial volume of air and A the area of the section of the cylindrical chamber that contains the air. The effect of the compressed air increases with the decrease of the initial air volume contained in the fork and the increase in wheel travel y.

The value of the reduced fork stiffness varies according to the weight of the motorcycle and its use. Values (for one fork leg) range from about 10 N/mm for light motorcycles to values of about 20 N/mm for heavy motorcycles. The progressive behavior due to the use of variable springs, or to the use of several springs placed in series and having different rigidities, and/or to the influence of air, causes an increase in the stiffness at the end of the stroke, which can be evaluated as 30 to 50%.

Example 2

Consider a motorcycle with a total mass (including the rider) of 200 kg and with a distribution of the loads at 50% - 50%. Let the unsprung mass in front be 18 kg and the angle of inclination of the steering head 24°.

Calculate the stiffness of the reduced spring so that with a load equal to the weight of the motorcycle the fork is compressed by 80 mm.

The maximum weight on the spring is equal to the weight force of the motorcycle minus the weight force of the front unsprung mass. The equivalent vertical stiffness is:

$$k_f = \frac{(100-18)\cdot 9.81}{80 \cos 24°} = 11 \text{ N/mm}$$

The overall stiffness of the fork must therefore be equal to:

$$k = k_f \cos^2 \varepsilon = 9.18 \text{ N/mm}$$

Finally, let us calculate the deflection of the spring, under conditions of static equilibrium, supposing that the spring of the fork is preloaded by 20 mm:

$$y = \frac{(100-18) 9.8 \cos 24°}{k} - \Delta y = 67.5 \text{ mm}$$

Example 3

Now let us evaluate the stiffness of a fork taking into account the influence of air.

- initial volume of air: case a) $V_{1_a} = 270\ \text{cm}^3$;
- initial volume of air: case b) $V_{1_b} = 300\ \text{cm}^3$;
- cross section area of the cylindrical chamber: $A = 10\ \text{cm}^2$;
- initial pressure: $p_1 = 1$ bar;
- linear stiffness of the metal spring: $k = 8$ N/mm;
- preload: $\Delta y = 10$ mm.

Figure 5-17 shows the increase in elastic forces and stiffness, under the variation of the deformation of the spring for two different initial volumes of air.

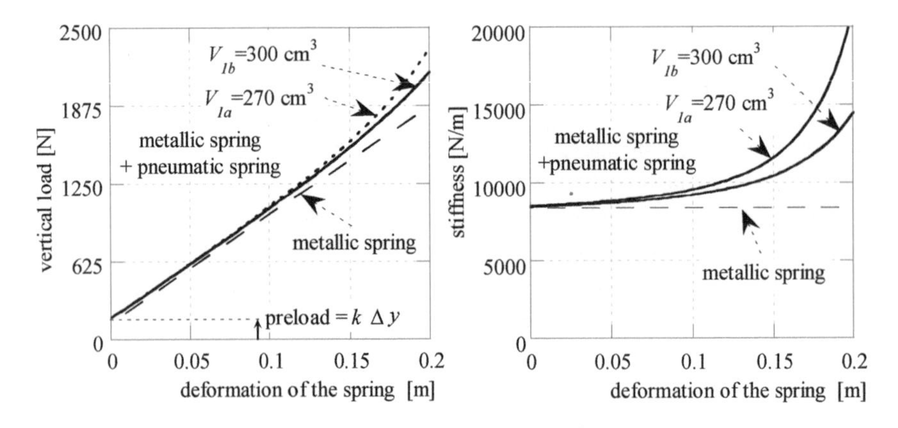

Fig. 5-17 Influence of the pneumatic spring on overall stiffness.

The influence of the pneumatic spring becomes important when the air volume is strongly compressed. For high wheel travel the value of the stiffness can even be doubled. The initial volume of air in the sleeve depends on the quantity of oil contained in the sleeve. A large quantity of oil corresponds to a lower initial volume and therefore to an increase in the influence of the pneumatic spring. Stiffness values and the loads presented in Fig. 5-17 refer to only one spring; therefore the fork as a whole will have a stiffness twice that indicated.

5.3.6 Rear suspension stiffness

Having the curve of desired progressive behavior, the synthesis of the mechanisms can be carried out by means of numerical optimization algorithms. Generally, the ratio between the vertical velocity of the wheel and the deformation velocity of the spring-shock absorber group (the inverse of the ratio τ_{m,y_C}) has values varying from 2 to 4. The higher values, with equal wheel travel rate, cause small deformation velocities of the shock absorber and there is the need to use shock absorbers of larger dimensions.

Figure 5-18 shows the kinematic scheme and the movement of a rear four-bar linkage suspension, with the spring-shock absorber group fastened between the chassis and the rocker. This mechanism makes it possible to generate curves with a significant progressive rate. As illustrated in the curves of the three examples shown on the right in the figure, the degree of the stiffness rising-rate depends on the positions of the spring attachment point. The greatest value of the velocity ratio, and therefore the greatest raising rate, is attained when the spring is orthogonal to the rocker.

Figure 5-19 represents a kinematic diagram with the spring fastened between the chassis and the connecting link. With this arrangement we attain very different curves of stiffness, as the position of the point to which the spring is attached varies. The right-hand figure shows how it is possible to obtain degressive, progressive and approximately constant stiffness.

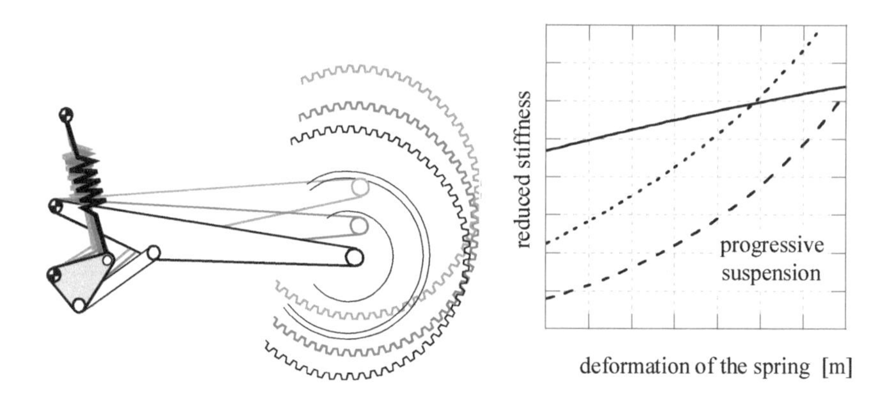

Fig. 5-18 Rear suspension with the spring attached to the rocker arm.

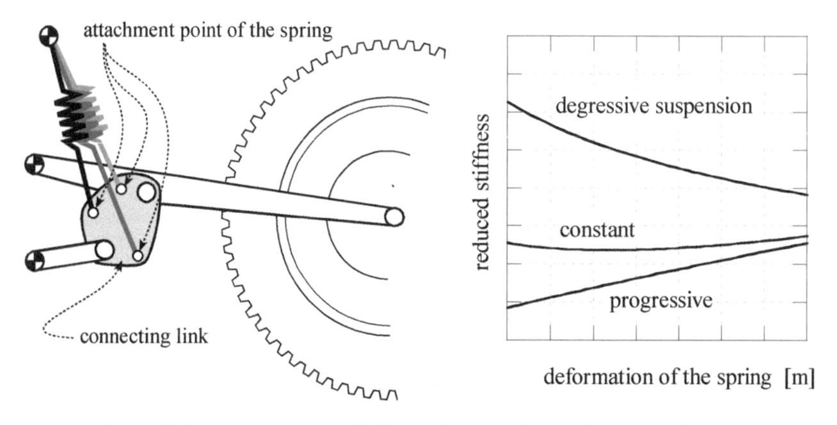

Fig. 5-19 Rear suspension with the spring attached to the connecting link.

Example 4

Consider a classic rear suspension with a swing-arm having the following characteristics:

- length of swinging arm: $L = 0.6$ m;
- distance from pivot to spring: $L_1 = 0.4$ m;
- length of undeformed spring: $L_{m_o} = 0.5$ m.

In the initial position, corresponding to $y = 0$ the axis of the wheel is lowered by 100 mm in relation to the pivot of the swinging arm.

In the graph in Fig. 5-20, the variation of the reduced vertical stiffness (made dimensionless with respect to the initial value calculated in correspondence to $y = 0$) under variation of the vertical displacement of the wheel pin y, is shown for various values of the initial inclination angle of the spring.

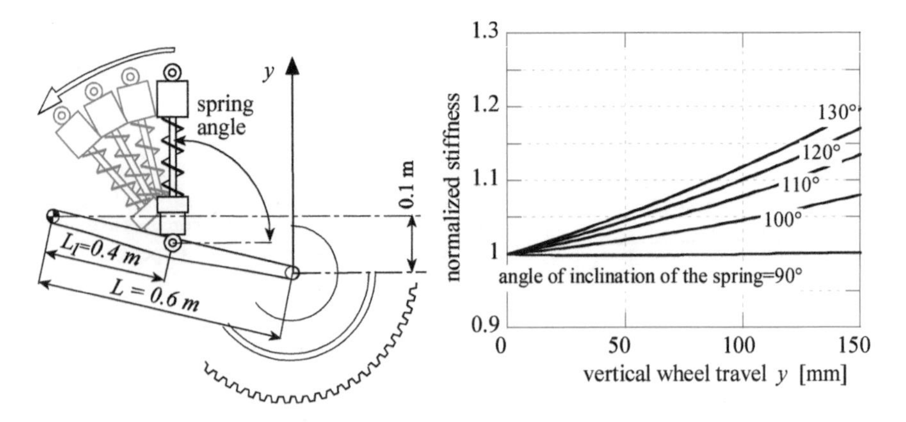

Fig. 5-20 Stiffness curve for various inclination values of the spring-shock absorber group.

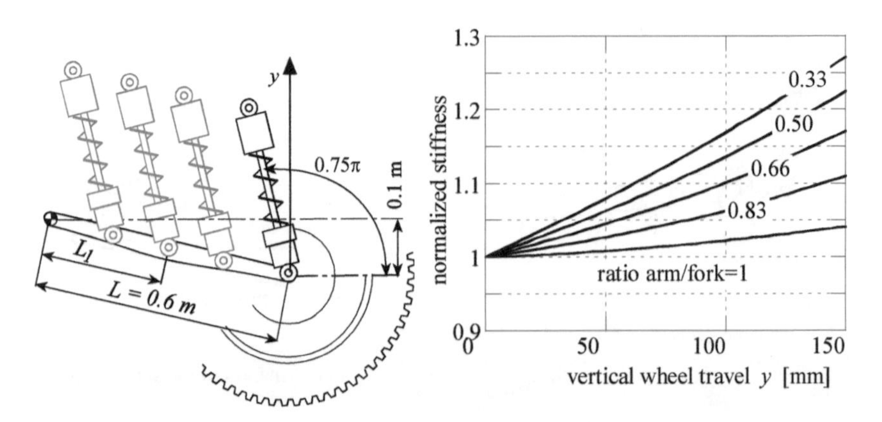

Fig. 5-21 Stiffness curve for various values of the ratio between the spring distance and the length of the swinging arm.

One can observe that when increasing the initial inclination of the spring, the curves obtained are characterized by a greater stiffness rising-rate.

Figure 5-21 shows the influence of the position of the spring attachment point, expressed by the ratio between the arm L_1 and the length of the swinging arm L. It can be observed that positioning the spring attachment point close to the swinging arm pivot causes an increase in the stiffness rate. The progressive behavior attainable with the classic suspension, with long springs that are greatly inclined and fastened near the swinging arm pivot, can reach 50-60% at most.

Example 5

Consider the suspension in the preceding example and evaluate the torsional stiffness when the springs are inclined 135° to the horizontal. Take the stiffness of the spring as $k = 80$ kN/m.

The reduced torsional stiffness is equal to:

- $k_\vartheta = 4.30$ kNm/rad for $y = 0$:
- $k_\vartheta = 5.67$ kNm/rad for $y = 150$ mm.

In the latter case, the principal term $k\,\tau^2_{m,\vartheta}$ gives a value of 5.31 kNm/rad, while the secondary term $k\,(L_m - L_{m_0})\partial\tau_{m,\vartheta}/\partial\vartheta$ gives a lower value of 0.46 kNm/rad.

5.4 Considerations on climbing a step

Consider a motorcycle traveling at constant velocity V which at a certain instant encounters a step with height h. To climb the step, the wheel must advance by a distance s:

$$s = \sqrt{h(2R-h)}$$

Consider first the case of a motorcycle with no suspension systems. The time the wheel takes to climb the step is equal to the ratio:

$$\Delta t = \frac{s}{V}$$

In that case, the advancement s of the wheel corresponds exactly to the advancement of the motorcycle. The velocity and vertical acceleration of the wheel are:

$$\dot{y} = V\sqrt{\left(\frac{R}{R-(h-y)}\right)^2 - 1} \qquad \ddot{y} = V^2\frac{R^2}{[R-(h-y)]^3} \qquad 0 < y < h$$

Example 6

At a velocity of 10 m/s, a wheel of radius $R = 0.307$ m travels over an obstacle 50 mm high in an interval equal to 0.0168 s. In that time the wheel climbs to a height equal to the step so that the average vertical lift speed is 3.3 m/s. The maximum value of the velocity is reached at the initial instant and is equal to 6.5 m/s.

The maximum vertical acceleration also occurs at the initial instant (assuming that the tire is infinitely rigid), and is 555 m/s^2, which is 55 times the acceleration due to gravity. Actually, the acceleration is much lower, since the tire is deformed and partially absorbs the impulse.

This simple example shows the importance of the suspension.

In the case of a motorcycle with suspension, the time taken by the wheel to climb the step depends on the type of suspension and no longer just on the dimensions of the obstacle.

When the front wheel provided with suspension climbs a step the force transmitted to the chassis depends on both the elastic and the damping characteristics of the suspension (Fig. 5-22). At low velocities, the elastic force prevails while at medium and high velocities the damping force is dominant.

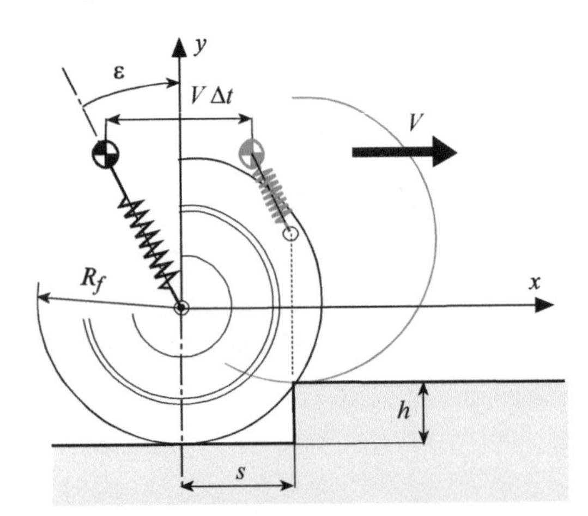

Fig. 5-22 The front wheel at the beginning and end of climbing the step.

Example 7

Let us consider a front suspension with the following properties: $k = 8000$ N/m and $c = 500$ Ns/m, and suppose that while climbing the same step of example 6 the chassis (sprung mass) does not raise. Determine the elastic and damping forces at a forward velocity of 1 m/s and 10 m/s.

Case 1: $\varepsilon = 27°$

The elastic force depends only on the deformation of the spring. Its maximum value is 356 N. If the forward velocity is low (1 m/s), the elastic force is greater than the damping force which is equal to 291 m/s. At a velocity of 10 m/s the damping force prevails (2916 N) and is at a maximum at the beginning of climbing the step.

Case 2: $\varepsilon = 33°$.

As the angle of inclination of the fork increases, both the maximum value of the elastic force (335 N) and that of the viscous force (2740 N) diminish.

Let us now consider the rear wheel provided with a suspension. Let us assume that the vertical position of the vehicle chassis does not change while overcoming the obstacle, or that the swinging arm pivot moves along the x-axis (Fig. 5-23).

The time necessary for the rear wheel to pass from the initial to the final position depends on the velocity of the vehicle and the geometry of the suspension (length of the fork, radius of the wheel, initial angle of inclination of the fork). The time needed to run over the obstacle is given by the following expression:

$$\Delta t = \frac{(\cos\vartheta_2 - \cos\vartheta_1)L + s}{V}$$

where:

ϑ_1 indicates the initial angle of inclination;

ϑ_2 indicates the final angle of inclination;

s indicates the distance between the pin of the wheel and the step.

The time Δt employed by the wheel to climb the step can differ from the time used by the vehicle to travel the distance s: when the wheel climbs the step and advances by distance s in the same time used by the vehicle to advance by the same distance s, the behavior of the suspension is defined as neutral. In the case of climbing times Δt that exceed those of the neutral suspension, the behavior of the suspension is referred to as positive.

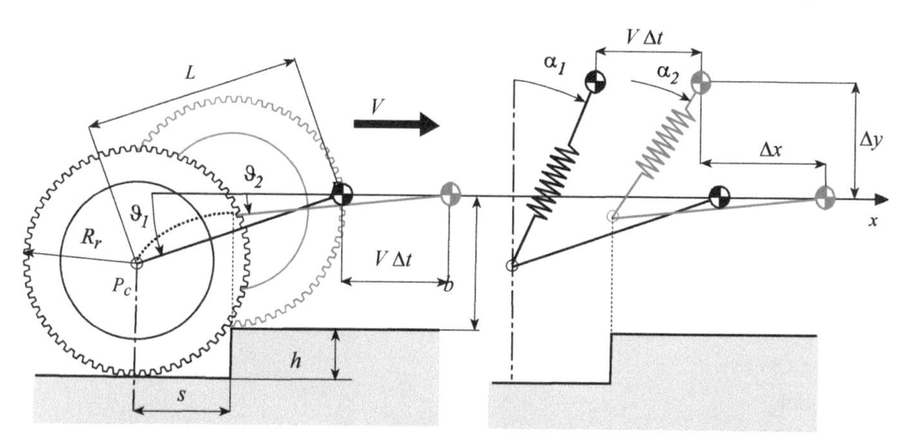

Fig. 5-23 The rear wheel at the beginning and end of climbing the step.

The suspension has positive behavior when the fork, in both the initial and final positions, remains below the x axis (Fig. 5-24); in the case of climbing times Δt that are lower than the neutral suspension, the behavior of the suspension is referred to as negative. The suspension behaves negatively when the fork remains above the x axis in both the initial and the final position (Fig. 5-25).

The circular trajectory described by the center of the wheel in climbing the step is therefore covered in differing times according to the type of behavior of the suspension. In the case of suspension with positive behavior, the trajectory is traversed in a greater time interval. This subjects the wheel to lower vertical accelerations. In particular, the deformation law of the spring-shock absorber group shows lower acceleration.

Fig. 5-24 Rear suspension with positive behavior ($s < V\Delta t$).

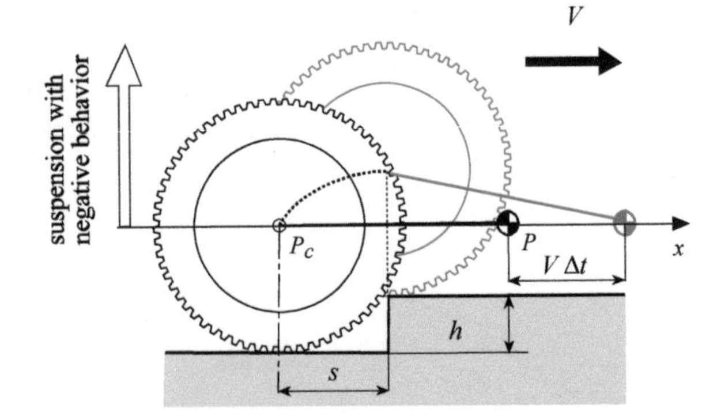

Fig. 5-25 Rear suspension with negative behavior ($s > V\Delta t$).

Example 8

Let us consider a rear wheel with the following properties:

- wheel radius: $R_r = 0.307$ m;
- swinging arm length: $L = 0.6$ m;
- obstacle height: $h = 0.05$ m;
- velocity: $V = 10$ m/s.

The peak value of vertical acceleration depends on the behavior of the rear suspension:

Suspension	ϑ_1	ϑ_2	Acceleration peak value
positive	4.8°	0°	448 m/s²
negative	0°	-4.8°	527 m/s²

As mentioned earlier, peak acceleration is lower in the case of positive suspension. The peak value increases as the radius of the wheel decreases. A diminution of 10% in the wheel radius causes a 5% increase in the peak acceleration value.

A diminution of the deformation velocity of the shock absorber is obtained by increasing the inclination of the spring-shock absorber unit. For example, passing from the vertical position to an angle of 45° reduces the viscous force maximum by 35%.

The same considerations can also be extended to the front suspension. The classic telescopic fork suspension is positive while the pulled or pushed arm suspension can be positive or negative according to the arm's angle.

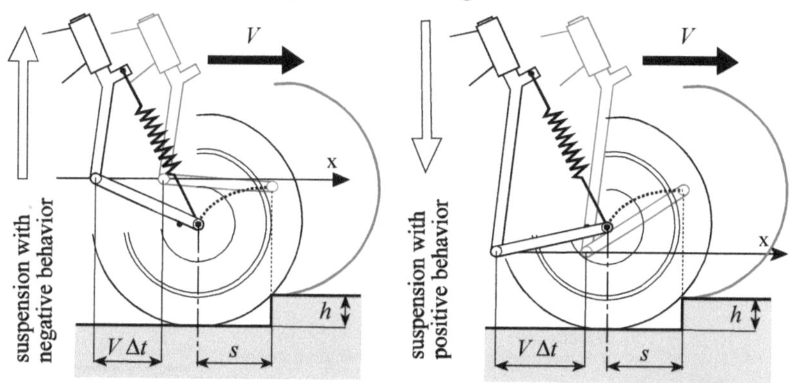

Fig. 5-26 Front suspension with forward arms with negative and positive behavior.

5.5 Slipping of the rear wheel contact point

Consider a motorcycle traveling at constant velocity under a thrust that is also constant.

Actually, the transmission of the thrust force always occurs in the presence of a relative slip between the tire and the road (the peripheral velocity of the tire is greater than the forward velocity of the motorcycle). Suppose the thrust force is constant and therefore the relative slip is also constant.

If there is relative motion between the chassis and the swinging arm (for example, because of irregularities in the road plane), there will be extra slip that will be added to, or subtracted from, the necessary slip due to the thrust.

To evaluate the amount of slip due to the oscillations of the swinging arm, we need to concentrate our attention on the relative motion of the rear arm with respect to the chassis.

The model can be simplified by assuming that the driving sprocket concentric with the axis of the swinging arm and that the motorcycle is stopped with its engine off (Fig. 5-27).

Let us suppose that initially the sprocket is lock to ground with the swinging arm and wheel free to rotate (Fig 5-27 b). If the swinging arm is rotated by the angle $\Delta\vartheta$, the point of contact of the chain on the drive sprocket is rotated by the same angle while a portion of the chain of length $\Delta\vartheta \, r_p$, equal to the product of the angle of rotation by the radius of the sprocket, does not rotate.

In this situation, the wheel is subject to:

- a counterclockwise rotation $\Delta\vartheta$, due to the rotation of the swinging arm (Fig. 5-27 b),
- a clockwise rotation $\Delta\beta$, caused by the extension of a portion of the chain from the larger wheel sprocket (Fig. 5-27 c).

The clockwise rotation $\Delta\beta$ is equal to:

$$\Delta\beta = \frac{\Delta\vartheta\, r_p}{r_c} = \tau_{cp} \cdot \Delta\vartheta \quad \text{where the ratio:} \quad \tau_{cp} = \frac{\omega_c}{\omega_p} = \frac{r_p}{r_c}$$

indicates the transmission ratio between the driving sprocket and the rear sprocket.

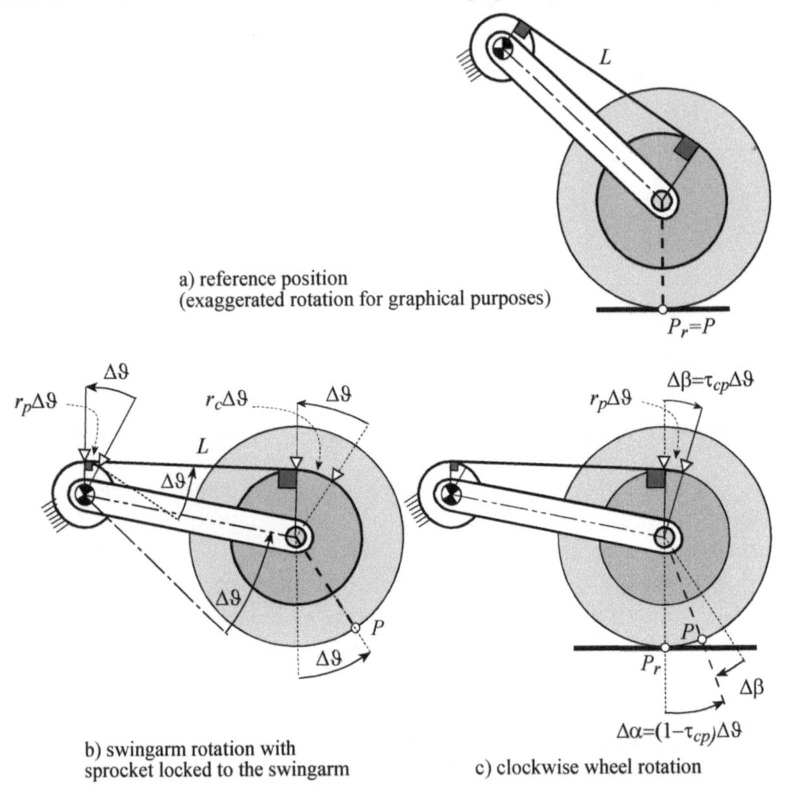

Fig. 5-27 Slippage of the contact point of the rear wheel.

Let P be the point belonging to the tire which at the initial instant represents the contact point between the tire and the ground. Following the oscillation of the fork, it is rotated through the angle:

$$\Delta\alpha = (1 - \tau_{cp})\Delta\vartheta$$

α therefore represents the slide angle due to the rotation of the swinging arm. A slide length of αR_r corresponds to the slippage angle (R_r indicates the radius of the wheel).

In conclusion, giving a counterclockwise rotation on the swinging arm, i.e., compressing the rear suspension, causes a slip that is added to that needed to transmit the thrust. Therefore, it is correct to state that the slip between the wheel and the ground increases during the compression of the suspension and diminishes during its extension.

If we suppose that the swinging arm is subject to an oscillatory motion:

$$\Delta\vartheta(t) = \vartheta_o + \Delta\vartheta \sin\omega t$$

the angular velocity of slippage of the contact point is:

$$\Delta\dot{\alpha} = (1-\tau_{cp})\Delta\vartheta\,\omega\cos\omega t$$

This means that during forward motion at constant velocity V, a fluctuating component is superimposed on the constant angular velocity of rotation of the wheel $\Omega = V/R_r$. Actually, the slippage due to the swinging arm oscillations transmits fluctuations to the sprocket motion as well, so that the irregularity of the motor's rotation is increased. In the same way, the pull on the chain is composed of a constant tension, to which a fluctuating tension generated by swinging arm oscillation is added.

In the more general case in which the sprocket is not concentric with the swinging arm pivot (Fig. 5-28), the expression for the slippage of point P_r is slightly modified.

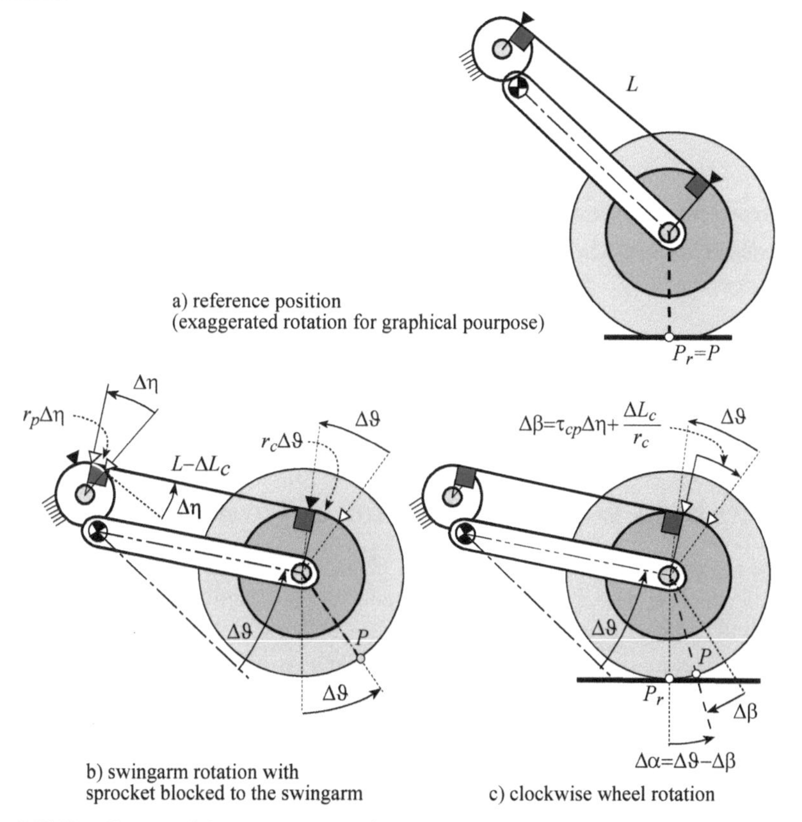

Fig. 5-28 The slippage of the contact point of the rear wheel with a sprocket keyed in a generic position.

In fact, the chain rotates through an angle $\Delta\eta$ which is smaller than the rotation angle $\Delta\vartheta$ imposed on the swinging arm. Furthermore, the distance between the points of tangency of the chain with the rear sprocket and the driving sprocket varies with the variation of the swinging arm rotation angle.

The length of the upper branch of the chain is at a maximum when the axis of the driving sprocket is aligned with the swinging arm axis (Fig. 5-29).

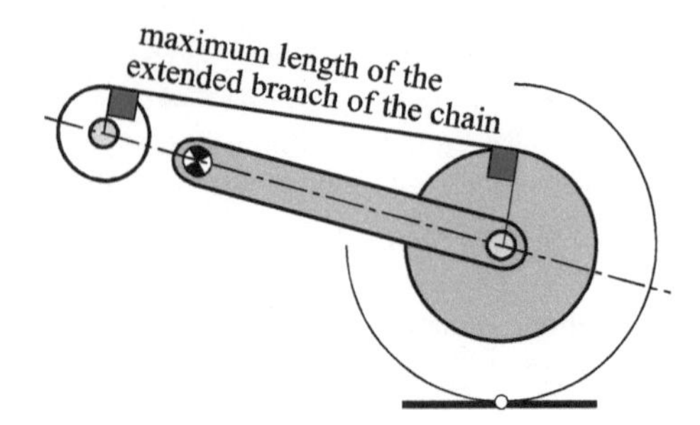

Fig. 5-29 Maximum length of the upper branch of the chain.

The wheel rotates in a counterclockwise direction equal to $\Delta\vartheta$ and a simultaneous clockwise rotation equal to $\Delta\beta$:

$$\Delta\beta = \tau_{cp}\Delta\eta + \frac{\Delta L_c}{r_c}$$

ΔL_c represents the variation in length of the upper branch of the chain (a reduction has a negative sign). The point P of the tire therefore rotates through the angle:

$$\Delta\alpha = \Delta\vartheta - \Delta\beta$$

Since the term ΔL_c is negligible with respect to the product $\Delta\eta\, r_p$, the sliding velocity depends substantially on the transmission ratio between rear sprocket and driving sprocket and on the angle $\Delta\vartheta$ which depends on the length of the swinging arm.

5.6 Models with one degree of freedom

The motorcycle, in straight running, is characterized by five degrees of freedom (one of which is associated with the vehicle forward motion), and is made up of three rigid bodies:

- the sprung mass (chassis, engine and rider);
- the rear unsprung mass (wheel, brake and part of the swinging arm);
- the front unsprung mass (wheel, brake and part of the fork).

In general the in-plane motion of the motorcycle can be considered as the combination of a vertical motion (bounce) and a rotating motion (pitch). These two motions correspond to the vibration modes of the motorcycle in the plane.

Only by ignoring the unsprung masses, and considering the system of two degrees of freedom as uncoupled, we can treat the motorcycle in the plane with two models each having one degree of freedom: one for the vertical bounce and the other for the pitching motion (Fig. 5-30). Considering the motorcycle as an uncoupled system is the same as assuming that, imposing a vertical displacement of the chassis, the consequent movement is composed only of vertical oscillations or that, imposing a rotation of the chassis around an axis passing through the center of gravity, the resulting motion is one of pure pitch.

We, therefore, consider that the motorcycle in the plane is composed of only one sprung mass, sustained by two springs, which represent the action of the suspension and the tires. The reduced stiffness of the suspension is connected in series with the stiffness of the tires, so that the equivalent elastic constants K_f and K_r for the front and rear sections respectively, are:

$$K_f = \frac{k_f k_{p_f}}{k_f + k_{p_f}} \qquad K_r = \frac{k_r k_{p_r}}{k_r + k_{p_r}}$$

where:

k_f is the reduced stiffness of the front suspension;

k_r is the reduced stiffness of the rear suspension;

k_{p_f} is the radial stiffness of the front tire;

k_{p_r} is the radial stiffness of the rear tire.

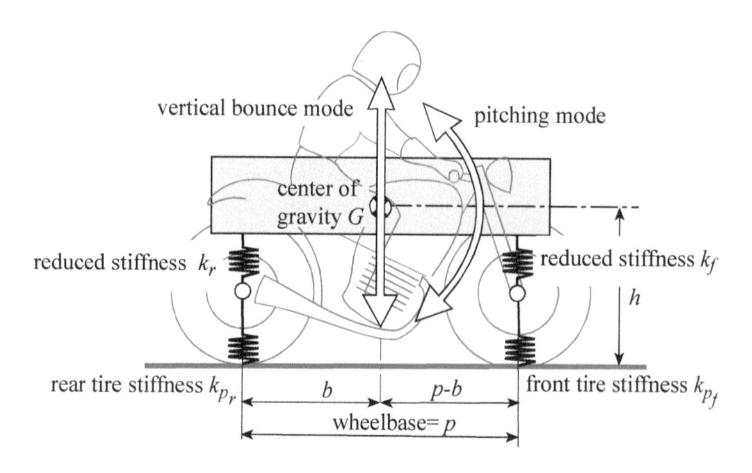

Fig. 5-30 Principal vibration modes in the plane.

Example 9

Consider a motorcycle with a reduced rear suspension stiffness of k_r = 20 N/mm. Compare the overall stiffness of the system with a rather stiff (k_{p_r} = 250 N/mm) and a rather deformable tire (k_{p_r} = 120 N/mm) rear tire.

With a radially stiff rear tire the overall stiffness is K_r = 18.5 N/mm (-7.4 %), while with a rather deformable rear tire the result is K_r = 17.1 N/mm (-14.3 %).

5.6.1 Bounce and pitch motion

The equilibrium equations of vertical forces and the moment about the horizontal axis:

$$m\ddot{z} + (K_r + K_f)z = 0$$

$$I_{y_G}\ddot{\theta} + [K_r b^2 + K_f (p-b)^2]\theta = 0$$

provide the expressions for natural frequencies of the vertical bounce mode, and for the pitching mode, where I_{y_G} is the polar moment of inertia around the y axis.

The natural frequencies ν_b, ν_p respectively, for the vertical bounce motion and the pitch motion are:

$$\nu_b = \frac{1}{2\pi}\sqrt{\frac{K_f + K_r}{m}} \qquad \nu_p = \frac{1}{2\pi}\sqrt{\frac{K_f (p-b)^2 + K_r b^2}{I_{y_G}}} = \frac{1}{2\pi}\sqrt{\frac{K_f \frac{(p-b)^2}{\rho^2} + K_r \frac{b^2}{\rho^2}}{m}}$$

In general, the radius of inertia ρ is less than the distances b and $(p-b)$ so that the frequency of the pitching motion ν_p is greater than that of the vertical bounce motion ν_b.

In this section, we have assumed that the two main vibration modes are uncoupled. This is an ideal condition that is not met in reality because these two modes are generally coupled to each other. The coupling of the modes can be experienced by uniformly lowering the entire motorcycle, without allowing it to rotate and then leaving it free to vibrate. It is easy to verify that both the vertical and the pitch motions are excited. In the same way, by imposing a rotation of the motorcycle around the center of gravity both the pitch and the vertical motions are excited.

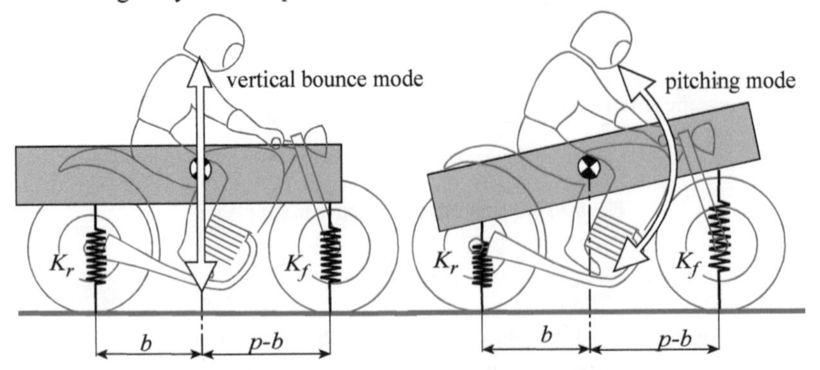

Fig. 5-31 Modes of vibration in the plane.

In conclusion, because of the road irregularities, the motorcycle oscillates with a motion that is a combination of two vibration modes; i.e., the motorcycle oscillates vertically and pitches at the same time.

Example 10

Consider a motorcycle with the following characteristics:

- wheelbase: $p = 1.4$ m;
- distance from the center of gravity to the rear wheel : $b = 0.7$ m;
- sprung mass $m = 200$ kg;
- pitch moment of inertia: $I_{yG} = 38$ kgm^2;
- reduced stiffness of the front suspension: $k_f = 15$ kN/m;
- reduced stiffness of the rear suspension: $k_r = 24$ kN/m;
- radial stiffness of the tires: $k_{p_f} = k_{p_r} = k_p = 180$ kN/m.

Determine the equivalent stiffness of the front and rear sections. Then determine the bounce and pitch frequencies.
The equivalent stiffnesses are:

- front stiffness: $K_f = 13.85$ kN/m;
- rear stiffness: $K_r = 21.18$ kN/m.

The natural frequencies of the two modes of vibration are:

- bounce motion: $\nu_b = 2.11$ Hz;
- pitch motion: $\nu_p = 3.38$ Hz.

The frequency of the vertical bounce motion is less than that of the pitch motion ($\nu_b < \nu_p$). Resonance occurs in correspondence with the following values of the vehicle velocity in the presence of an irregularity on the road plane with a wavelength of 6 m,

- bounce motion: critical velocity = 12.64 m/s;
- pitch motion: critical velocity = 20.29 m/s.

5.6.2 Wheel hop resonance

The suspension stiffness values are significantly lower than those for tire stiffness (tire stiffness is approximately 6-12 times suspension stiffness). The unsprung mass m is, therefore, connected to the ground with a hard spring and to the sprung mass with a soft spring. In a first approximation, the influence of the connection to the suspended mass can be ignored. In this way, the unsprung mass, elastically supported by the vertical stiffness of the tire alone, can be represented by a simple system with one degree of freedom. Therefore the natural frequency of the vertical motion of the unsprung mass is (Fig. 5-32):

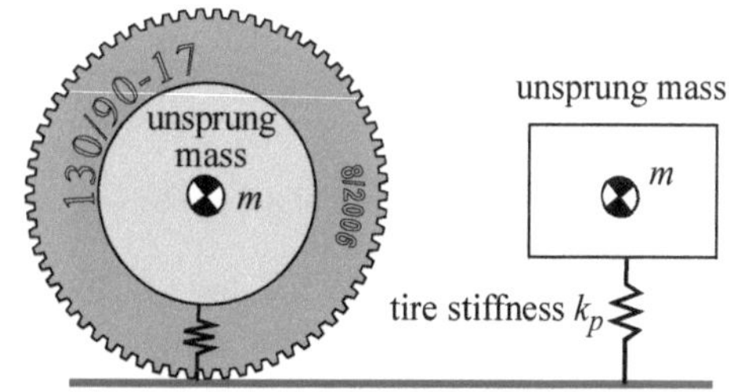

Fig. 5-32 Model with one degree of freedom of the vertical movement of the wheel.

$$\nu_t = \frac{1}{2\pi}\sqrt{\frac{k_p}{m}}$$

Example 11

Consider the motorcycle of the previous example.
With the unsprung masses, 15 kg on the front and 18 kg on the rear. Determine the natural frequencies.

The natural frequencies are:

- vertical motion of the front tire: $\nu_{t_f} = 17.43$ Hz;
- vertical motion of the rear tire: $\nu_{t_r} = 15.92$ Hz.

With an irregularity in the road plane with a wavelength of 3 m, the condition of resonance occurs in correspondence with a vehicle forward velocity of 52.3 and 47.5 m/s, respectively, for the front and rear wheels.

5.7 Two degree of freedom model

If we ignore the unsprung masses, the system has two degrees of freedom. They can be associated with the vertical displacement of the motorcycle center of gravity (sprung mass), and with its pitch rotation about a horizontal axis (Fig. 5-33).

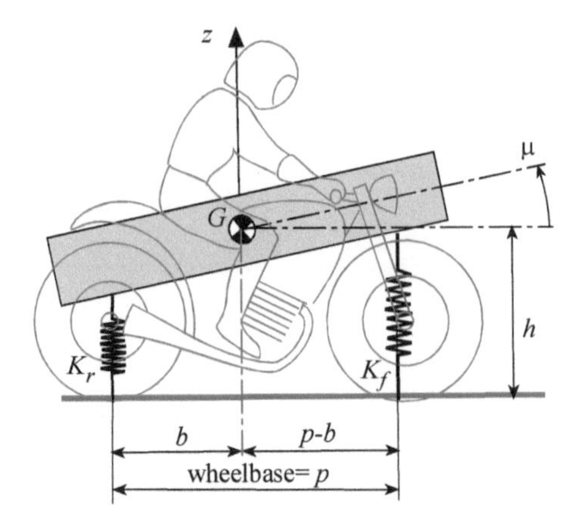

Fig. 5-33 Model of the motorcycle in the plane with two degrees of freedom.

When studying the modes of vibration, having ignored the unsprung masses, does not lead to significant inaccuracies since the stiffness of the suspension is 6 to 12 times smaller than the vertical stiffness of the tires. The influence of the unsprung masses becomes important at medium and high frequencies of the irregularities in the road plane, i.e., at high velocities and short wavelengths.

The free oscillations, ignoring the damping effect, are described by the following equations:

$$\begin{bmatrix} m & 0 \\ 0 & I_{yG} \end{bmatrix} \begin{Bmatrix} \ddot{z} \\ \ddot{\mu} \end{Bmatrix} + \begin{bmatrix} K_f + K_r & K_f(p-b) - K_r b \\ K_f(p-b) - K_r b & K_f(p-b)^2 + K_r b^2 \end{bmatrix} \begin{Bmatrix} z \\ \mu \end{Bmatrix} = 0$$

The frequency equation is then:

$$-m I_{yG}\,\omega^4 + \left[(I_{yG} + m b^2)K_r + (I_{yG} + m(p-b)^2)K_f\right]\omega^2 - p^2 K_r K_f = 0$$

The two roots of the equation are the undamped system's two natural frequencies. The ratio between the amplitudes of the vertical oscillation and the pitch oscillation is given by the equation

$$s_i = \frac{z_{0_i}}{\mu_{0_i}} = \frac{-K_f(p-b) + K_r b}{K_f + K_r - m\omega_i^2} = \frac{K_f(p-b)^2 + K_r b^2 - I_{yG}\,\omega_i^2}{-K_f(p-b) + K_r b} \qquad i = 1,2$$

ω_i indicates the natural frequency of the vibration mode considered.

These expressions represent the distance from the center of rotation to the center of gravity.

Example 12

Consider the motorcycle in example 10.

The natural frequencies calculated with the two degrees of freedom model, are:

- vertical bounce motion: $\nu_b = 2.02$ Hz;
- pitching motion: $\nu_p = 3.42$ Hz.

These values differ at most by 4% with respect to the corresponding values calculated when considering two uncoupled systems with one degree of freedom each.

The first mode of vibration, represented in Fig. 5-34, is basically a vertical bounce motion of the motorcycle. The instantaneous rotation center is located behind the motorcycle at a distance $s_1 = 2.08$ m from the center of gravity.

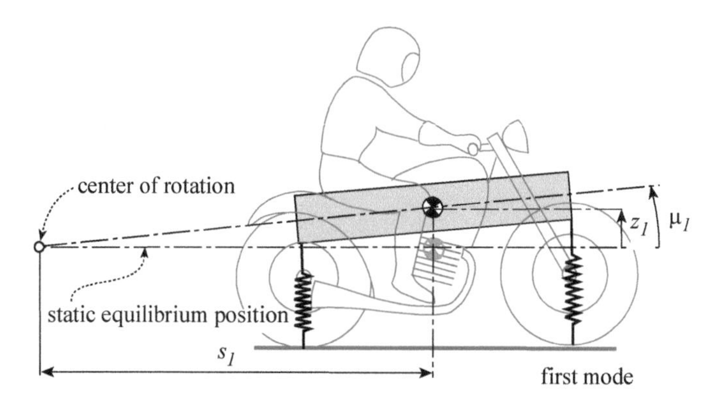

Fig. 5-34 First mode of motorcycle vibration in the plane (vertical bounce mode, $i = 1$).

The second mode, represented in Fig. 5-35, is essentially a pitch mode, its rotation center is located immediately in front of the center of gravity, at a distance of only $s_2 = 0.09$m.

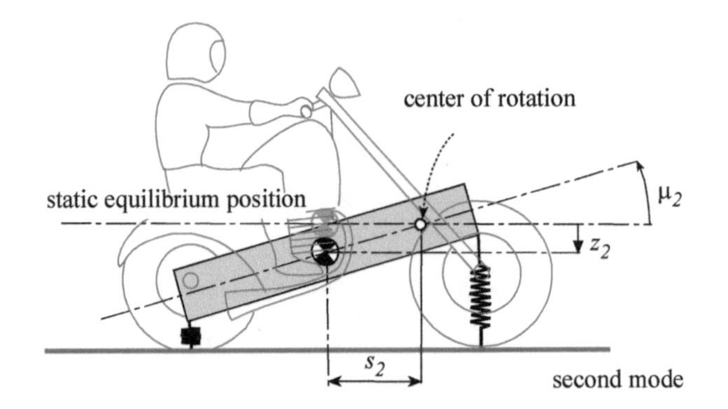

Fig. 5-35 Second mode of motorcycle vibration in the plane (pitching mode $i = 2$).

* * *

Only if the following condition between stiffnesses and distances is fulfilled can the two motion equations, for z and μ respectively (p. 174), be uncoupled:

$$-K_r b + K_f (p-b) = 0$$

Therefore, the natural frequencies will be equal to those previously determined in the models with one degree of freedom.

In this case the first mode, called the bounce mode, now becomes a pure vertical translation while the second mode, called the pitch mode, becomes a pure rotation around the center of gravity. In this case where the modes are not coupled, the distance s_1 from the rotation center of the first mode tends towards infinity, while the distance s_2 from the center of rotation of the second mode is zero.

Example 13

Consider the motorcycle of the previous example. Determine the value of b needed to uncouple the bounce and pitch modes.

The uncoupling of the pitch motion from the bounce motion can be accomplished by reducing the distance b to a value of 0.55 m. Consequently the distribution of the weights becomes 40% on the front section and 60% on the rear.

The uncoupling can also be obtained by modifying the stiffness of the suspensions. Since the distances $(p-b)$ and b are equal, the stiffness of the front suspension needs to be equal to that of the rear suspension, i.e., equal to 24 kN/m.

5.8 Four degrees of freedom model

As mentioned earlier, the motorcycle in its plane of symmetry can be represented as three rigid bodies whose vibrating motion is described by four independent coordinates, as illustrated in Fig. 5-36:

- the vertical displacement of the sprung mass center;
- the pitching rotation of the sprung mass;
- the vertical displacements of the two unsprung masses (the equivalent masses at the center of the wheels).

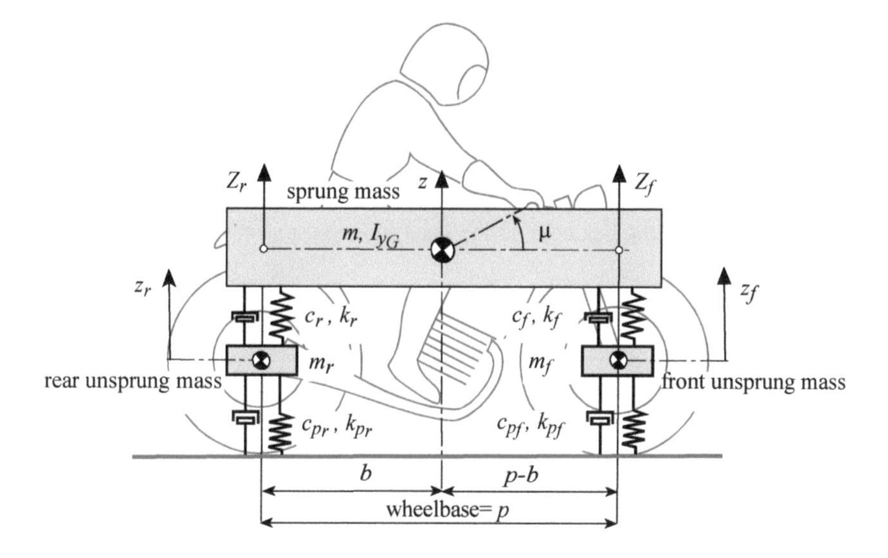

Fig. 5-36 Diagram of the motorcycle with four degrees of freedom.

The equations of free, undamped motion are summed up in the following matrix equation:

$$\begin{bmatrix} m & 0 & 0 & 0 \\ 0 & I_{yG} & 0 & 0 \\ 0 & 0 & m_f & 0 \\ 0 & 0 & 0 & m_r \end{bmatrix} \begin{Bmatrix} \ddot{z} \\ \ddot{\mu} \\ \ddot{z}_f \\ \ddot{z}_r \end{Bmatrix} + \begin{bmatrix} k_f + k_r & (p-b)k_f - bk_r & -k_f & -k_r \\ (p-b)k_f - bk_r & (p-b)^2 k_f + b^2 k_r & -(p-b)\ k_f & bk_r \\ -k_f & -(p-b)\ k_f & k_f + k_{p_f} & 0 \\ -k_r & b\ k_r & 0 & k_r + k_{p_r} \end{bmatrix} \begin{Bmatrix} z \\ \mu \\ z_f \\ z_r \end{Bmatrix} = 0$$

The natural frequencies are calculated by solving the related eigenvalue problem numerically. Rewriting the equations of free motion, using coordinates z_f, Z_f, z_r, Z_r, brings some interesting points to light.

$$\begin{bmatrix} m\frac{b^2}{p^2}+\frac{I_{y_G}}{p^2} & 0 & m\frac{(p-b)b}{p^2}-\frac{I_{y_G}}{p^2} & 0 \\ 0 & m_f & 0 & 0 \\ m\frac{(p-b)b}{p^2}-\frac{I_{y_G}}{p^2} & 0 & m\frac{(p-b)^2}{p^2}+\frac{I_{y_G}}{p^2} & 0 \\ 0 & 0 & 0 & m_r \end{bmatrix} \begin{Bmatrix} \ddot{Z}_f \\ \ddot{z}_f \\ \ddot{Z}_r \\ \ddot{z}_r \end{Bmatrix} + \begin{bmatrix} k_f & -k_f & 0 & 0 \\ -k_f & k_f+kp_f & 0 & 0 \\ 0 & 0 & k_r & -k_r \\ 0 & 0 & -k_r & k_r+kp_r \end{bmatrix} \begin{Bmatrix} Z_f \\ z_f \\ Z_r \\ z_r \end{Bmatrix} = 0$$

Observing the mass matrix, we note that the first two equations are uncoupled from the other two if the following term is zero:

$$m(p-b)b - I_{y_G} = 0$$

That is, the product of the distances $(p-b)$ and b equals the square of the inertia radius $\rho^2 = I_{y_G}/m$.

In this case, the four equations represent two mono-suspensions with two degrees of freedom that are independent of each other, as shown in Fig. 5-37. The first two expressions describe the behavior of the front mono-suspension, while the last two describe the rear one.

$$\begin{bmatrix} mb/p & 0 \\ 0 & m_f \end{bmatrix} \begin{Bmatrix} \ddot{Z}_f \\ \ddot{z}_f \end{Bmatrix} + \begin{bmatrix} k_f & -k_f \\ -k_f & k_f+kp_f \end{bmatrix} \begin{Bmatrix} \ddot{Z}_f \\ \ddot{z}_f \end{Bmatrix} = 0$$

$$\begin{bmatrix} m(p-b)/p & 0 \\ 0 & m_r \end{bmatrix} \begin{Bmatrix} \ddot{Z}_r \\ \ddot{z}_r \end{Bmatrix} + \begin{bmatrix} k_r & -k_r \\ -k_r & k_r+kp_r \end{bmatrix} \begin{Bmatrix} \ddot{Z}_r \\ \ddot{z}_r \end{Bmatrix} = 0$$

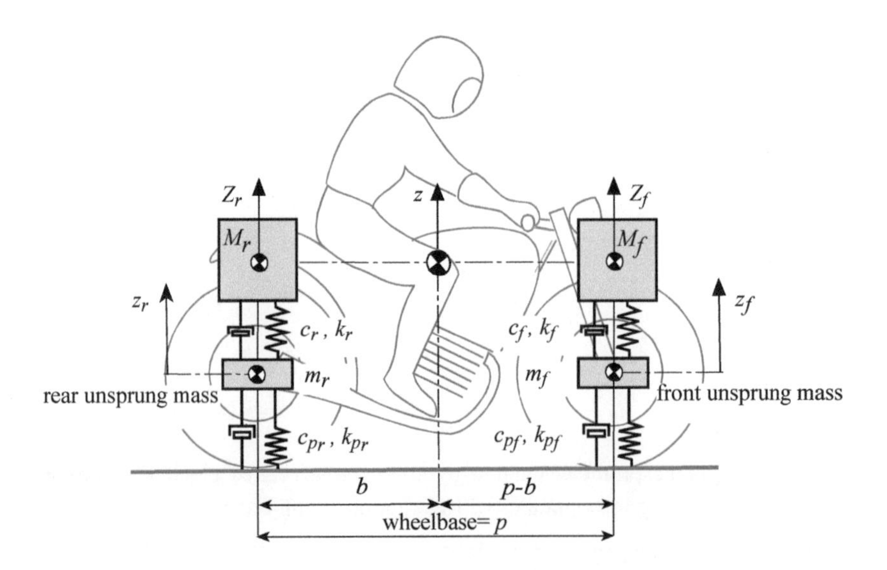

Fig. 5-37 Mono-suspensions of the front and rear sections.

From a physics point of view, this means that the suspended mass can be represented by an equivalent dynamic system composed of two masses placed at the extremities in correspondence with the two wheels:

$$M_f = m\frac{b}{p} \qquad M_r = m\frac{(p-b)}{p}$$

The suspended mass m is distributed between the front section (M_f) and the rear section (M_r) in the same proportion in which the static loads on the wheels are distributed.

It is vital to specify that this condition is difficult to accomplish under real conditions (with reference to the suggested example, the product of the distances $(p-b)b$ is equal to 0.49, while the square of the inertia radius is 0.44), nevertheless it is important from a physics point of view. Therefore, in assuming that this condition does apply, it is possible to conclude that the vertical displacements of the front and rear sections will occur independently of each other, and that, therefore, the pitching motion depends on the value of the phase between the two vertical motions.

Ignoring the unsprung masses, the natural frequencies of the two systems can be easily calculated. For the front and rear sections, we have:

$$\nu_f = \frac{1}{2\pi}\sqrt{\frac{K_f}{M_f}} = \frac{1}{2\pi}\sqrt{\frac{K_f p}{mb}}$$

$$\nu_r = \frac{1}{2\pi}\sqrt{\frac{K_r}{M_r}} = \frac{1}{2\pi}\sqrt{\frac{K_r p}{m(p-b)}}$$

These frequencies are equal if the necessary condition for the uncoupling of the vertical motion from the pitching $K_r b = K_f(p-b)$ is satisfied.

In this case, the bounce has the same frequency as the pitch motion, so that each free motion is made up of a combination of vertical and rotational oscillations with the same frequency.

Generally, the front suspension has a relatively lower stiffness than the rear suspension and, therefore, the frequency of the front section is lower than the frequency of the rear section. In percentage terms ν_f is equal to 70 to 80% of ν_r.

The suspension stiffness requirement can be evaluated on the basis of considerations on the frequencies of the front and rear sections. For good riding comfort, the two frequencies should have values around 1.5 Hz, and the pitch rotation center should be located in the area of the rider's seat.

In racing vehicles, known for their rather rigid suspensions, the natural frequencies vary from 2 Hz to 2.6 Hz.

Usually in street motorcycles and in scooters, the distance ($p-b$) is greater than b, and therefore in order to obtain the same natural frequency at both the front and rear sections, the stiffness at the front should be assumed to be less than the rear.

To evaluate the approximate values of the frequencies of the front and rear sections experimentally, the static deflections Δ_f, Δ_r, due to the weight pressing on the two wheels need to be measured. As such the corresponding frequencies are given by the equations:

$$\nu_f = \frac{1}{2\pi}\sqrt{\frac{g}{\Delta_f}} \qquad \nu_r = \frac{1}{2\pi}\sqrt{\frac{g}{\Delta_r}}$$

Hence, for a frequency of 2 Hz a static deflection of 60 mm is required.

Example 14

Consider the vehicle described previously and suppose that the unsprung masses have the following values:

- front mass: $m_f = 15$ kg;
- rear mass: $m_r = 18$ kg.

Determine the four natural frequencies of the system.

The four natural frequencies are: 2.03, 3.42, 16.98, 18.18 Hz; it is useful to recall that the natural frequencies of the sprung mass, calculated while the unsprung masses are ignored, are equal to 2.11, 3.38 Hz, while those of the unsprung masses are equal to 15.91 and 17.43 Hz.

Example 15

The sprung masses on the front and rear sections are, in the case of load distribution 50% - 50%, equal to $M_f = M_r = 100$ kg. Determine the natural frequencies of the front and rear sections.

Ignoring the unsprung masses, the natural frequencies are:

- front section frequency: $\nu_f = 1.87$ Hz;
- rear section frequency: $\nu_r = 2.32$ Hz.

In percentage terms, ν_f is equal to 81% of ν_r.

Example 16

A motorcycle with a weight (including the rider) of 1970 N has a front unsprung mass of 14 kg and a rear one of 16 kg. With a 50% - 50% distribution of the loads, calculate the front and rear reduced stiffnesses in such a way that the front frequency is equal to 1.9 Hz and that of the rear 2.3 Hz.

The masses pressing down on the suspensions are:

$$M_f = \frac{1}{2}\frac{1970}{9.81} - 14 = 86.4\,\text{kg}$$

$$M_r = \frac{1}{2}\frac{1970}{9.81} - 16 = 84.4\,\text{kg}$$

The overall front and rear stiffnesses are:

$$K_f = (2\pi\nu_f)^2 M_f = 12.32\,\text{kN/m}$$

$$K_r = (2\pi\nu_r)^2 M_r = 17.65\ \text{kN/m}$$

The reduced stiffnesses of the suspensions, taking into account the tire stiffnesses, have the values:

$$k_f = \frac{K_f \cdot kp_f}{kp_f - K_f} = 13.14\ \text{kN/m}$$

$$k_r = \frac{K_r \cdot kp_r}{kp_r - K_r} = 19.36\ \text{kN/m}$$

In Fig. 5-38 the in-plane modes of vibration of the motorcycle are given, calculated with a mathematical model that takes into account the motorcycle geometry. The meaning of the damping ratio will be explained in the next section.

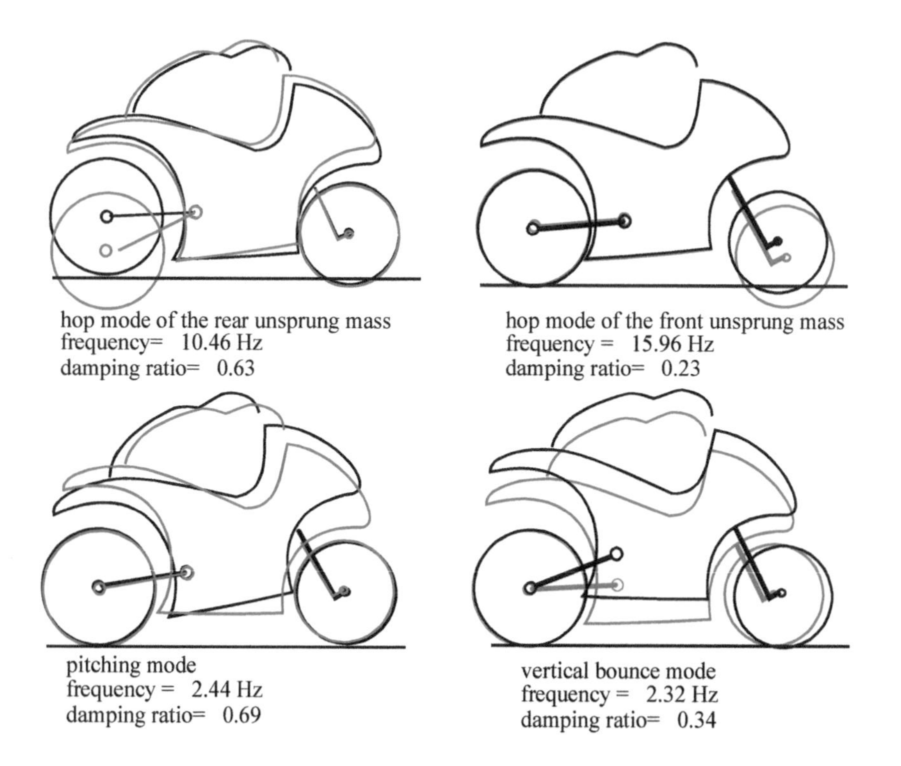

Fig. 5-38 Modes of vibration in the plane.

5.9 One degree of freedom mono-suspension

5.9.1 Oscillatory motion imposed by road irregularities

Consider a motorcycle running on a sinusoidal profile road at constant velocity V and suppose that the motorcycle can be represented by two separate mono-suspensions. Suppose furthermore that the unsprung masses are negligible. The model of the mono-suspension with one degree of freedom can represent either the front or rear suspensions.

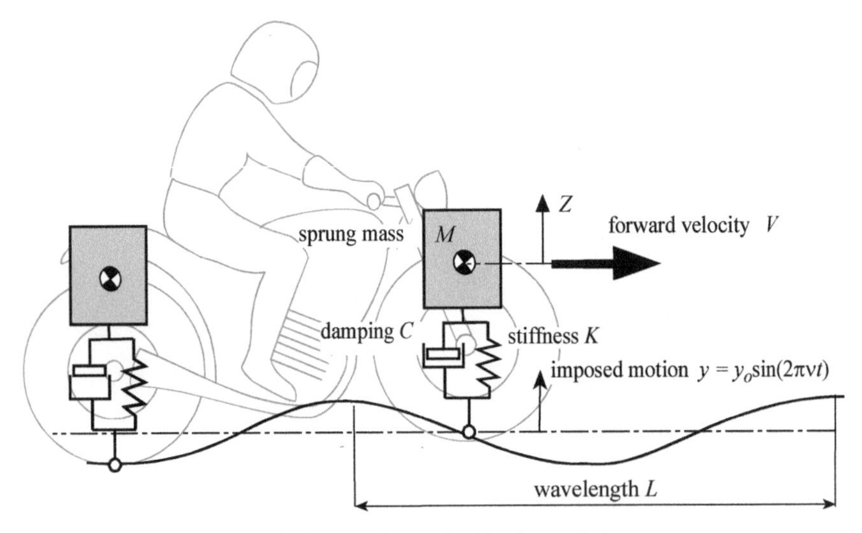

Fig. 5-39 Suspension excited by the road plane.

Consider the front suspension affected by the motion caused by road irregularities, represented in Fig. 5-39. The contact point of the wheel with the road profile moves in harmonic motion according to the law:

$$y = y_o \sin 2\pi \nu t$$

$\nu = V / L$ represents the frequency of the motion imposed on the system by the road irregularities:

It can easily be demonstrated that in steady state (periodic response) the ratio between the amplitude Z_o of the vertical motion of the sprung mass and the amplitude y_o of the imposed motion is:

$$T = \frac{Z_o}{y_o} = \sqrt{\frac{1 + \left(2\zeta \frac{\nu}{\nu_n}\right)^2}{\left(1 - \left(\frac{\nu}{\nu_n}\right)^2\right)^2 + \left(2\zeta \frac{\nu}{\nu_n}\right)^2}}$$

ν_n represents the frequency of the mono-suspension. The ratio T is called transmissibility. In Fig. 5-40 the transmissibility curves T for various values of the damping ratio ζ are given. It is useful to recall that the damping ratio is given by the equation:

$$\zeta = \frac{C}{2\sqrt{K M}}$$

From the graph it is clear that the transmissibility, for any value of the damping, is always equal to one at the value $\sqrt{2}$ of the frequency ratio. This value appears when the forward velocity satisfies the relationship:

$$V = L\, \nu_n \sqrt{2}$$

If the wavelength is equal to 6 m and the natural frequency is equal to 2 Hz, the transmissibility has a value of unity when the forward velocity is equal to approximately 17 m/s.

The plot clearly highlights that:

- for values of the frequency ratio less than $\sqrt{2}$ (velocity $V < 17$ m/s), the introduction of suspension increases the oscillation amplitude ($T > 1$). Therefore, the application of suspension is useful ($T < 1$) only for values of the frequency ratio greater than $\sqrt{2}$ (velocity $V > 17$ m/s);
- high values of the damping ratio attenuate the increase in transmissibility for ratios of the frequencies less than $\sqrt{2}$ (velocity $V < 17$ m/s), but they worsen the responsiveness of the system at high velocities (velocity $V > 17$ m/s).

For the study of riding comfort, the vertical acceleration graph of the motorcycle is more interesting. The rider (which in this model is assumed to be fixed to the motorcycle and forced to move along the vertical axis), perceives a sensation of comfort, which is related to the accelerations to which his body is subjected.

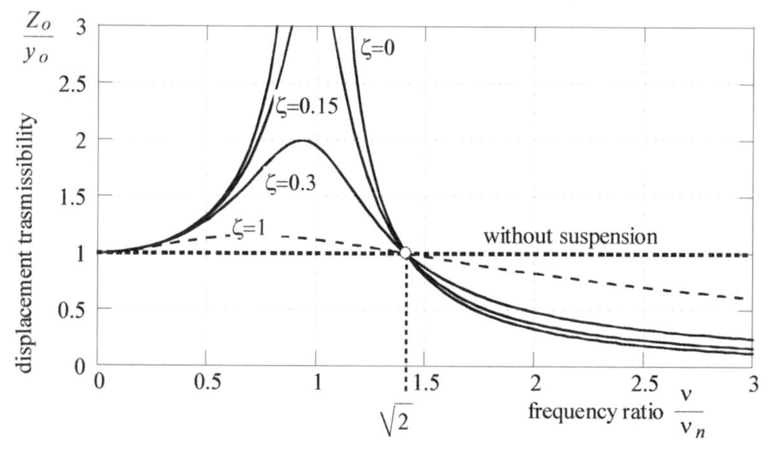

Fig. 5-40 Displacement transmissibility versus the frequency ratio.

Figure 5-41 represents the transmissibility of vertical acceleration, as a function of the ratio of the frequency for various values of the suspension damping ratio.

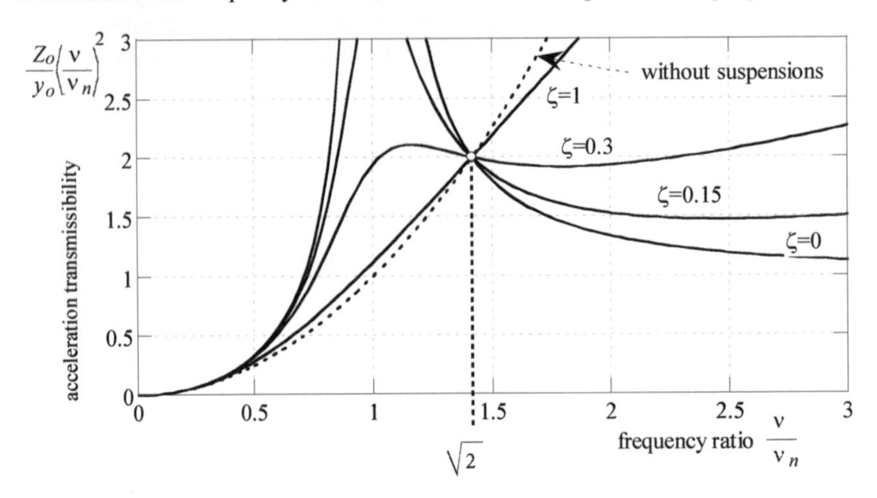

Fig. 5-41 Transmissibility of accelerations versus the frequency ratio.

All the curves assume the value of 2, when the frequency ratio is equal to $\sqrt{2}$, but with different slopes. The slope is zero when the value of the damping ratio is equal to: $\zeta = \sqrt{2}/4 = 0.354$.

The curve characterized by such a slope ensures minimum accelerations around the point $\nu/\nu_n = \sqrt{2}$. The ratio ζ = 0.354 represents the optimal value around the point considered. The graph allows us to draw the following important conclusions:

- the suspension behaves like a filter that cuts the high frequencies and amplifies those found in a narrow band around the condition of resonance.
- a significant increase in riding comfort can be obtained by reducing the motorcycle's natural frequency values. This reduction can be obtained by diminishing suspension stiffness, that is by using softer springs. It must be noted, however, that excessively soft springs can compromise the vehicle trim especially in phases of rapid acceleration or sudden stops.

5.9.2 Optimal value of the damping ratio

Consider the mono-suspension illustrated in Fig. 5-42, and suppose that the mass is oscillating freely and that, at the initial instant, it passes through the position of static equilibrium at velocity $\dot{Z}_o$.

The law of harmonic motion, assuming zero damping, is:

$$Z(t) = Z_o \sin 2\pi\nu_n t = \frac{\dot{Z}_o}{2\pi\nu_n} \sin 2\pi\nu_n t$$

The maximum value of acceleration is:

$$\ddot{Z}_o = (2\pi\nu_n)^2 Z_o$$

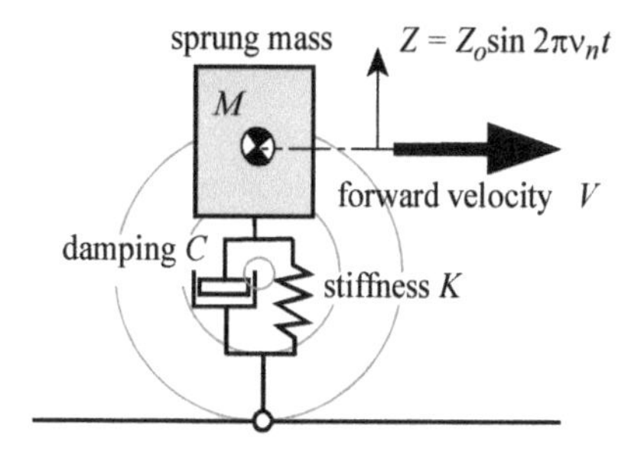

Fig. 5-42 Mono-suspension.

In order to reduce maximum acceleration, it is necessary to attach a viscous shock absorber to the spring; hence, the mass oscillates freely after a disturbance according to the following law:

$$Z(t) = Z_o e^{-\zeta(2\pi\nu_n)t} \sin q_n t = \frac{\dot{Z}_o}{q_n} e^{-\zeta(2\pi\nu_n)t} \sin q_n t$$

where q_n indicates the natural frequency of the damped mono-suspension.

The evaluation of riding comfort can be associated with the maximum peak of vertical acceleration of the sprung mass. The best comfort occurs when peak acceleration is at a minimum.

The graph of Fig. 5-43 shows the course of the vertical acceleration of the sprung mass for several values of the damping ratio. The ideal condition to provide a comfortable ride is with a damping ratio value of 0.35, at which the acceleration becomes minimal.

It is important to note that the optimal value of the damping ratio coincides with that derived previously on the basis of the periodic response in steady state.

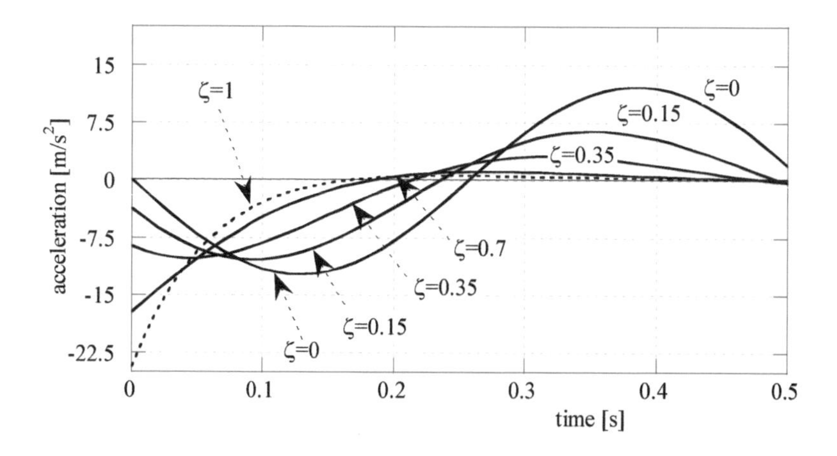

Fig. 5-43 Acceleration after a road bump for various values of the damping ratio.

5.9.3 Considerations on single and double effect shock absorbers

Suppose that the mono-suspension moves along a cosinusoidal shaped bump and assume that the time necessary for the vehicle to transit the irregularity is small compared to the inverse of the natural frequency of the system (Fig. 5-44).

The viscous shock absorber affects the motion of the sprung mass during the passage over the road irregularity. We evaluate its influence in three cases:

- a shock absorber with constant c acting in both the compression and extension phases (double effect);
- a shock absorber with constant $2c$ acting only in the compression phase (single effect);
- a shock absorber with constant $2c$ acting only in the extension phase (single effect).

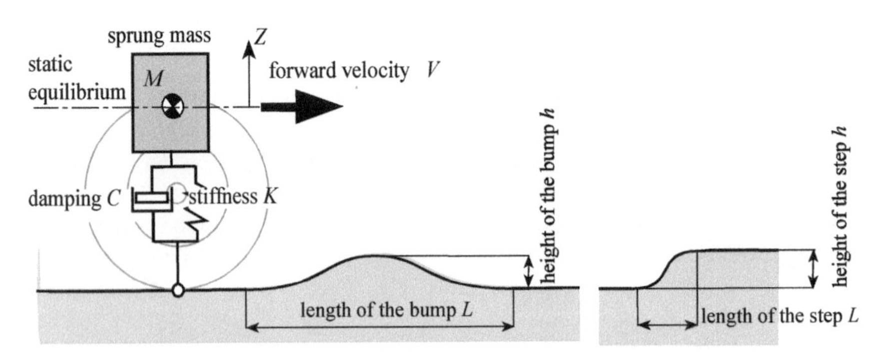

Fig. 5-44 Mono-suspension traveling over a bump and over a step.

Example 17

Consider a mono-suspension with the following properties:

- sprung mass: $M = 140$ kg;
- stiffness: $K = 20$ kN/m;
- damping in extension and compression $C = 1171$ Ns/m, $\zeta = 0.35$;
- damping only in extension: $C = 2342$ Ns/m;
- damping only in compression: $C = 2342$ Ns/m.

Determine the natural frequency.

The natural frequency ν_n is equal to 1.9 Hz.

Passing over a bump

The bump has a height of 0.008 m and a length of 0.6 m. With a forward velocity of 15 m/s, the bump is passed over in 0.04 s; it is assumed that the wheel never departs from the road profile.

Figure 5.45 shows the course of the displacement of the sprung mass versus the time. It can be noted that a double effect damper behaves optimally, since it has less displacement from the position of static equilibrium.

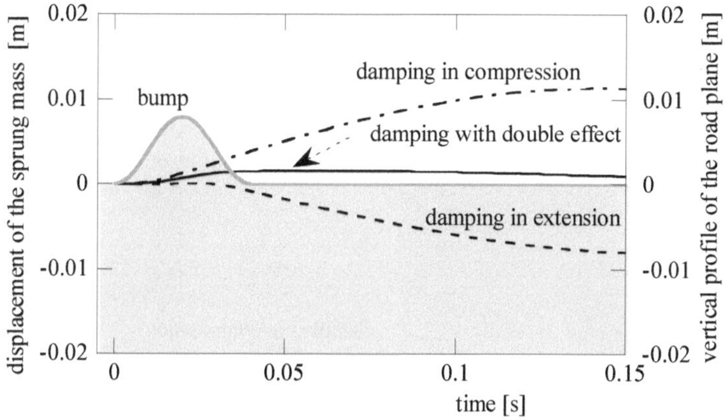

Fig. 5-45 Traveling over a bump: evolution of the displacement of the sprung mass.

The behavior of the double effect shock absorber can be easily explained if the time employed in driving over the bump is compared with the natural period of the suspension. Since the natural period is of a larger order of magnitude, an almost impulsive force on the sprung mass is exerted on the shock absorber, first with a positive sign and then with a negative sign, which produces overall equal effects of opposite sign.

However, in the case of a shock absorber acting only in the compression or extension phase, the sprung mass undergoes, respectively, a positive or negative impulse.

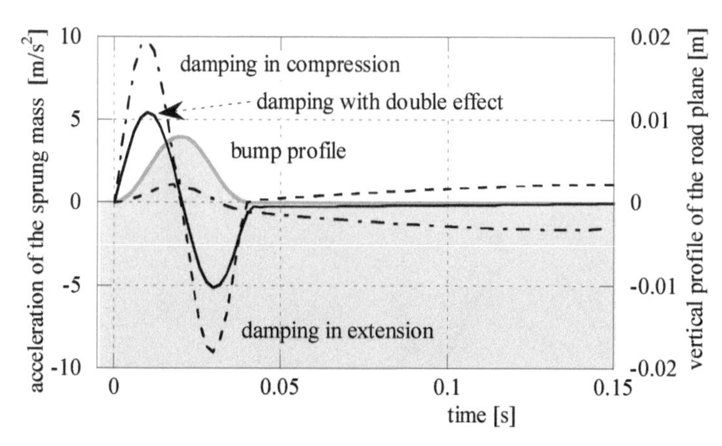

Fig. 5-46 Traveling over a bump: evolution of the acceleration of the sprung mass.

The acceleration graph (Fig. 5-46) highlights an important aspect. The shock absorber with damping only in extension has the characteristic of drastically reducing the positive acceleration of the sprung mass, so that the rider is not subjected to annoying accelerations upward that could throw him off the saddle.

Passage over a step

Now consider the mono-suspension passing over a step. Figure 5-47 highlights the different behavior of the mono-suspension in going over the step compared to the passage over a bump. The suspension with the shock absorber, which operates only in extension, enables a passage over the step without harmful oscillations of the sprung mass.

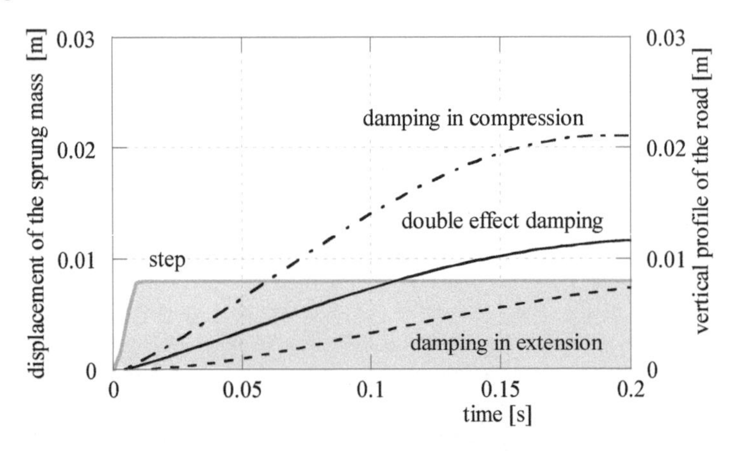

Fig. 5-47 Passage over a step: evolution of the displacement of the sprung mass.

The shock absorber that is active only in compression is stressed (the force is proportional to velocity) at a level such as to compromise its integrity.

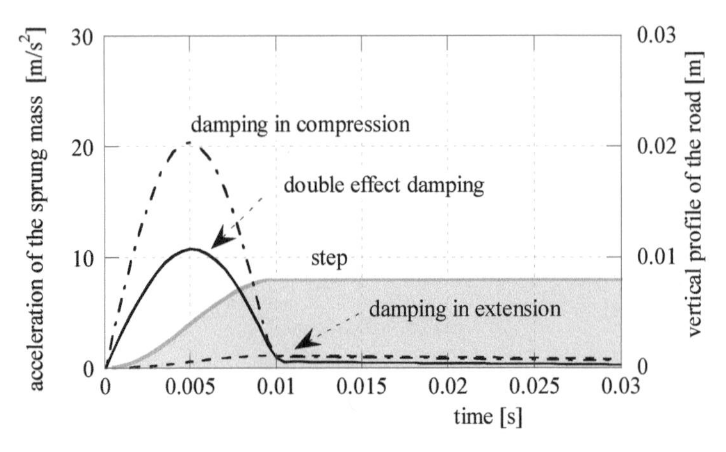

Fig. 5-48 Passage over a step: evolution of acceleration of the suspended mass.

For the purpose of riding comfort, the graph representing the acceleration of the sprung mass against time (Fig. 5-48) clearly shows that the shock absorber must act primarily in extension.

5.10 Characteristics of shock absorbers

We have seen that the damping coefficient in compression should be lower than that in extension, because when a wheel encounters a step or a bump, it must follow the profile of the obstacle without generating too much opposing force. While if it encounters a rut or a pothole, it can jump over it with only a temporary loss of wheel contact with the road plane.

The shock absorber characteristics are represented in a typical graph that shows the force on the ordinate and the displacement of the imposed harmonic motion on the abscissa, as illustrated in Fig. 5-49.

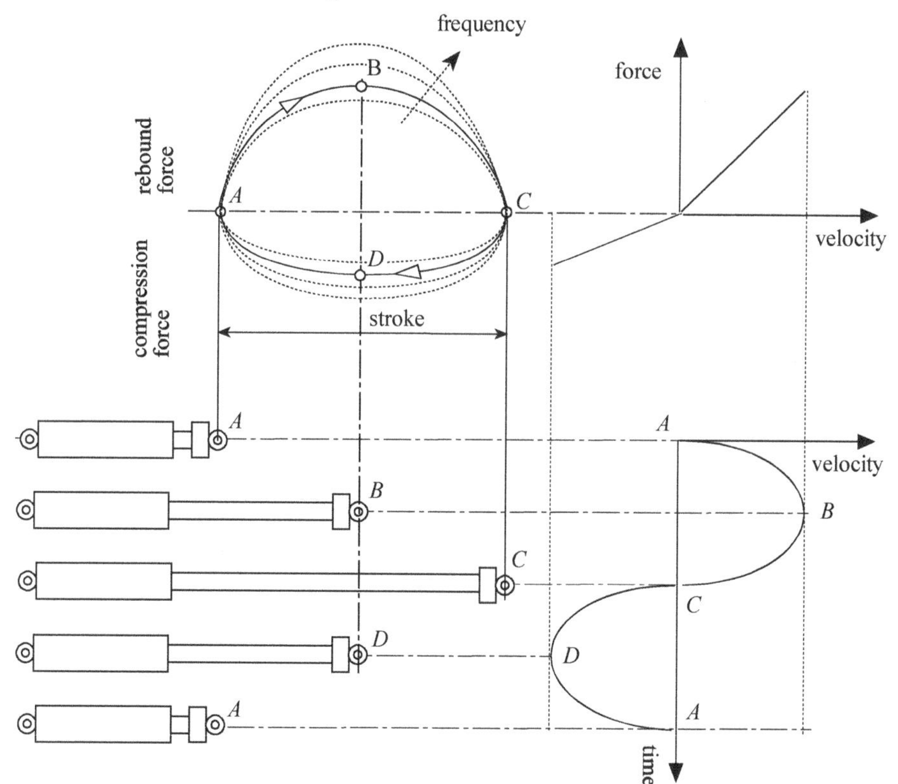

Fig. 5-49 Diagram of the force generated by a shock absorber.

As the frequency of the imposed motion increases, wide closed curves arise, the area of which represents the energy dissipated by the damping. The energy dissipated by the double-effect shock absorber is proportional to the frequency and to the square of the amplitude Δ of the imposed harmonic motion:

$$E = \frac{1}{2}\pi \cdot 2\pi\nu\Delta^2(c_e + c_c)$$

c_e and c_c indicate the constant damping coefficients respectively in extension and compression.

In addition to its non-linearity, due to the asymmetry of the damping coefficient in rebound (deformation velocity>0) and in compression (deformation velocity<0), the shock absorber will have a damping coefficient that varies with the velocity. Depending on how much the damping coefficient depends on the velocity, we can have progressive or degressive behavior.

$$F = \begin{cases} c_e\, \dot{y} \\ c_c\, \dot{y} \end{cases} \qquad \text{linear shock absorber}$$

$$F = \begin{cases} (c_e + \Delta c_e \left| \dot{y}^n \right|)\dot{y} \\ (c_c + \Delta c_c \left| \dot{y}^n \right|)\dot{y} \end{cases} \qquad \text{progressive shock absorber}$$

$$F = \begin{cases} (c_{est} - \Delta c_e \left| \dot{y}^n \right|)\dot{y} \\ (c_{comp} - \Delta c_c \left| \dot{y}^n \right|)\dot{y} \end{cases} \qquad \text{degressive shock absorber}$$

The exponent n gives the degree of dependence of the damping coefficient on the velocity.

The area of the force-displacement graph represents the energy dissipated. It can be observed that with equal maximum force, the degressive shock absorber dissipates more energy than the linear and progressive ones.

Fig. 5-50 gives a comparison of the characteristic graphs for three cases:

- linear damping;
- progressive damping;
- degressive damping.

The choice of the shock absorber and its calibration are made with the following features in mind:

- the overall energy to be dissipated in a cycle;
- the distribution of the energy to be dissipated in the two phases of extension and compression;
- the value of the progressive or degressive property of the shock absorber.

As far as the quantity of energy to be dissipated is concerned, we have seen that, for the sake of a comfortable ride, the average of the compression and extension coefficients must generally have a value of around 30-35% with respect to the critical damping.

In general, the damping in the compression phase is less than half of that in extension. The distribution depends on the motorcycle type as well as suspension stiffness. Some riders prefer somewhat rigid suspension with little damping in compression; others prefer additional damping in compression, and softer suspension.

Degressive damping has the advantage that it dissipates a greater quantity of energy at an equal maximum force level. It can be noted in Fig. 5-50 that the area of the force-stroke graph is greater in the case of degressive damping.

We have seen that the reduced damping is equal to the product of the actual damping coefficient and the square of the velocity ratio. Therefore, if we use a linear shock absorber (constant coefficient) and a progressive suspension (τ_{m,y_C} increasing with the amplitude of the wheel), the reduced damping is progressive because of the general progressive behavior of the suspension mechanism.

The choice of the shock absorber's degree of degressive rate must, therefore, be made by taking into account the contrary effect generated by any progressive rate of the suspension.

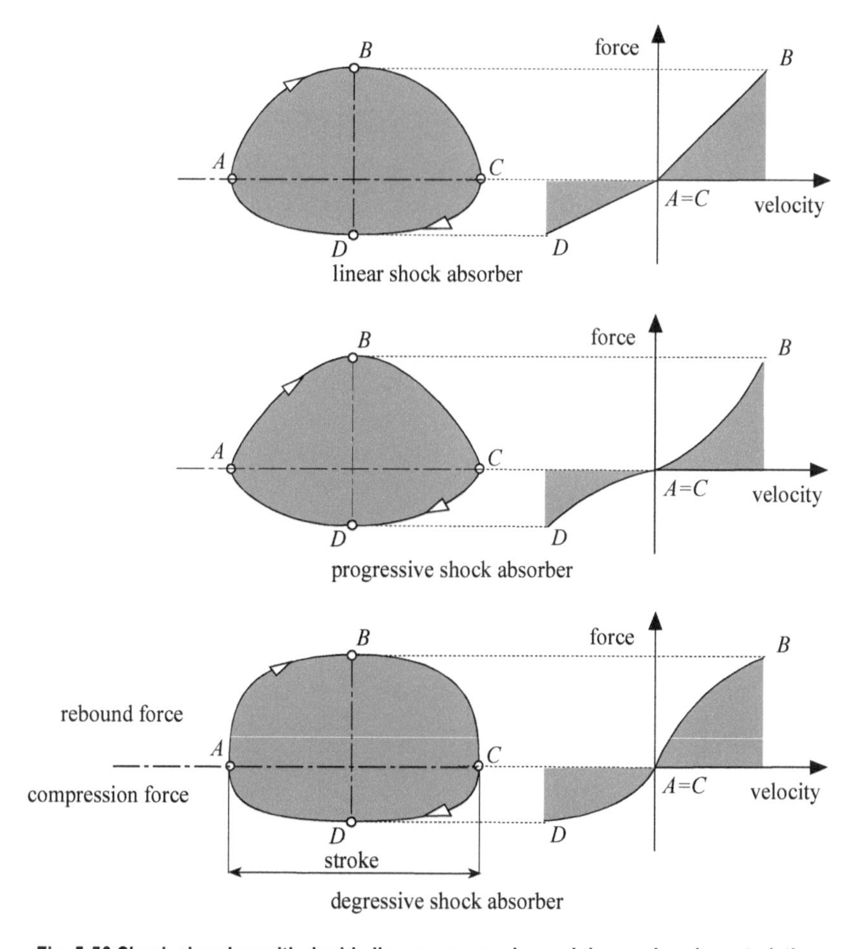

Fig. 5-50 Shock absorber with double linear, progressive and degressive characteristics.

5.11 The influence of the unsprung mass

The one degree freedom, mono-suspension model has allowed us to bring to light some interesting considerations regarding the most appropriate value of the damping ratio. However, it must be highlighted that the model with one degree of freedom disregards the influence of the unsprung mass. We have seen that the hop frequency of the system composed of only the unsprung mass and the tire radial spring is in the range 12-18 Hz.

Now for example, let us consider a mono-suspension with two degrees of freedom, representing the rear section, and observe the influence of the value of the unsprung mass on comfort and road adherence.

Suppose a motorcycle proceeds at constant velocity along a road with a profile that imposes a harmonic motion on the wheel, as shown in Fig. 5-51.

The equations of the motion of the sprung mass and unsprung mass are:

$$\begin{bmatrix} M & 0 \\ 0 & m \end{bmatrix} \begin{Bmatrix} \ddot{Z} \\ \ddot{z} \end{Bmatrix} + \begin{bmatrix} c & -c \\ -c & c+c_p \end{bmatrix} \begin{Bmatrix} \dot{Z} \\ \dot{z} \end{Bmatrix} + \begin{bmatrix} k & -k \\ -k & k+k_p \end{bmatrix} \begin{Bmatrix} Z \\ z \end{Bmatrix} = \begin{Bmatrix} 0 \\ k_p y + c_p \dot{y} \end{Bmatrix}$$

where M represents the sprung mass, m the unsprung mass, c the shock absorber damping, c_p the tire damping, k the suspension spring stiffness, and k_p the tire radial stiffness.

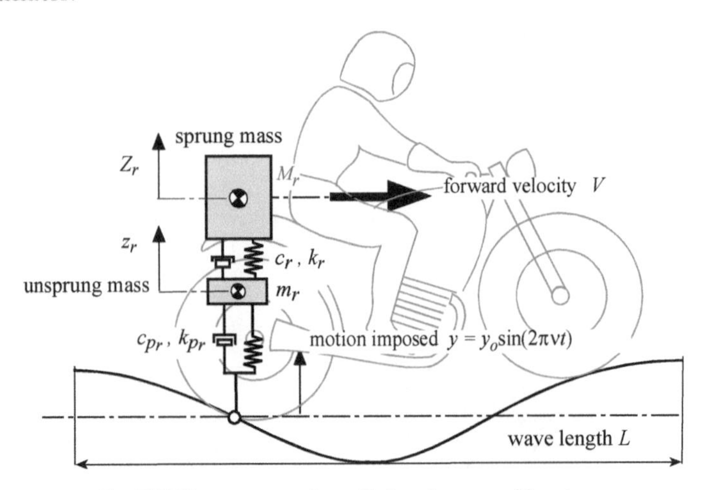

Fig. 5-51 Mono-suspension with two degrees of freedom.

The imposed motion can be described as:

$$y = y_o e^{i2\pi vt} \qquad \dot{y} = i\omega y_o e^{i2\pi vt}$$

where $v = V/L$ represents the frequency of the imposed motion, depending both on the forward velocity and the road profile wavelength.

In a steady state periodic motion, the masses oscillate with the same frequency as the imposed motion. The complex amplitudes of the sprung and unsprung masses normalized, in relation to the amplitude of the imposed motion, are given by the following equation:

$$\begin{Bmatrix} \dfrac{Z}{y_o} \\ \dfrac{z}{y_o} \end{Bmatrix} = \left\| \left[-(2\pi\nu)^2 \begin{bmatrix} M & 0 \\ 0 & m \end{bmatrix} + i2\pi\nu \begin{bmatrix} c & -c \\ -c & c+c_p \end{bmatrix} + \begin{bmatrix} k & -k \\ -k & k+k_p \end{bmatrix} \right]^{-1} \begin{Bmatrix} 0 \\ (k_p + i2\pi\nu c_p)e^{i2\pi\nu t} \end{Bmatrix} \right\|$$

The dynamic load on the wheel is composed of a constant component, equal to the static load, and of a fluctuating component. The minimum dynamic load on the wheel, normalized in relation to the static load, is given by:

$$N_{a_{\min}} = \frac{N_{\min}}{g(M+m)} = g - \frac{(z_r - y_o)(k_p + i2\pi\nu c_p)e^{i2\pi\nu t}}{g(M+m)}$$

The maximum normalized dynamic load is given by:

$$N_{a_{\max}} = \frac{N_{\max}}{g(M+m)} = g + \frac{(z_r - y_o)(k_p + i2\pi\nu c_p)e^{i2\pi\nu t}}{g(M+m)}$$

Example 18

Consider a suspension with the following characteristics:

- sprung mass: $M = 110$ kg;
- tire stiffness: $k_p = 130$ kN/m;
- reduced stiffness of the suspension: $k = 30$ kN/m;
- unsprung mass: $m = 15$ kg.

Determine the natural frequencies of the suspension, considered as a system with two degrees of freedom:

- 1st mode (displacement of the sprung and unsprung masses, in phase):
  $\nu_1 = 2.29$ Hz;
- 2nd mode (displacement of the sprung and unsprung masses, in opposite phase):
  $\nu_2 = 14.88$ Hz.

By way of example, compare these two values with the frequencies calculated with the one degree freedom models:

- a system composed of only sprung mass M and suspension stiffness k:
  $\nu_s = 2.63$ Hz (see example 15);
- a system composed of unsprung mass m and tire stiffness k_p:
  $\nu_t = 15.92$ Hz (see example 11).

It can be observed that:

- the first frequency ($\nu_1 = 2.29$ Hz), is near the natural frequency of the sprung mass ($\nu_s = 2.63$ Hz);
- the second frequency ($\nu_2 = 14.88$ Hz), is near the value of the natural frequency

of the unsprung mass (ν_t = 15.92 Hz).

* * *

Now let us introduce the non-dimensional ratio of the frequencies:

$$\Omega = \frac{\nu}{\nu_s} = \frac{\nu}{\dfrac{1}{2\pi}\sqrt{\dfrac{k}{M}}}$$

where:

ν represents the frequency of the imposed motion;

ν_s represents the frequency of the system composed of only sprung mass M and suspension stiffness k .

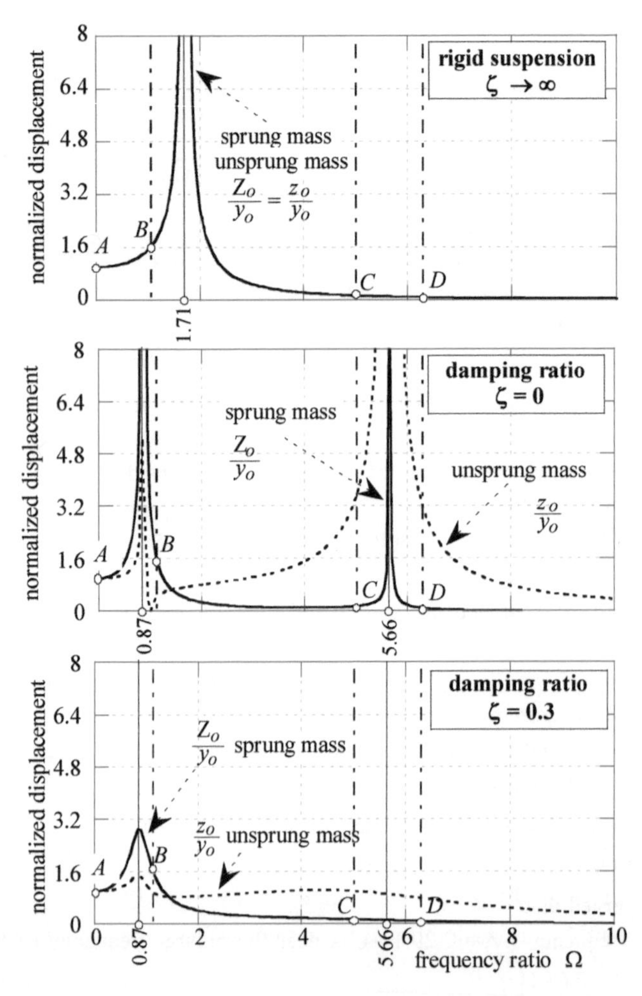

Fig. 5-52 Periodic response amplitude of the sprung (solid line) and unsprung masses (dotted line) versus the frequency ratio.

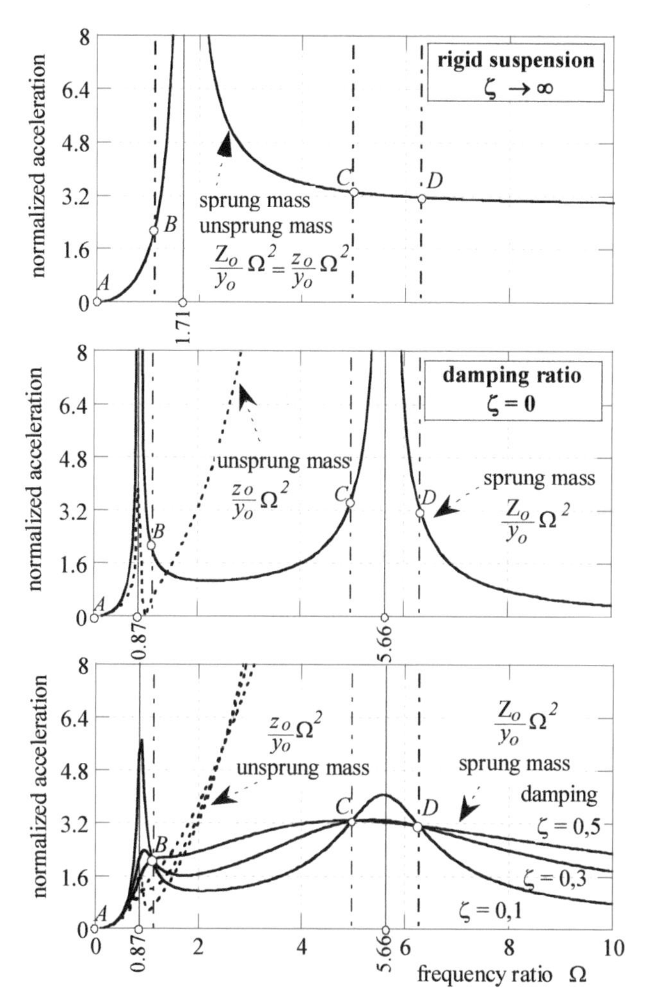

Fig. 5-53 Acceleration of the sprung (solid line) and unsprung masses (dotted line) versus the frequency ratio.

Fig. 5-52 shows the system's response as a function of the frequency ratio Ω, in the three following cases:

- rigid suspension;
- suspension without a shock absorber;
- suspension with damping ratio of 0.3.

The calculated natural frequencies of the two vibration modes of the suspension ($\nu_1 = 2.29$ Hz, $\nu_2 = 14.88$ Hz) correspond, respectively, to $\Omega = 0.87$ and $\Omega = 5.66$. Without suspension, i.e., with infinite suspension stiffness, the natural frequency depends only on the tire stiffness and the sum of the masses:

$$\nu_n = \frac{1}{2\pi}\sqrt{\frac{k_p}{M+m}}$$

Its value, with the data of the preceding exercise, is equal to 4.5 Hz (Ω = 1.71).

The damping ratio is calculated in relation to the critical damping $\sqrt{4kM}$ of the system with one degree of freedom, which is obtained by assuming that the unsprung mass is zero and the tire stiffness infinite.

For the purposes of riding comfort, the acceleration graphs can be obtained from the previous ones by multiplying the displacement by the square of the frequency ratio. The following interesting conclusions can be drawn:

- all the curves for the sprung mass displacement amplitude pass through the points A, B, C and D;
- in the frequency ranges $A-B$ and $C-D$ the maximum acceleration of the sprung mass diminishes with an increase in damping, while for values of the frequency ratio, between B and C, or higher than D, the increase in damping causes an increase in acceleration of the sprung mass;
- in the range of low frequencies, the optimal curve is the one that gives the minimum value of acceleration in correspondence with the first resonance. Since the curve must pass through point B, the optimal curve is the one that has its maximum at B. It can be shown that this curve is obtained with a value of the damping ratio equal to about ζ = 0.35. This value of ζ also makes the acceleration approximately minimal in the range $C-D$;
- in the intermediate field of the frequency ratio $A-B$ and for values beyond point D, for comfort purposes it would be appropriate to adopt lower damping values, but that would mean an increase in the maximum values of accelerations under resonance conditions;
- the acceleration to which the unsprung mass is subjected is not significantly influenced by the value of the damping ratio.

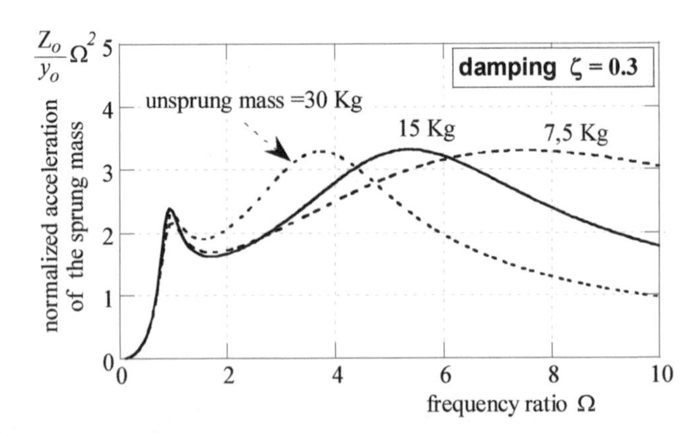

Fig. 5-54 Sprung mass acceleration for various values of the unsprung mass versus the frequency ratio.

Figure 5-54 shows the influence of the unsprung mass on the amplitude of acceleration of the sprung mass. Dividing the unsprung mass in half ($m = 7.5$ kg), the acceleration of the suspended mass diminishes at low frequencies, but increases at high ones. Alternatively, in doubling the unsprung mass ($m = 30$ kg), there is a net acceleration decrease at high frequencies and an increase at the low ones.

Now let us analyze the fluctuations of the loads on the wheels. As mentioned earlier, the load is composed of a constant term to which a fluctuating harmonic component, with frequency equal to that of the imposed motion, is added. Its amplitude normalized with respect to the static load is given by:

$$\Delta N_a = \left| \frac{(z_o - y_o)(k_p + i2\pi\nu c_p)}{g(M+m)} \right|$$

It is clear that the fluctuating component works against road-holding. Furthermore, if the fluctuating term exceeds the static load, the tire will lose contact with the road plane.

In Fig. 5-55, the fluctuating component is plotted against the frequency ratio. The fluctuating component is normalized in relation to the static load and divided by the amplitude of the sinusoidal profile of the road. It can be observed that the value of the damping ratio equal to 0.3 ensures good grip at low frequencies; at high frequencies, it would be appropriate to have a greater value of ζ which, however, could reduce riding comfort and adherence in the B-C field of the low frequencies. The loss of adherence happens when ΔN_a is equal to one. For example, at a frequency of 5 Hz and with a damping ratio ζ=0.1 separation occurs for values of the profile height greater than y_o=1/142 m.

The reduction of the unsprung masses benefits road-holding especially at low frequencies; moreover the value of the second frequency of the mono-suspension increases as the value of the unsprung mass decreases as shown in Fig. 5-56.

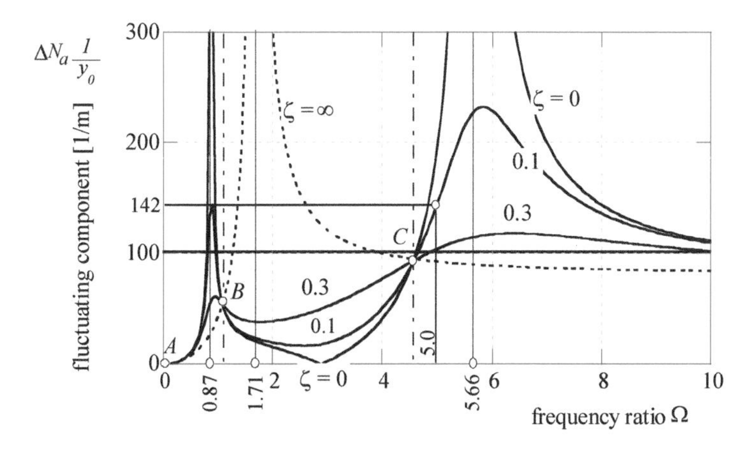

Fig. 5-55 Amplitude of the fluctuating part of the vertical load on the wheel for various values of the damping ratio.

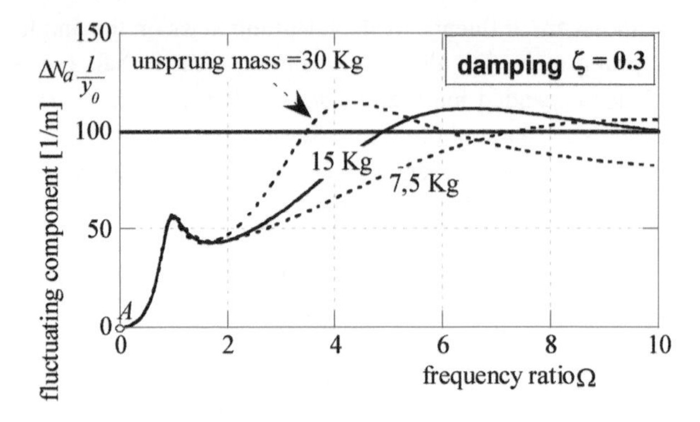

Fig. 5-56 Amplitude of the fluctuating part of the vertical load on the wheel for various values of the unsprung mass.

5.12 The rear suspension of the scooter

The scooter can be regarded as a special case of the motorcycle in which the engine is an integral part of the swing-arm. This construction solution facilitates the transmission of motion from the engine to the rear wheel, but interferes with its dynamic behavior in the vertical plane for the following reasons:

- the unsprung rear mass of the vehicle (mass of the wheel and part of the mass of the engine), with respect to the sprung mass (chassis + rider) has a significantly higher value than that of a conventional motorcycle;
- the unbalanced alternating forces (the engines of scooters are not usually provided with equilibrating countershafts) generated by the engine are transmitted to the chassis and are the cause of unwanted vibrations, which are felt by the rider in the handlebars, saddle and footrest.

5.12.1 Considerations on the position of the attachment point of the engine

Consider a scooter with the engine connected directly to the chassis with a pivot, as shown in Fig. 5-57.

We set a goal of seeking the best position for the attachment point P of the engine with the chassis, in order to minimize the forces transmitted to the chassis, generated by road unevenness. For this purpose, we substitute the "engine" with an equivalent system from a dynamic point of view, composed of a moment of inertia and two masses placed in correspondence:

- to the intersection point O of the horizontal line passing through the center of gravity with the vertical line passing through the wheel contact point;
- to the point P where the engine is attached to the chassis, whose position is unknown. The reduced masses at the points O and P and the moment of inertia are:

$$m_o = m\frac{b}{c} \qquad m_P = m\frac{a}{c} \qquad I_o = I_G - m\,ab$$

where: I_G represents the moment of inertia around the center of gravity;
m indicates the mass of the engine;
a,b are the distances indicated in Fig. 5-57.

The moment of inertia I_o is zero when the distance b satisfies the equation:

$$b = \frac{I_G}{ma}$$

The distance between the two points O and P is therefore equal to:

$$c = a + b = \frac{I_G + ma^2}{ma}$$

In this case the point O is the center of percussion in relation to point P. The vertical forces acting on O do not generate reaction forces at P.
It is interesting to note that the engine suspended at the percussion point presents interesting behavior from the dynamic point of view.

The swinging arm-engine system, attached at P and elastically supported at the point O, constitutes a vibrating system with one degree of freedom. Given k_r, which is the reduced vertical suspension stiffness at the point O, the natural frequency of the system is:

$$\omega = \sqrt{\frac{k_r\, c^2}{I_G + mb^2}} = \sqrt{\frac{k_r}{(I_G + mb^2)/c^2}}$$

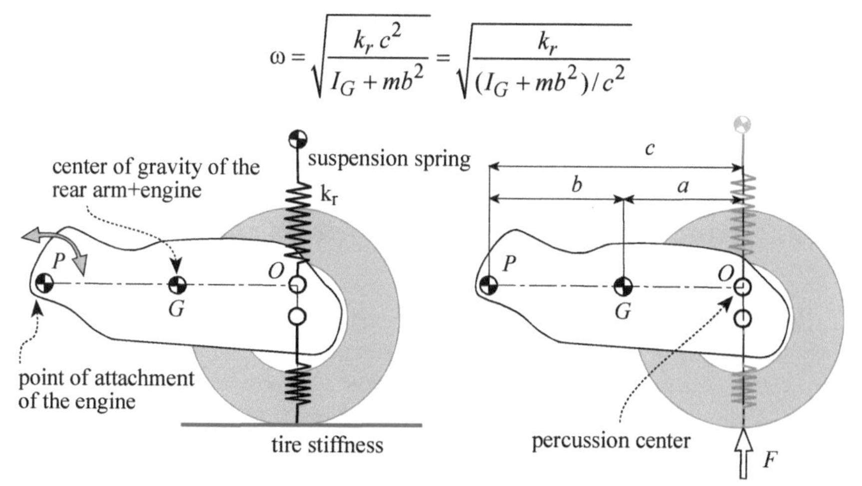

Fig. 5-57 Classic suspension of the scooter.

If the distance c is varied with a constant, the natural frequency changes. It can be shown that it reaches its maximum value when the distance c satisfies the necessary condition for making the point O the center of percussion with respect to the point of attachment of the engine P.

The frequency in this case is:

$$\omega = \sqrt{\frac{k_r\,(I_G + ma^2)}{I_G m}}$$

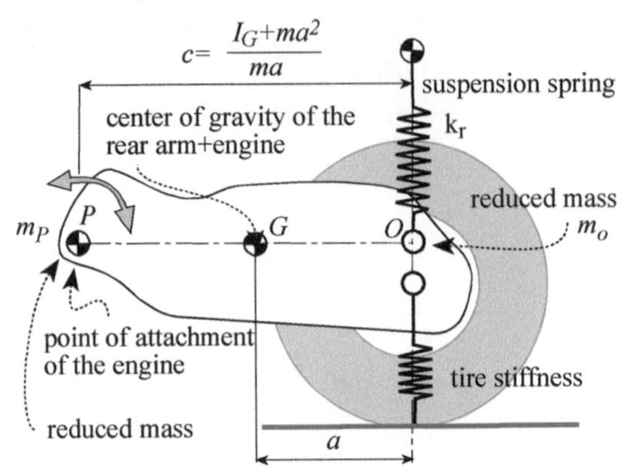

Fig. 5-58. Reduced system.

At the maximum value of the frequency with the same reduced stiffness, the equivalent unsprung mass corresponds to that of the minimum frequency; thus providing benefits for both riding comfort and road-holding.

The direct attachment of the engine to the chassis has some disadvantages for the isolation of vibrations. In fact, the engine generates alternating forces of imbalance, with frequency equal to and double that of the engine's rotation, which vary in the 50 to 400 Hz range. The unbalancing forces are transmitted to the chassis through the attachment point of the engine and also through the spring-shock absorber group. These forces generate vibrations that are felt by the rider on the handlebars, footrest and saddle. To reduce the vibrations transmitted to the chassis through the mounts, the junction with the chassis can be accomplished elastically with a system consisting of a simple rocker arm or a rocker arm with a link rod.

5.12.2 Attachment of the engine with a rocker arm (two degrees of freedom)

The engine connected to the chassis (assumed locked), with a simple rocker arm constitutes a system with two degrees of freedom, as shown in Fig. 5-59.

The two vibration modes of the system have their centers of rotation aligned along the axis of the rocker arm. Their position depends on the inertial characteristics of the engine, the tire radial stiffness, the spring stiffness and the mount stiffness of the rocker arm.

For the purposes of operating as a suspension, point *P*, where the rocker arm is attached, should be the center of percussion with respect to the point *O*, so that the vertical forces generated by the irregularities in the road do not stress the mounts of the rocker arm, but are opposed by the spring-shock absorber group.

Since the shock absorber behaves at high frequencies almost like a strut, the engine with the spring-shock absorber group and with the rocker arm makes up a four-bar linkage. The center of rotation of the engine in relation to the chassis can easily be found. It is the intersection point of the axes of the rocker arm and the shock absorber.

The largest component of the unbalancing force should be normal to the axis of the rocker arm.

The component of the unbalancing force parallel to the axis, however, is completely transmitted to the chassis and this is the disadvantage of such a suspension. For this reason, the rocker arm is joined elastically to the chassis by an elastomer plug, to reduce the transmission of vibrations.

If the balancing of the engine is accomplished in such a way as to have a normal component equal to 100% of the alternate unbalancing force, this type of suspension assures an excellent isolation of the vibrations. Such a distribution of the alternate force can generate excessively high loads on the drive shaft bearings.

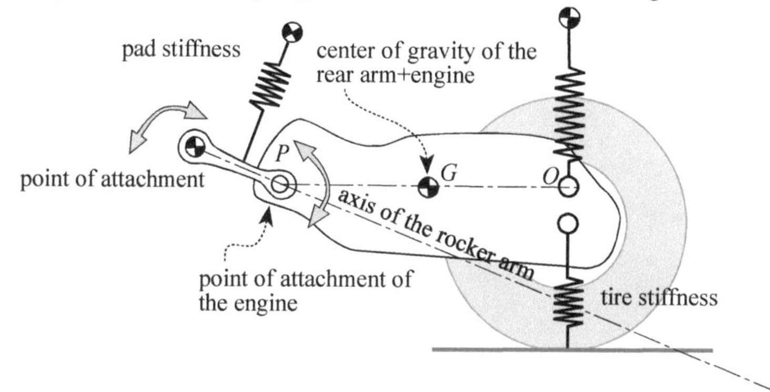

Fig. 5-59 Rear suspension of the scooter with a rocker arm.

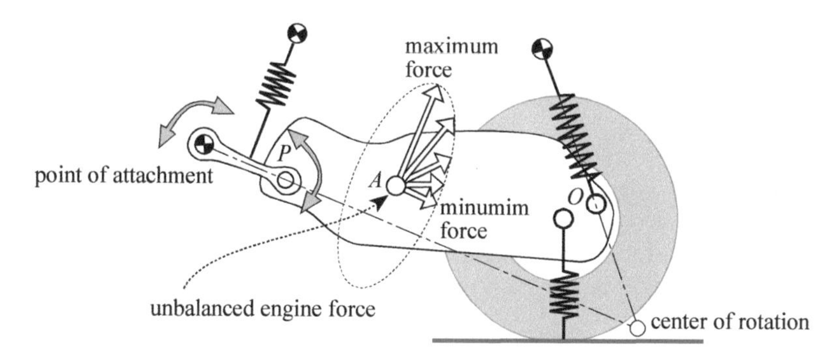

Fig. 5-60 Centers of instantaneous rotation of the scooter with a rocker arm.

From a theoretical point of view, for an optimal dynamic performance, this suspension arrangement should have the following properties:

- the distribution of the unbalancing force equal to: 100% normal component in relation to the rocker arm and 0% in the tangential direction;
- the point of attachment P of the engine coinciding with the center of percussion with respect to O;
- the center of rotation in correspondence with the contact point;
- the center of rotation of one vibration mode in correspondence with point P (mode excited by road irregularity);
- the center of the second vibration mode in correspondence with the contact point (mode excited by unbalanced force).

5.12.3 Rocker arm and link rod attachment of the engine (three degrees of freedom)

The swinging arm-engine assembly attached to the chassis with a rocker arm and with a link rod, represented in Fig. 5-61, has three degrees of freedom and therefore three modes of vibration. The introduction of an additional degree of freedom responds to the need to isolate both components of the alternate unbalancing force.

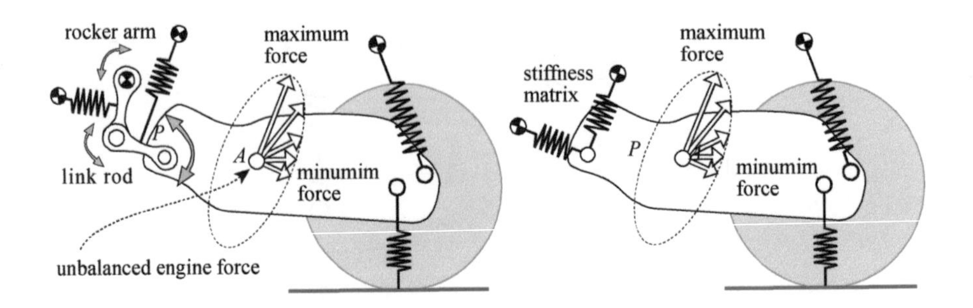

Fig. 5-61 Suspension of the scooter with rocker arm and link rod.

The position of the rotation centers of the three vibrating modes depends on the inertial characteristics of the engine, the stiffness characteristics of the mounts, the tire and the suspension spring.

The engine behavior can be studied by substituting the system, made up of the rocker arm, link rod and the relative elastic plugs, with an appropriate stiffness matrix as shown in Fig. 5-61. This matrix is generally not diagonal: a force applied to the swinging arm-engine system along the x direction brings on a displacement along the y direction and vice versa.

In principle, the location of the centers of rotation of the three modes is set up in the ideal diagram, illustrated in Fig. 5-62.

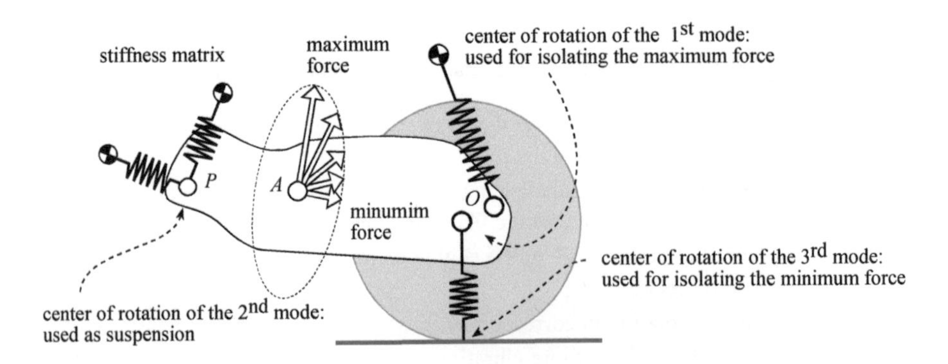

Fig. 5-62 Suspension of the scooter with rocker arm and link rod.

5.13 Road excitation

The motorcycle's in-plane response, excited by an uneven road, is important for human perception (rider comfort) and also for tire adherence with the road. Such excitation is random and is described by the road profile's statistical properties.

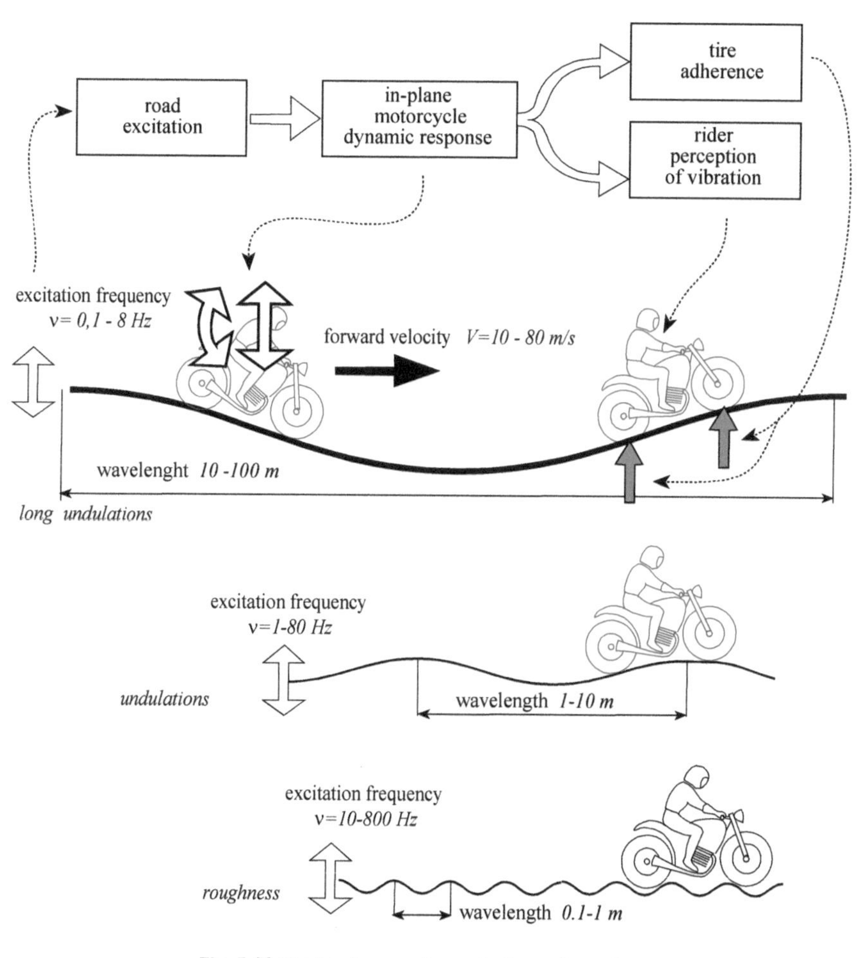

Fig. 5-63 The in-plane motorcycle dynamic system.

Consider a motorcycle with a forward velocity equal to 20 m/s. For long undulations exceeding 40 m the equilibrium of the motorcycle may be considered with a static analysis. In fact the frequency of the excitations (less than 0.5 Hz) are low enough in relation to the pitch and bounce frequencies to consider the motorcycle a quasi-static system. The excitation frequency range 0.5-25 Hz may be defined as the ride range. In this range the dynamic response depends primarily on the suspension. With a forward velocity of 20 m/s this frequency range corresponds to wavelengths from 40 m to 0.8 m. The excitations above 25 Hz are in the noise field.

To evaluate the rider's index of comfort it is important to concentrate on accelerations rather than displacements: in fact the human body is especially sensitive to the RMS value of accelerations. Discomfort is felt more in a range of frequencies that lie between 4 Hz and 8 Hz as shown in Fig. 5-64. This figure shows the lines of maximum tolerable levels for different durations of the vibration exposure established by the ISO 2631 Standards [Mechanical vibration and shock, *Evaluation of human exposure to whole-body vibration*, International Organization for Standardization, 1997].

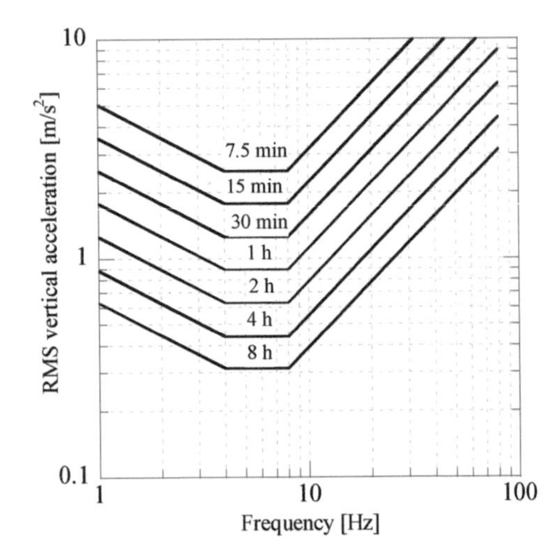

Fig. 5-64 Discomfort threshold for different exposure times.

Of course acceleration limits are related to the time of exposure: high accelerations are tolerated for a shorter time while low accelerations are tolerated for a longer time.

5.13.1 Power spectral density of the road

The elevation profile measured over a length of road can be decomposed into a series of sine waves varying in amplitude and phase relationship by means of the Power Spectral Density function. This function represents the amplitude density versus path frequency. Path frequency is the inverse of the wavelength and is expressed in cycles/m and thus the PSD function of the road elevation profile is expressed in s.

From experimental measurements of the road profile some laws regarding the PSD function have been proposed.

According to ISO standard [ISO 2631, ISO 5349, ISO, Draft Standard ISO/TC 108/WG9] the power spectral density of a road profile is described by the following equation:

$$PSD_{road} = PSD_o \left(\frac{\upsilon_o}{\upsilon} \right)^n \qquad \begin{cases} n = n_1 & \upsilon \le \upsilon_0 \\ n = n_2 & \upsilon > \upsilon_0 \end{cases}$$

where:

- $n_1 = 2, n_2 = 1.5$ are exponents;
- PSD_0 is the roughness magnitude parameter that depends on the quality of the road ($\mathrm{m^2/(cycle/m)}$);
- $\upsilon = 1/L_w = v/V$ is the path frequency (cycle/m);
- υ_0 is the cutoff path frequency (cycle/m).

PSD_0 values lie between $4*10^{-6}$ and $1024*10^{-6}$ ($\mathrm{m^2/(cycle/m)}$). Some examples of PSD functions are plotted as a function of path frequency υ in Fig. 5-65.

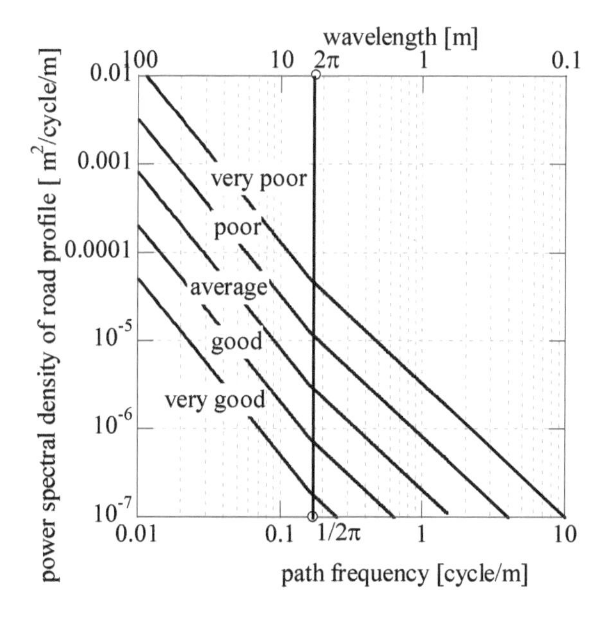

Fig. 5-65 Power spectral density of some road [ISO, Draft Standard ISO/TC 108/WG9].

5.13.2 In-plane frequency response function

The in-plane front and rear transfer function of the motorcycle is the ratio between the bounce or pitch acceleration amplitude of the motorcycle center of mass and the displacement of the contact point of the wheels versus frequency.

Figure 5.66 shows an example of transfer functions. It can be seen that the bounce resonance is more visible in relation to the pitch resonance, which is more damped. Also the wheel hop resonances are clearly visible; it should be highlighted that the radial damping of the tires are negligible in relation to the damping of the shock absorbers.

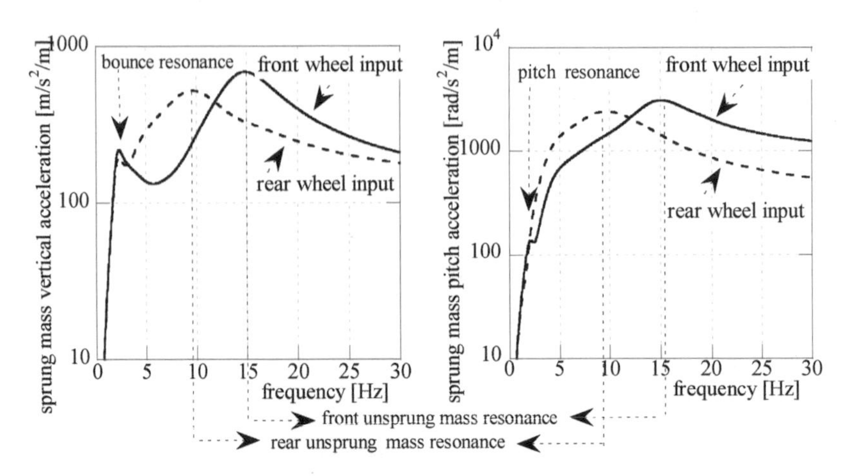

Fig. 5-66 Example of front and rear transfer function.

5.13.3 Motorcycle response

Once the power spectral density of the road profile is known it is possible to calculate the sprung mass acceleration spectrum by multiplying the road spectrum by the square of the motorcycle transfer function. The power spectral density is simply:

$$PSD_{sprung\ mass} = PSD_{road} \left|H(\omega)\right|^2_{sprung\ mass}$$

where:

$PSD_{sprung\ mass}$ is the PSD of the acceleration of the sprung mass;

PSD_{road} is the PSD of the road input;

$\left|H(\omega)\right|_{sprung\ mass}$ is the magnitude of the in-plane complex transfer function of the motorcycle.

The isolation properties of the suspension generate an acceleration spectrum of the sprung mass with high amplitudes at the sprung mass resonances, with a moderate attenuation in the range of the wheels' resonances and a rapid attenuation thereafter.

The motorcycle responses to the road excitation with two different forward velocities are represented in Fig. 5-67 for an average road profile. In this case the bounce and pitch modes of vibration are coupled. The bounce response is more pronounced at low frequencies while the pitch acceleration of the sprung mass is more pronounced at middle frequencies. The figures also highlight the wheelbase filtering phenomenon (dips in the curves).

Wheelbase filtering is closely related to the ratio p/V between the vehicle wheelbase and forward velocity. In fact the rear wheel sees the same road profile as the front wheel, only with a time delay which is equal to the ratio between the wheelbase and the forward velocity. Wheelbase filtering causes characteristic lobes (in between the dips) in these plots: when the speed increases (as the ratio p/V decreases) the number of lobes diminishes and only the peak characteristics of the system response to the road surface PSD remain in the plots. The first frequency of the minimum bounce acceleration response is equal to the velocity divided by twice the wheelbase while the pitch acceleration presents the first minimum when the ratio between the velocity and the wheelbase is equal to one. The coupling of the bounce and pitch mode makes the filtering phenomenon less clear. The motorcycle has a 1.5 m wheelbase, with a speed of 20 m/s null response (dip) occurs at approximately 6.5 Hz for the bounce and 11 Hz for the pitch.

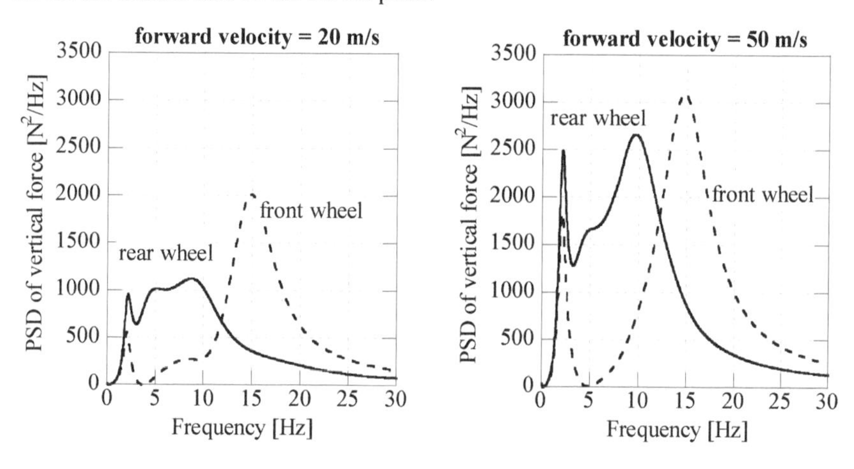

Fig. 5-67 Example of sprung mass response.

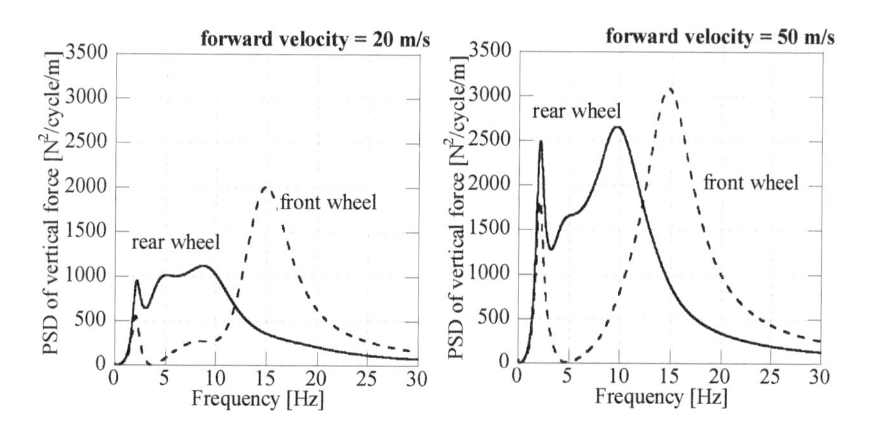

Fig. 5-68 Example of vertical load response.

The suspensions are also important for tire adherence, which diminishes with the increase of the vertical load fluctuations. The PSD of the vertical loads induced on the wheel are shown in Fig. 5-68. At low speeds the front wheel is subjected to higher load variations compared with the rear wheel, while as the speed increases the rear wheel load fluctuations become more important and are distributed over a large frequency range.

The load fluctuation rises with the increase of the unsprung mass in relation to the sprung mass and with the increase of the tires' radial stiffness.

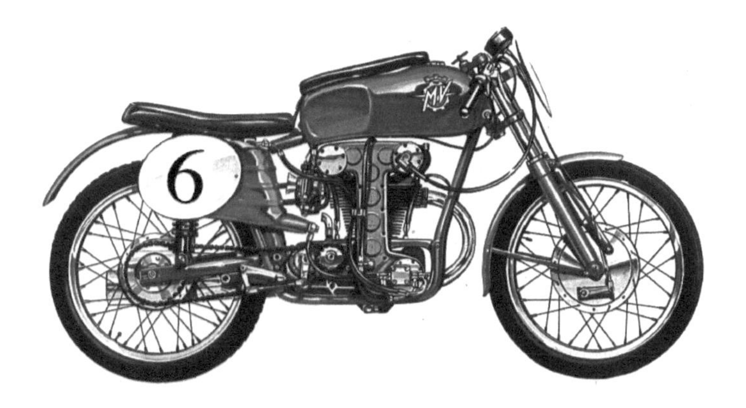

MV Grand Prix 125 cc of 1953

6 Motorcycle Trim

In the previous chapters, the forces acting on the motorcycle were calculated: resistance forces, driving force and dynamic loads on the wheels, in different conditions of both stationary and non-stationary motion, in acceleration and in braking. In this chapter, variations in the trim exhibited by the motorcycle under various driving conditions will be studied, and the importance of the chain force will be highlighted.

The term vehicle trim implies the geometric configuration that the motorcycle acquires in different conditions during transient and steady motion, in acceleration and in braking.

As shown below, the motorcycle trim depends on the stiffness characteristics of the front and rear suspensions, on the forces operating on the motorcycle, and on the inclination angle of the chain and the swinging arm.

6.1 Motorcycle trim in steady state motion

Figure 6-1 illustrates the system of forces operating on the motorcycle in steady state conditions. In this case, the study of the trim is specifically designed to determine the attitude and configuration of the motorcycle, and in particular, that of the rear suspension.

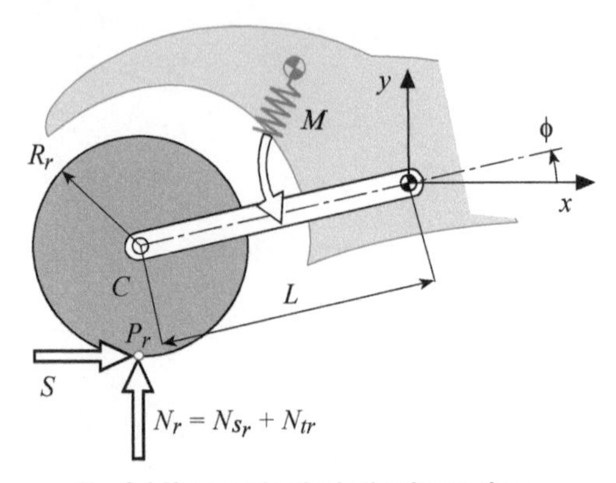

Fig. 6-1 Motorcycle trim in the thrust phase.

6.1.1 Rear suspension equilibrium

Now let us consider the rear swinging arm with its wheel represented in Fig. 6-2, assuming the forward motion of the vehicle to be at constant speed. The following forces are applied to the rear swinging arm and wheel system:

- the thrust force S;
- the vertical dynamic load N_r;
- the chain force T;
- the elastic torque M.

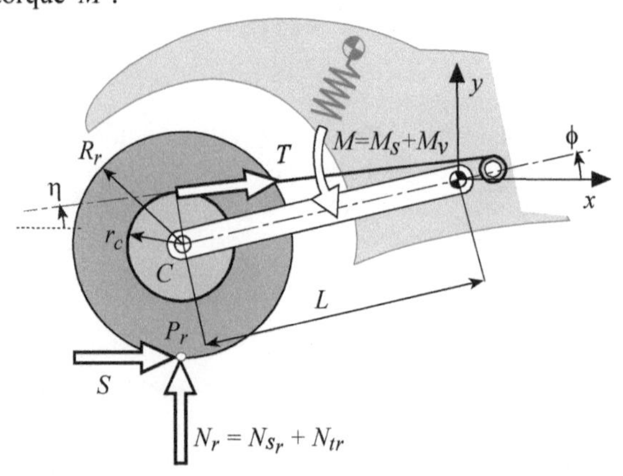

Fig. 6-2 Rear suspension balance with chain transmission.

The balance of the moments on the swinging arm pivot gives the following expression:

$$M_v = N_{tr} L \cos\phi - S(R_r + L\sin\phi) + T[r_c - L\sin(\phi - \eta)]$$

The equilibrium equation does not report the static elastic moment M_s exerted by the suspension spring and the moment generated by the static vertical load N_{S_r}

which is equal to $-M_S$.

The four moments acting on the swinging arm are:

- the moment generated by the load transfer N_{tr} that compresses the suspension;
- the moment generated by the driving force S that tends to extend the suspension;
- the moment generated by the chain force T that compresses the suspension.
- the additional elastic moment generated by the suspension movement M_v that can be positive or negative.

The driving force is assumed to be constant and is related to the chain force, exerted by the chain on the wheel, through the equation:

$$S = \frac{T \cdot r_c}{R_r}$$

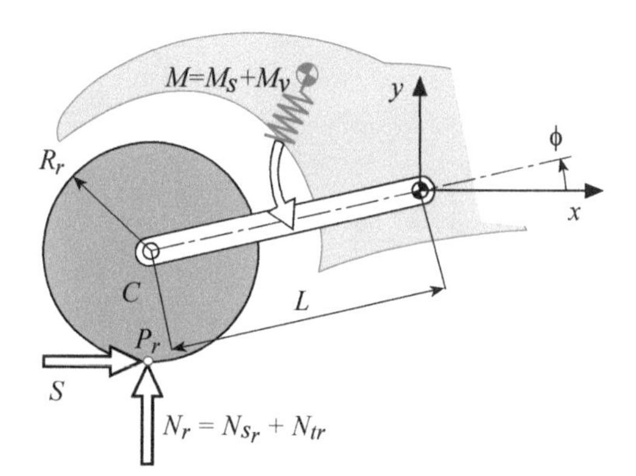

Fig. 6-3 Rear suspension balance with transmission shaft.

If there is no thrust, the chain force and the load transfer are null, so the moment exerted the static load balances the elastic moment:

$$M_S = N_{S_r} L \cos\phi$$

Conversely, if there is a thrust force, the trim of the rear suspension (arm position with respect to the frame), depends on the values of the three aforementioned components. Expressing the driving force as a function of the chain force, the equilibrium equation, with respect to the swinging arm pivot, can be rearranged as follows:

$$M_v = N_{tr} L \cos\phi - T L[\frac{r_c}{R_r} \sin\phi + \sin(\phi - \eta)]$$

where M_v is the part of the elastic moment, necessary to balance the moments generated by the load transfer, chain force and driving force.

If the term due to the load transfer is larger than that due to the chain force and driving force, the suspension is further compressed, with respect to the deflection caused by only the static load ($M_v > 0$). Vice versa, if the component due to the chain force and driving force prevails over the load transfer component, then the suspension is extended ($M_v < 0$).

It is interesting to note that the chain force disappears for both the shaft-drive transmission, as shown in Fig. 6-3, and in the case of scooters (with engine integrated on the swinging arm). Therefore, only the load transfer force and the thrust force exert a moment on the swinging arm.

In this case, the balance of the moments results in:

$$M_v = N_{tr} L\cos\phi - S(R_r + L\sin\phi)$$

6.1.2 Inclination angle of chain

In order to study the balance of the moments, it is necessary to express the inclination angle of chain η, as a function of the angular inclination ϕ of the swinging arm.

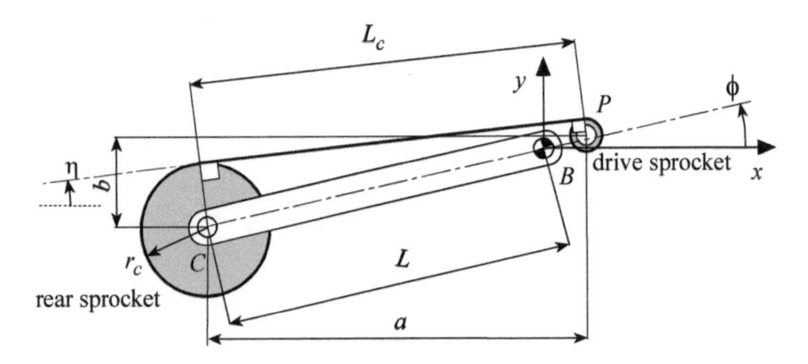

Fig. 6-4 Geometry of the rear swinging arm with chain transmission.

With reference to Fig. 6-4, we may write the following equation:

$$b - (r_c - r_p)\cos\eta = L_c \sin\eta$$

where r_p and r_c represent the radii of the drive sprocket and the rear sprocket, b represents the vertical distance between the axes of the sprockets, L_c is the length of the straight line section of the chain. The equation gives the inclination angle as a function of the swinging arm angle.

For practical purposes, the following approximate expression (assuming $\cos\eta \cong 1$) is sufficient:

$$\eta = \arcsin\frac{b - (r_c - r_p)}{L_c} = \arcsin\frac{L\sin\phi + y_P - (r_c - r_p)}{L_c}$$

where y_P is the vertical coordinate of the drive sprocket shaft.

6.1.3 Squat ratio and squat angle

Chain transmission

Consider the intersection point A, between the axis of the upper chain branch and the straight line passing through the center of the wheel and through the swinging arm pivot, shown in Fig. 6-5. The straight line that connects the point of contact between the rear wheel P_r and point A is called the squat line. Its inclination to the horizontal plane is called the squat angle σ.

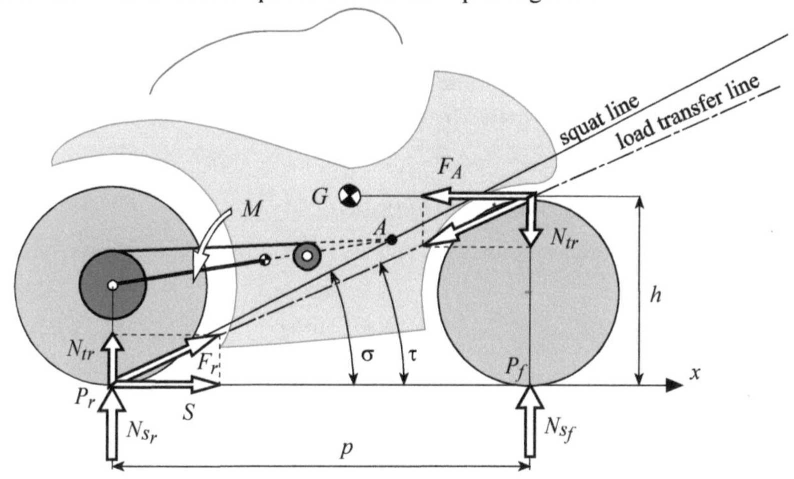

Fig. 6-5 Squat and load transfer lines.

The resultant F_r, from the sum of the load transfer force and the driving force, applied to the point of contact of the rear wheel, is inclined with respect to horizontal by an angle named the angle of load transfer τ. The line of action of the resultant F_r is called the straight line of the load transfer.

We define the squat ratio $\Re$ as the ratio between the moment generated by the load transfer and the moment generated by the sum of the chain force and the driving force:

$$\Re = \frac{N_{tr} L \cos\phi}{S L \sin\phi + T L \sin(\phi - \eta)}$$

Expressing the load transfer as a function of the driving force, the ratio is a function of only the geometric characteristics, and in particular, it is equal to the ratio between the tangent of the load transfer angle and the tangent of the squat angle:

$$\Re = \frac{h\cos\phi}{p\left[\sin\phi + \dfrac{R_r}{r_c}\sin(\phi - \eta)\right]} = \frac{\tan\tau}{\tan\sigma}$$

The ratio varies according to the variation of the swinging arm inclination angle and

depends on the difference between the arm inclination angle and the chain inclination angle. Such a difference is sensitive to the position of the axis of the drive sprocket in relation to the position of the swinging arm pivot.
Three cases can occur:

- Point A lies on the straight line of the load transfer, that is $\sigma = \tau$; in this case $\Re = 1$. During the thrust phase there are no additional moments operating on the swinging arm, so the suspension spring is no longer stressed compared with the static condition;
- Point A lies under the straight line of the load transfer, that is $\sigma < \tau$; in this case $\Re > 1$: the moment generated by the resultant F_r causes a compression of the spring in addition to the one created by the static load;
- Point A lies above the straight line of the load transfer, that is $\sigma > \tau$; in this case $\Re < 1$. The moment generated by the resultant F_r causes the extension of the spring.

Transmission shaft with universal joints

In the case of final transmission shaft and also in scooters, the squat ratio is:

$$\Re = \frac{N_{tr} L \cos\phi}{S(R_r + L \sin\phi)}$$

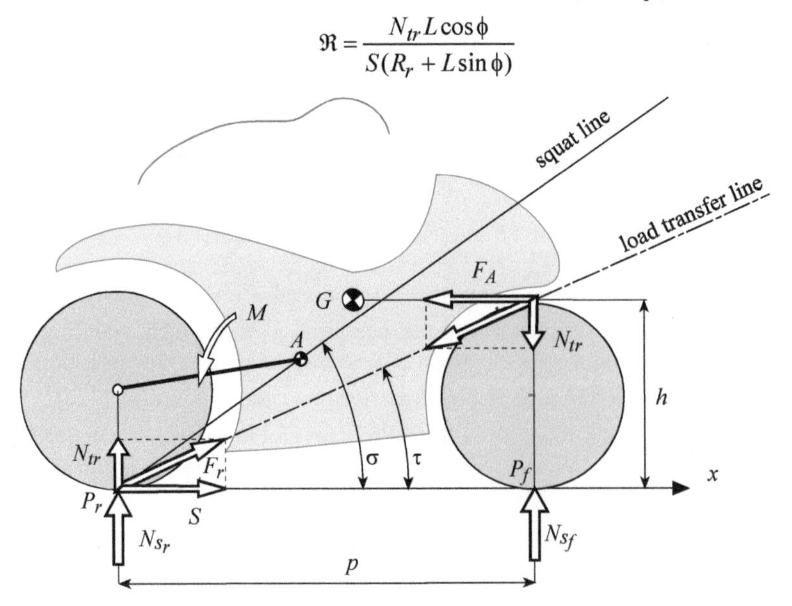

Fig. 6-6 Swinging arm balance with transmission shaft.

The ratio can be expressed as a function of the geometric characteristics of the suspension and also as a ratio between the tangent of the load transfer angle and the tangent of the squat angle. In this case the squat line passes through the point of contact and the swinging arm pivot as can be seen in Fig. 6-6:

$$\Re = \frac{h L \cos\phi}{p\,(R_r + L \sin\phi)} = \frac{\tan\tau}{\tan\sigma}$$

In general the load transfer angle τ with the transmission shaft, is smaller than the angle σ, thus the ratio $\Re$ is less than one, i.e., the suspension is always extended in the thrust phase. In order to obtain ratios close to unitary values swinging arms of great length should be used.

Four-bar suspension with transmission shaft

As shown in Fig. 6-7, in this case the ratio $\Re$ can be expressed as a function of both the load transfer angle τ and the squat angle. In this case the squat line is the straight line passing through the center of rotation with respect to the frame (point A) of the suspension connecting links and the point of contact of the rear wheel.

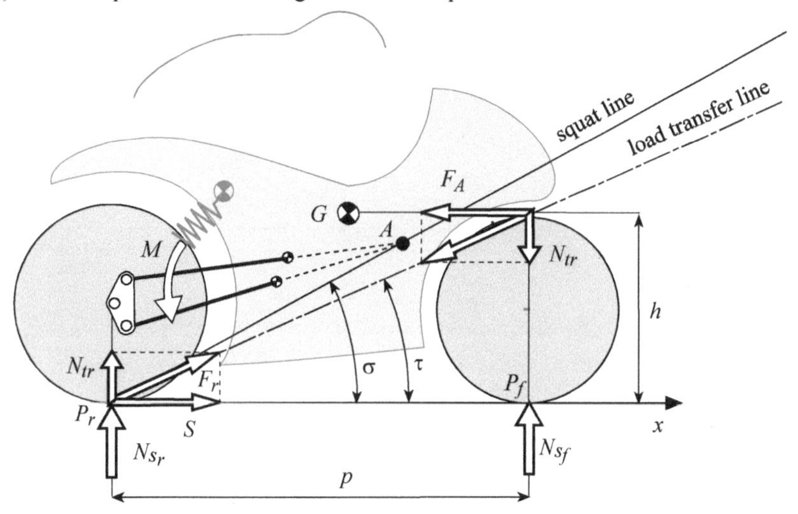

Fig. 6-7 Motorcycle balance with four-bar rear suspension.

Example 1

We will now make some remarks regarding a motorcycle with the following properties:

- motorcycle wheelbase: $p = 1370$ mm;
- height of gravity center: $h = 600$ mm;
- swinging arm length: $L = 590$ mm;
- rear wheel radius: $R_r = 317$ mm;
- rear sprocket radius: $r_c = 111.3$ mm;
- drive sprocket radius of: $r_p = 43.2$ mm;
- sprocket horizontal position compared with swinging arm pivot: $x_P = 75$ mm;
- sprocket vertical position compared with swinging arm pivot: $y_P = 0$ mm.

Let us examine the motion of the swinging arm in relation to the frame (assumed to be fixed).

Figure 6-8 depicts the deviation of squat ratio versus the variation of the vertical position y_C of the rear axle with respect to the frame. It is worth pointing out that in this example the load transfer tends to compress the suspension spring, whereas the chain force and the driving force tend to extend it.

It is possible to observe that in the reference case, for those values regarding the wheel vertical positions that are lower than approximately -65 mm, the effect of the chain force and thrust prevails as the value of the squat ratio is less than one ($\Re < 1$), whereas for the higher y_C values, the effect of the load transfer force prevails since the value of the ratio is greater than one ($\Re > 1$).

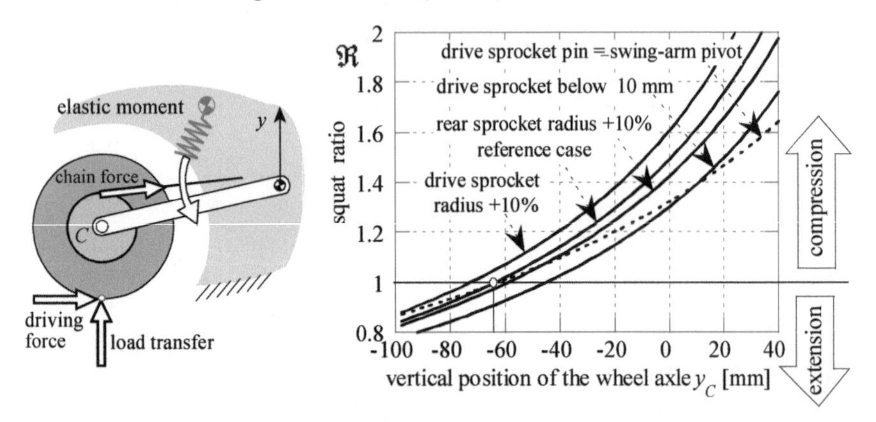

Fig. 6-8 Typical deviation of the squat ratio $\Re$ as y_C varies.

The figure also illustrates a comparison among the values of the ratio $\Re$, obtainable when only one single geometric parameter varies at a time: it is interesting to observe that as the geometry varies (radii of the drive sprocket and of the rear sprocket, position of the drive sprocket axis with respect to the swinging arm pivot) the values also vary, however, the curves still maintain an increasing deviation as the vertical position of the rear wheel axle varies.

6.1.4 Motorcycle trim as the squat ratio varies

Let us now continue with a few observations on the trim of the vehicle in motion at a constant speed and in the presence of a thrust (balanced by the aerodynamic resistant force).

With the increase of the chain force and, thus, of the driving force, the front axle lifts up because of the front wheel load decrease, while the back of the rear frame lifts up or lowers as a function of the squat ratio.

The vertical extension of the front suspension is equal to the ratio between the load transfer and the reduced vertical stiffness:

$$\Delta L_f = \frac{N_{tr}}{k_f}$$

As can be seen in Fig. 6-9, in the case of a unitary ratio ($\Re = 1$), as the chain force varies the force operating on the rear suspension spring does not undergo any variation as the action of the load transfer is perfectly balanced by the chain force. Under these specific conditions variations in the thrust affect only the front suspension.

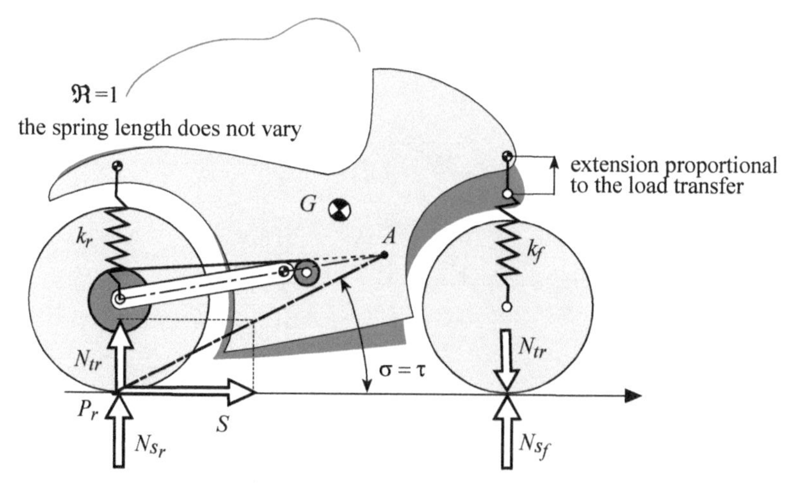

Fig. 6-9 Motorcycle trim with unitary squat ratio.

If the ratio is greater than one ($\Re > 1$), the rear suspension spring will be compressed compared to the condition of balance with a null chain force. With the increase of the chain force value, the front axle lifts up while the back of the rear frame lowers proportional to the value of the ratio. Furthermore, if the ratio increases with the increasing y_C (such as the case in example 1), with the increase of the chain force and, therefore, of the driving force, the movement becomes more appreciable as represented in Fig. 6-10.

If the ratio is less than one ($\Re < 1$) the rear suspension spring will be extended compared with the condition of balance with a null chain force. With the increase of the chain force, both the front axle and the rear axle are extended, causing the motorcycle center of gravity to raise, also illustrated in Fig. 6-11.

The deformation of the spring ΔL_r, reduced to the rear wheel axle (which becomes negative during compression), is provided by the following (approximate) expression:

$$\Delta L_r \cong \frac{\Delta\phi}{L} \approx \frac{M_v}{k_r L \cos\phi} = \frac{N_{tr}}{k_r}\left(\frac{1-\Re}{\Re}\right)$$

The previous observations allow us to conclude that the trim of the vehicle depends on the value of the squat ratio.

The variation in the trim, caused by the driving force, can also be expressed by the variation of the frame pitch angle:

$$\Delta\mu \approx \frac{\Delta L_f - \Delta L_r}{p}$$

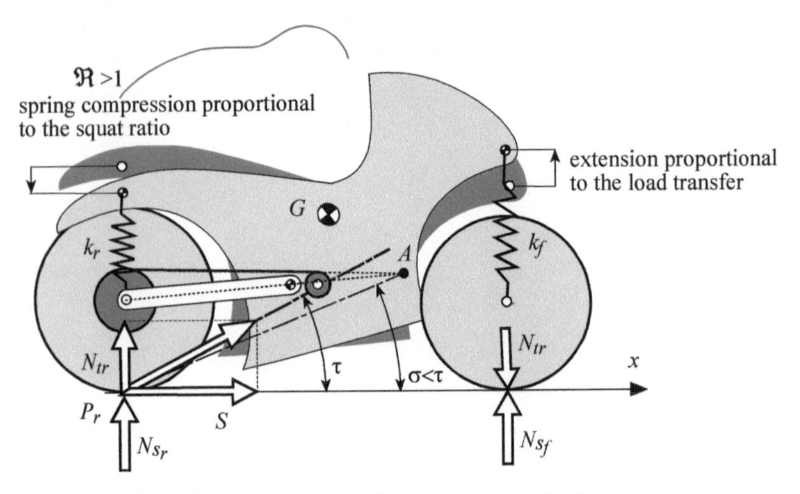

Fig. 6-10 Motorcycle trim with squat ratio greater than one.

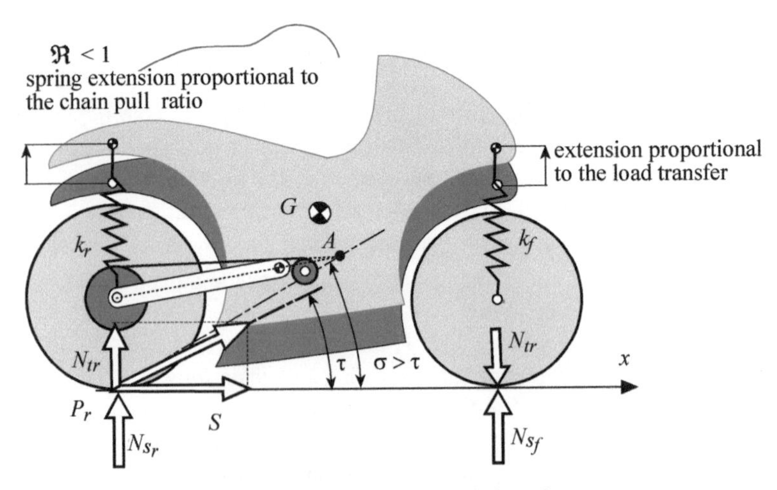

Fig. 6-11 Motorcycle trim with squat ratio less than one.

Example 2

Consider a motorcycle with the following characteristics:

- motorcycle wheelbase: $p = 1370$ mm;
- height of center of gravity: $h = 600$ mm;
- reduced stiffness of the rear suspension: $k_r = 20$ kN/m;
- reduced stiffness of the front suspension: $k_f = 13$ kN/m.

The aim is to determine the variations in the motorcycle trim after applying a chain force T equal to 4000 N, as the squat ratio value varies:

- $\mathfrak{R} = 0.7$;

- $\Re = 1.0$;
- $\Re = 1.3$.

The load transfer is equal to 615 N and, therefore, the variation $\Delta\mu$ of the pitch angle is:

- 1.42°, for a value of the ratio $\Re = 0.7$;
- 1.98°, for a value of the ratio $\Re = 1.0$;
- 2.27°, for a value of the ratio $\Re = 1.3$.

We can see that, as the squat ratio $\Re$ increases, the variation in the trim also increases. This is represented by the variation of the frame pitch angle. It is worth pointing out that the positive direction of the pitch is counter-clockwise.

6.2 Motorcycle trim in a curve

As represented in Fig. 6-12, the motorcycle that runs at a constant speed from a straight line to a cornering motion lowers and also slightly pitches forward.

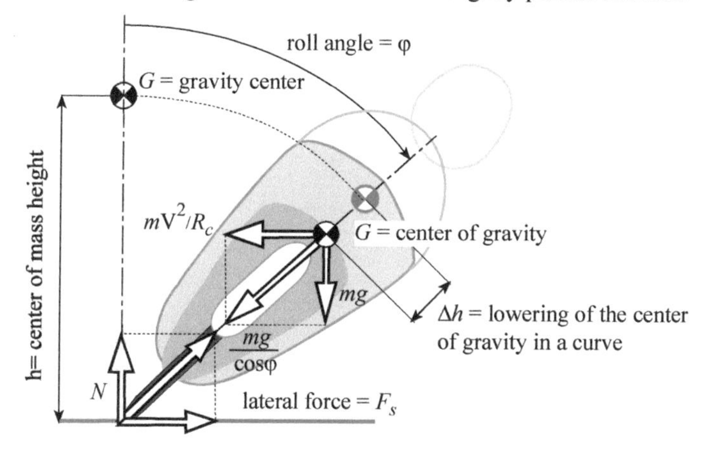

Fig. 6-12 Lowering of the motorcycle center of gravity in a curve.

The lowering is caused by the increase in load that acts in the motorcycle plane that increases in inverse proportion to the cosine of the roll angle φ. The pitch forward is due to the fact that the reduced stiffness of the front suspension is less than that of the rear suspension.

The lowering of the center of gravity Δh, and the variation of the pitch angle $\Delta\mu$, in the change from a straight line to a curve, is provided by the expressions (Fig. 6-13):

$$\Delta h = \frac{mg\left(\dfrac{1}{\cos\varphi} - 1\right)}{\dfrac{k_f \cdot k_r}{(p-b)^2 k_f + b^2 k_r} p^2} \qquad \Delta\mu = -\frac{mg\left(\dfrac{1}{\cos\varphi} - 1\right)}{\dfrac{k_f \cdot k_r}{b\ k_r - (p-b)k_f} p^2}$$

and it is possible to note that, if $b\ k_r > (p-b)k_f$, the pitch angle will be negative, and, thus, forward.

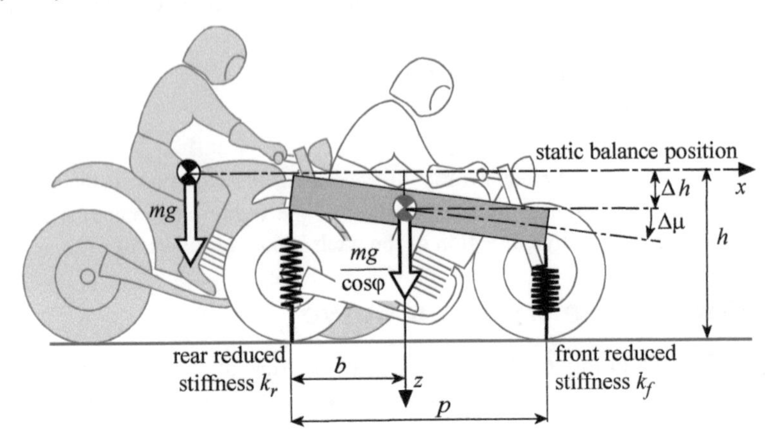

Fig. 6-13 Variations of the motorcycle trim in a curve.

6.2.1 Squat ratio in a curve

The squat ratio, in the change from a straight line to a curve, decreases because of the lowering of the center of gravity:

$$\Re = \frac{\tan\tau}{\tan\sigma} = \frac{(h-\Delta h)/p}{\tan\sigma}$$

Actually, in the change from straight running to cornering, the ratio decreases less than the chain force angle σ because point A is also lowered.

6.2.2 Trim in entering a curve

In order to evaluate the variations of a motorcycle trim in the change from a straight line to a curve we should consider the example shown below.

Example 3

Consider a motorcycle with the following characteristics:

- motorcycle mass: $m = 200$ kg;
- wheelbase: $p = 1.4$ m;
- longitudinal distance of the center of gravity: $b = 0.6$ m;
- reduced stiffness of the rear suspension: $k_r = 25$ kN/m;
- reduced stiffness of the front suspension: $k_f = 13$ kN/m;
- tire stiffness: $k_{p_f} = k_{p_r} = 150$ kN/m;
- roll angle: $\varphi = 45°$.

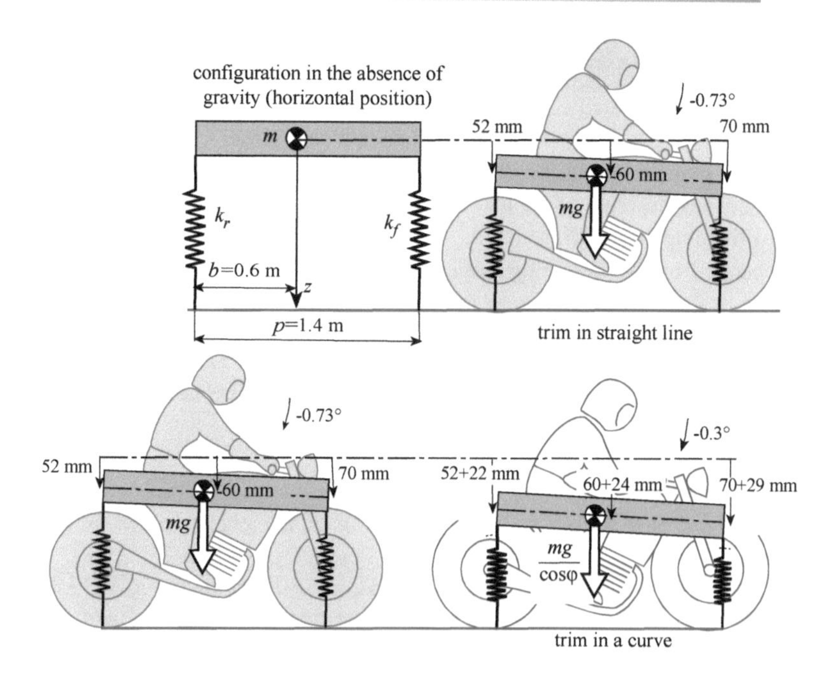

Fig. 6-14 Motorcycle trim in a straight line and in a curve.

Case 1: Straight line motion.

The lowering of the vehicle center of gravity, due to its weight, and in the hypothesis of no spring preload, is equal to 60 mm (lowering of 52 mm with respect to the rear wheel and 70 mm to the front wheel), while the clockwise pitch rotation is equal to - 0.73°. Obviously if the springs are preloaded the lowering will be less. For example if the rear suspension is preloaded at 28 mm the lowering of the rear frame should be 24 mm instead of 52 mm.

Case *2:* Change from straight line motion to cornering motion.

In the change from a straight line to a curve the load on the wheels increases gradually according to the increase in the motorcycle angle of inclination. At a roll angle φ that is equal to 45° the lowering Δh of the gravity center, caused by the increase of the wheel load, is equal to 24 mm.

The vehicle front frame lowers to a greater extent compared with the rear frame (29 and 22 mm respectively) because the front suspension spring is softer than the rear one, and the resulting forward pitch rotation $\Delta\mu$ is equal to an additional -0.3°.

The above example suggests some interesting considerations regarding the motorcycle pitch motion, in the change from a straight line to a curve:

- the motorcycle pitch is further increased in the braking phase while entering the curve,
- the forward pitch rotation causes a decrease in the front trail, which potentially helps entering a curve. In fact, the lateral force moment that tends to align the front frame decreases.

6.2.3 Trim in exiting a curve

If a thrust force is applied in a curve, and in the exiting phase, another variation of the trim appears due to several factors that come into play:

- the load transfer from the front wheel to the rear one, due to the thrust, determines a decrease of the load operating on the front wheel and hence, causes a positive rearward pitch rotation;
- the change from the cornering motion to the straight line motion determines a decrease in the load operating on the wheels and, therefore, the motorcycle, with the typical values ofsuspension stiffness, undergoes a positive counter-clockwise pitch rotation.

Let us now illustrate, with an example, another event related to the squat ratio.

Example 4

Consider the motorcycle from the previous example.
In the curve balance configuration, the lowering of the gravity center is equal to Δh=24 mm.

Let's now take a look at the effects of applying the thrust force. Fig. 6-15 depicts the deviation of the squat ratio, both in a straight line and in a curve (angle of roll φ= 45°) versus variation in the vertical position of the wheel compared with the frame.

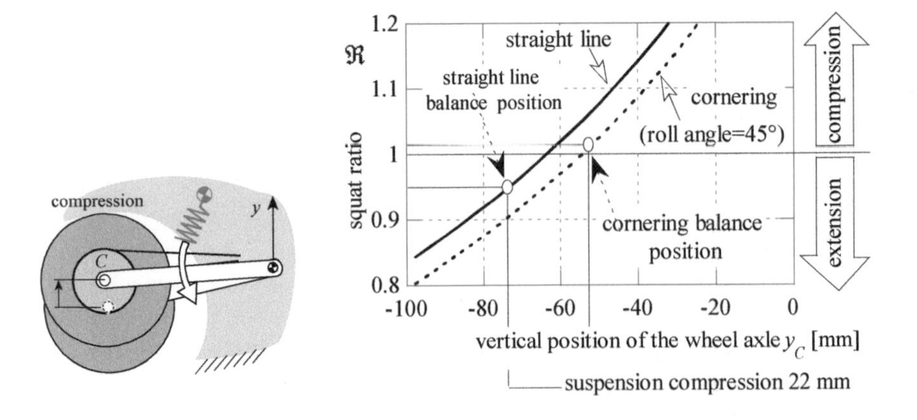

Fig. 6-15 Squat ratio in a straight line and in a curve.

It is possible to observe that in the balance positions, both in a straight running and in a curve, the squat ratio will be different.

Applying a thrust in a curve causes a variation in the trim, in the same way noted for the straight line:

- with $\Re = 1$ the rear suspension does not change trim;
- with $\Re < 1$ the rear suspension is extended;
- with $\Re > 1$ the rear suspension is compressed.

In this example, the squat ratio in the change from a curve to a straight line decreases from 1.02 to 0.95. This decrease, therefore, causes the rear suspension spring to extend.

Let us now examine the parameters that can be modified in order to limit the variations of the motorcycle trim in the thrust phase:

- the variation of the trim due to the difference in stiffness between the front and rear suspension is unavoidable because of the lower stiffness of the front suspension;
- the variation of the trim caused by the load transfer is also inevitable in the thrust phase;
- conversely, since in the thrust phase the trim varies depending on the value of the squat ratio, it is possible to modify this parameter by changing the suspension geometry.

6.3 Motorcycle trim in accelerated motion

One of the most important characteristics of motorcycle dynamic behavior in the acceleration phase, is its, more or less, ease to mount up or wheelie when it is subjected to a high driving force.

Aside from the amount of torque provided by the engine the motorcycle wheelie depends on the rear suspension characteristics and the transmission system that links the engine to the rear wheel.

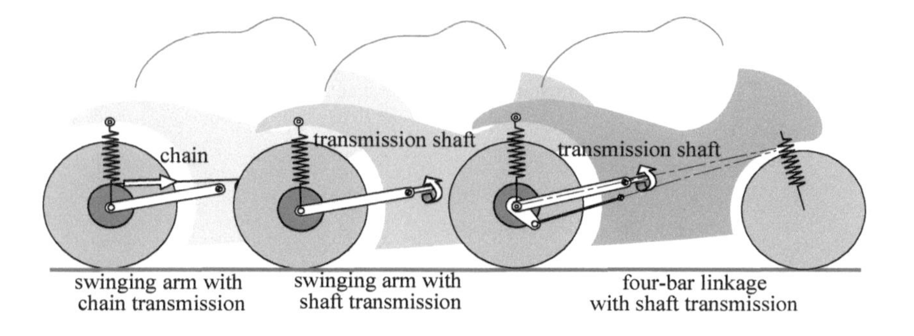

Fig. 6-16 Motorcycles with different typologies of rear suspensions.

Consider three motorcycles with equal inertial and geometric characteristics, subject to the same driving force, but featuring different rear suspensions and final drive systems:

- motorcycle with a classic swinging arm with chain;
- motorcycle with a classic swinging arm and transmission shaft;
- motorcycle with a four-bar linkage rear suspension and transmission shaft.

If in the initial instant, motorcycles have a constant speed equal to 100 km/h, and suddenly the engine generates a high torque transmitted to the rear wheel, the wheel accelerates and transfers the thrust force to the ground by means of longitudinal slip between the wheel and the ground.

First of all, consider three motorcycles with a classic swinging arm rear suspension and with chain transmission, but with different squat ratio values:

- $\Re = 1$ reference configuration;
- $\Re = 0.7$ obtained by moving the drive sprocket down;
- $\Re = 1.3$ obtained by moving the drive sprocket up.

High ratio values cause a compression of the rear suspension during the thrust phase and thus reduces the tendency for the vehicle to mount up, as shown in Fig. 6-17. A squat ratio less than one, while causing the rear suspension to extend, facilitates mount up. The figure demonstrates that, where $\Re = 1$ and $\Re = 1.3$, the front wheel lifts up and returns to the ground after a time interval of 0.8-1 seconds, whereas in the case of the ratio being $\Re = 0.7$, the vehicle travels with its front wheel skyward.

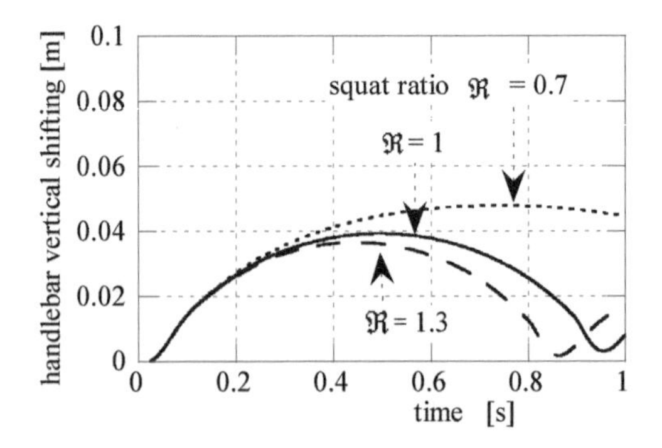

Fig. 6-17 Motorcycle mount up as the squat ratio varies.

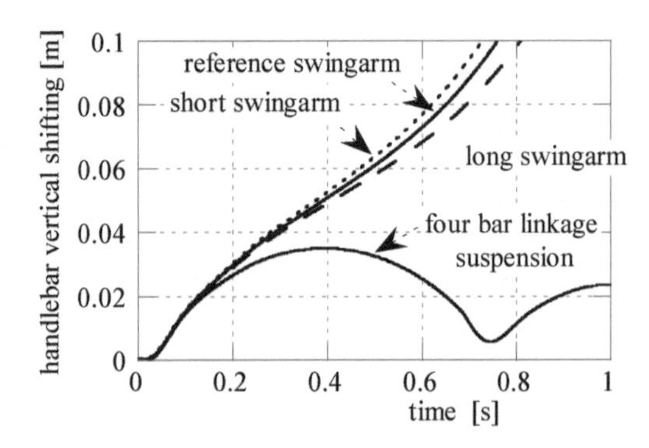

Fig. 6-18 Motorcycle mount up with final shaft drive.

Let's now see how motorcycles featuring different rear suspension systems behave. Figure 6-18 shows that the motorcycle with shaft transmission mounts up easily, and in the numerical simulation represented here, if the applied torque is not controlled it reaches the point where the vehicle overturns. The ease to mount up is proportional to the swinging arm length. The four-bar linkage suspension, in the case of shaft transmission, behaves much better. The figure illustrates that the configuration taken into consideration, that has the center of rotation of the connecting rod approximately above the front wheel, shows a behavior which is similar to that of the motorcycle with a classic swinging arm and with a high squat ratio. The BMW "Paralever" suspension is based on this kinematic scheme.

6.4 Influence of rear wheel slippage on the trim

Consider a motorcycle in straight line motion, with a constant speed, that suddenly loses its grip on an oily or icy spot.

In straight running a thrust force, equal and opposite to the resultant of the resistant forces, is applied to the rear wheel. The loss and the sudden pickup of grip create a violent variation of the driving force that impulsively excites both the suspension and the rear part of the motorcycle. The behavior of the motorcycle can be understood by considering this event as being constituted of five different sequential phases, represented in Fig. 6-19:

- steady state with the motorcycle traveling at a constant speed,
- sudden slippage of the rear wheel,
- rear wheel acceleration due to poor grip,
- sudden pickup of grip,
- damped transient oscillation.

Steady state with the motorcycle traveling at a constant speed.

The elastic moment M_v depends on the vertical transfer load N_{tr}, driving force S and chain force T:

$$M_v = N_{tr} L\cos\phi - S(R_r + L\sin\phi) + T[r_c - L\sin(\phi - \eta)]$$

It is important to remember that if $\Re$ is equal to 1, the moment exerted by the chain force T, the driving force S and the load transfer N_{tr} balance themselves out, therefore, the variable component M_v is equal to zero (the static elastic moment M_s balances the static load N_{s_r}). If $\Re < 1$, the spring is slightly extended, conversely if $\Re > 1$, it will be slightly compressed.

Sudden slippage of the driving wheel.

The loss of grip suddenly nullifies the driving force, so that the swinging arm and wheel system are no longer balanced.

$$I\ddot{\phi} = M_v - N_{tr} L\cos\phi - T[r_c - L\sin(\phi - \eta)]$$

The swinging arm is subject to a sudden angular acceleration in a clockwise direction (negative value), that is transmitted through the spring and shock absorber assembly and even the frame. The impulsive, upwards acceleration of the rear part of the motorcycle tends to throw the driver forward.

Rear wheel acceleration due to poor grip.

The rear wheel, subject to the torque generated by the driving force, accelerates and, furthermore, the motorcycle pitches forward until it nullifies the load transfer. This force, in steady state, is proportional to the driving force. On the other hand, in transient conditions it occurs with a delay compared with the driving force, as it is linked to the vehicle pitch. Only in the absence of suspension, the load transfer is perfectly in phase with the thrust.

In this phase, the rear suspension is extended because of the slow decrease in the load transfer force and, therefore, the rear frame tends to lift up while the front frame lowers.

Sudden pickup of grip.

The high value of the longitudinal slippage creates a large and sudden driving force, the swinging arm is subject to a sudden moment that causes an acceleration (which is positive) in a counter-clockwise direction (towards extension) that transmits through the spring and shock absorber assembly to the frame.

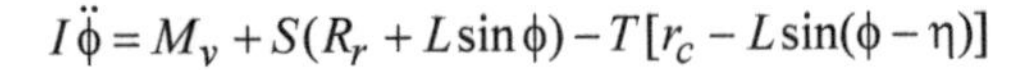

$$I\ddot{\phi} = M_v + S(R_r + L\sin\phi) - T[r_c - L\sin(\phi - \eta)]$$

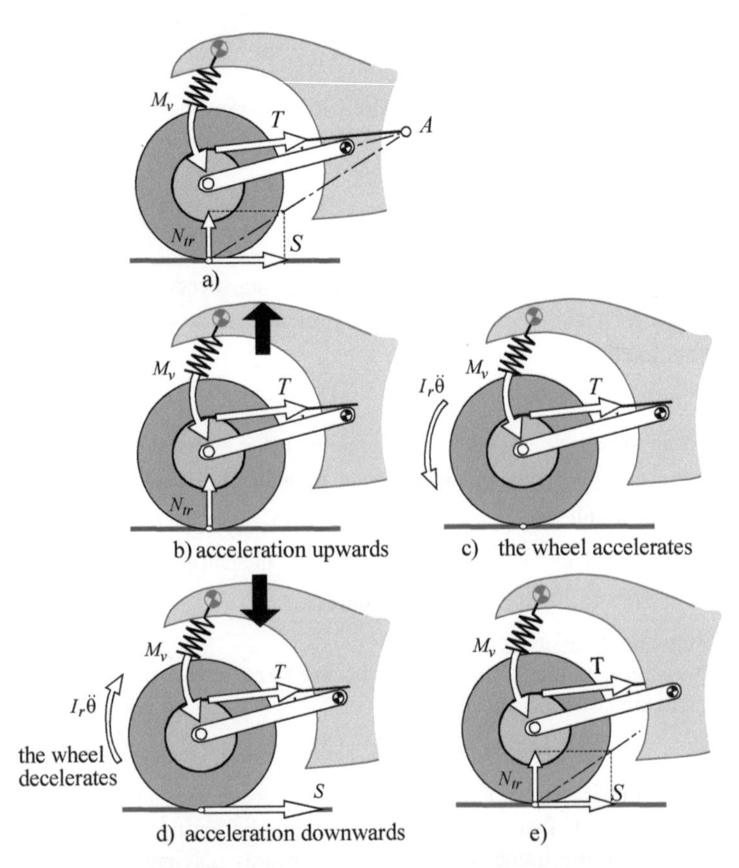

Fig. 6-19 Rear suspension trim during rear wheel slippage.

Damped transient oscillation.

A damped transient oscillation of the frame pitch occurs and after a certain time interval the initial steady condition is attained.

Figure 6-20 shows the deviation of the driving force and the vertical load on the rear wheel. As we can see, the load transfer force in the low grip zone slowly decreases and the driving force becomes very high when the pickup of the grip occurs.

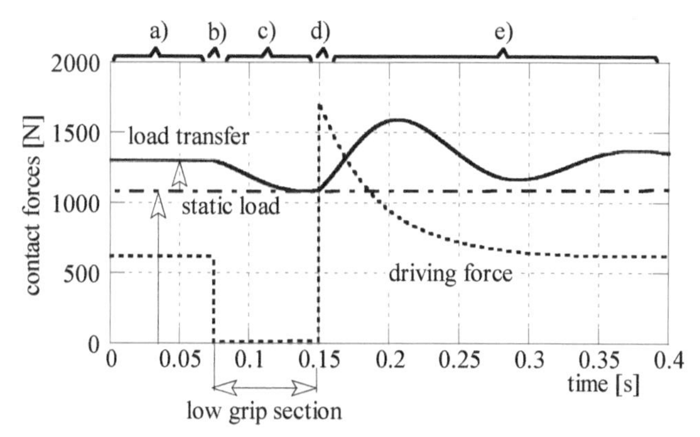

Fig. 6-20 Deviation of the vertical load and driving force.

Figure 6-21 illustrates the deviation of the vertical acceleration about the rear axle and the frame point to which the spring and shock absorber assembly is attached. The upward acceleration peak at the beginning of the low grip section, and the downward acceleration peak at the grip pickup point are quite noticeable.

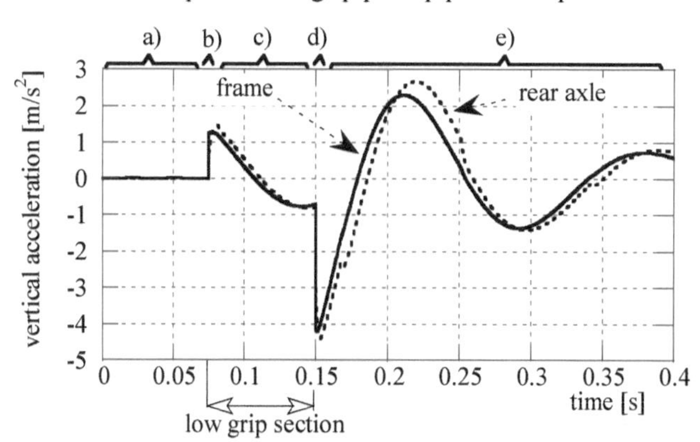

Fig. 6-21 Vertical acceleration of the rear part of motorcycle.

The moment balance for the swinging arm in steady conditions, as represented in Fig. 6-22:

$$0 = M_v - N_{tr} L \cos\phi + S(R_r + L\sin\phi) - T[r_c - L\sin(\phi - \eta)]$$

allows us to express a few observations regarding the chain force and the driving force. As we can see, the elastic moment becomes independent of the chain force and the driving force if the following four conditions are verified:

- the chain angle of inclination η is equal to the swinging arm angle of inclination

ϕ;

- the inclination angle of the swinging arm ϕ is equal to the load transfer angle $\tau = h/p$;
- the load transfer is proportional to the driving force, as it was for the stationary condition $N_{tr} = Sh/p = S\tan\tau$;
- the wheel angular speed is constant in order to have a direct proportionality between the driving force and the chain force $T = SR_r/r_c$.

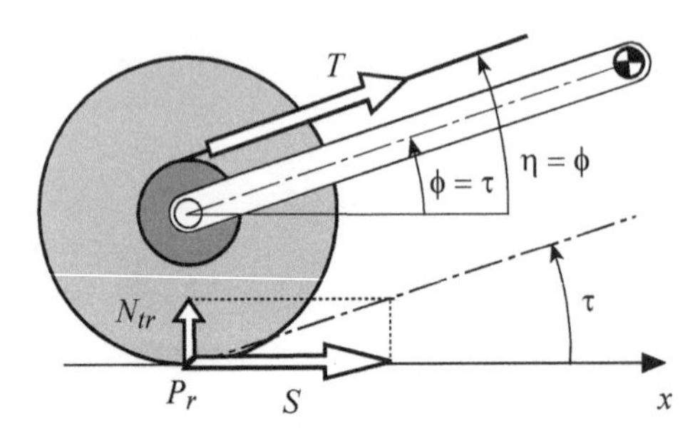

Fig. 6-22 Geometric configuration of the rear suspension that should ensure, in steady conditions, a trim independent of the forces applied.

In fact, the motion equation can be simplified in the following way:

$$M_v = N_{tr}\, L\cos\phi - S\,L\sin\phi$$

The variable elastic momentum M_v is null and therefore, as the chain force and then the driving force vary the swinging arm remains in equlibrium.

Two new architectures have been proposed [Romeva et al., 1993] that meet the conditions necessary to keep the squat ratio constant and unitary under steady conditions. The functional schemes of the two solutions, respectively named the "Bilever" and "Tracklever" systems, are shown in Fig. 6-23.

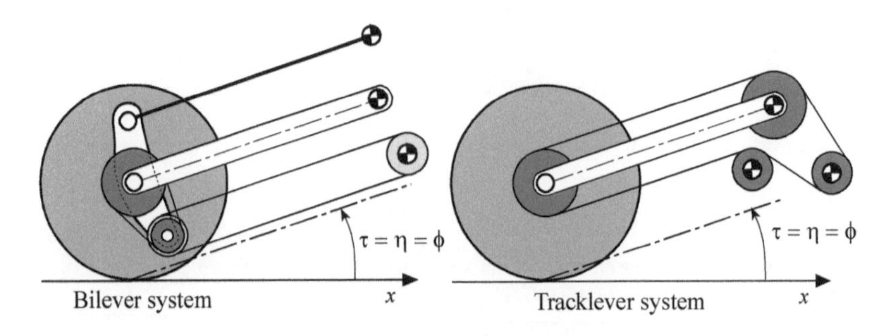

Fig. 6-23 Rear suspensions "Bilever" and "Tracklever".

Unfortunately, as we have already seen, in transient conditions, the load transfer (function of the driving force) occurs with a certain delay compared with the driving force and, therefore, these innovative suspension schemes do not present any advantages over the classic swinging arm suspension. In fact, Fig. 6-24 shows that even if the $\eta = \phi$ condition is met in the transient event both in the change from dry to wet ground and vice versa, the driving force, the load transfer and the chain force moments on the swinging arm are not in equilibrium. This behavior is thus like that of a traditional swinging arm suspension.

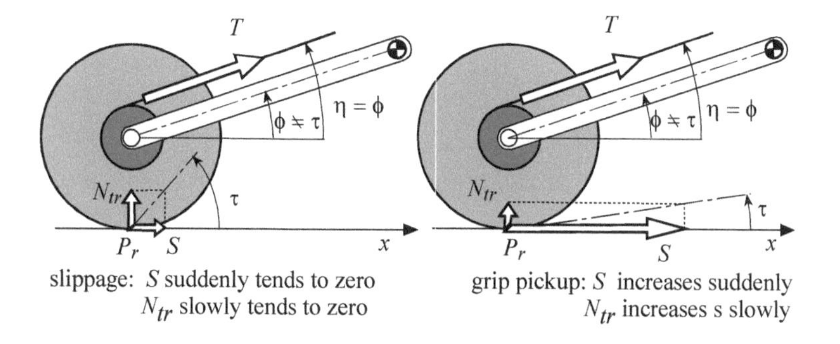

Fig. 6-24 Forces operating on the suspension during slippage and grip pickup.

6.4.1 Rear suspension with the pinion attached to the swinging arm

Another valid solution designed to improve the suspension dynamic behavior in the presence of the chain force is represented by the introduction of a second sprocket on the swinging arm (ATK System: Anti Tension Chain System).

In the original version, the chain inclination angle is equal to the angle of the swinging arm. The second sprocket, integral to the swing itself, is positioned so that the line of action of the reaction force exerted by the pinion on the swinging arm would pass through the swinging arm pivot without causing any additional moment.

Let us consider the more general case which presents a constant value between the chain inclination angle and the swinging arm inclination angle (Fig. 6-25):

$$\nu = \phi - \eta$$

If the reaction force generated by the second sprocket on the swinging arm does not generate a moment, due to the fact that the line of action passes through the swinging arm pivot, the equilibrium between the moments acting on the wheel-swinging arm system in relation to the swinging arm axle, will be:

$$I\ddot{\phi} = M_v - N_{tr} L \cos\phi + S L \sin\phi + T L \sin\nu$$

Let us analyze the above equilibrium equation:

- the moment caused by the load transfer compresses the suspension;
- the driving force compresses the suspension if $\phi < 0$, or extends it if $\phi > 0$;
- the moment generated by the chain force extends the suspension if $\nu > 0$, or compresses it if $\nu < 0$.

In order to obtain values of the chain ratio, very close to the unit, the chain must be more inclined in relation to the swinging arm, that is: $\nu = (\phi - \eta) > 0$.

The curves in Fig. 6-26 show the growing trend of the chain force ratio as the vertical wheel travel y_c becomes higher.

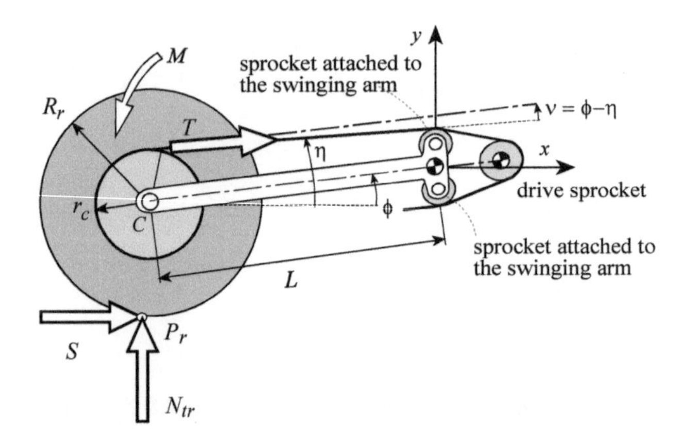

Fig. 6-25 Suspension with a second sprocket attached to the swinging arm.

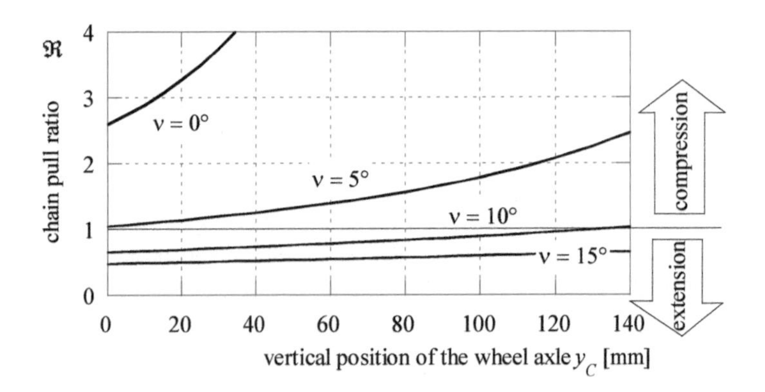

Fig. 6-26 Chain force ratio varying with the y_c displacement for several values of the ν angle.

The curve shape can be modified by varying the sprocket position in relation to the swinging arm pivot. To do this, placing the second pinion in various positions on the swinging arm (see Fig. 6-27), it is possible to take advantage of the reaction force thus generating a moment which will extend or compress the suspension.

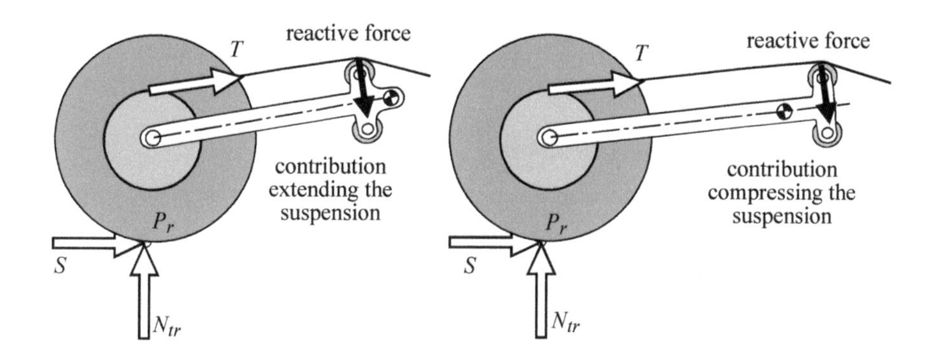

Fig. 6-27 Influence of the sprocket position on the chain force ratio.

6.5 The braking action

The load transfer during the braking action is directly proportional to the total force of the braking action, to the height of the mass center, and inversely proportional to the wheelbase (see Fig. 6-28):

$$N_{tr} = (F_r + F_f)\frac{h}{p} = F\frac{h}{p}$$

where F indicates the sum of the front and rear braking action.

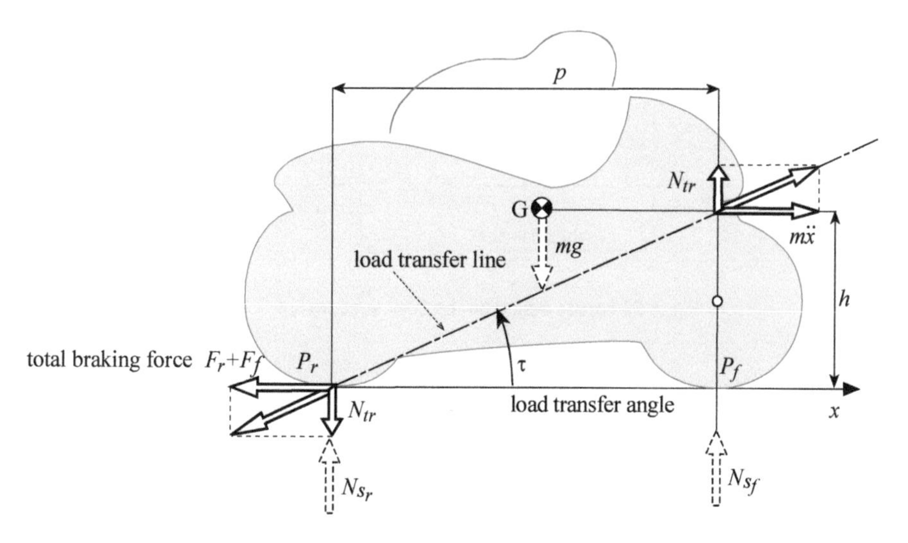

Fig. 6-28 Motorcycle equilibrium during braking.

The braking action generates a pitching motion in the motorcycle, especially at the beginning, when the braking force is suddenly applied. During the residual time of the braking action, assumed to be uniform, the front and rear suspensions take a different trim depending on the type of suspension.

In the following paragraph the effects of the braking force on the front and rear suspensions are examined.

6.5.1 The front suspension

During braking the front suspension is subject, in addition to the static vertical load, to two additional forces:

- the front braking force F_f,
- the load transfer N_{tr} generated by the total braking force F .

These two forces define the load transfer angle of the front wheel τ_f, i.e., the inclination angle of the line of action of the resulting force acting on the front wheel in relation to the road (Fig. 6-29).

$$\tau_f = \arctan\frac{N_{tr}}{F_f}$$

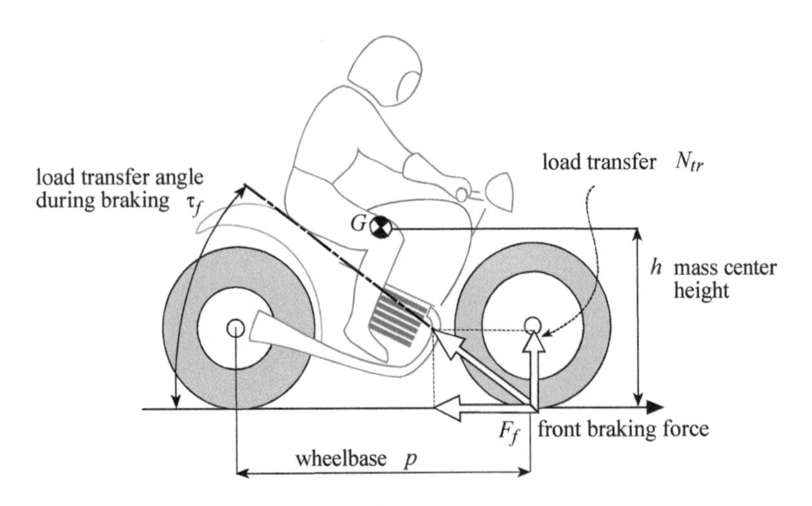

Fig. 6-29 Forces acting on the front suspension.

The telescopic forks

In the case of telescopic forks, the contact point trajectory of the front wheel in relation to the frame can be presumed to be straight and parallel to the steering axis. During the braking action, the fork is compressed, due to the load transfer component N_{tr} and the effect generated by the braking force component F_f, as shown in Fig. 6-30. The magnitude of the compression depends primarily on the fork inclination angle $\gamma = \pi/2 - \varepsilon$ (trajectory angle).

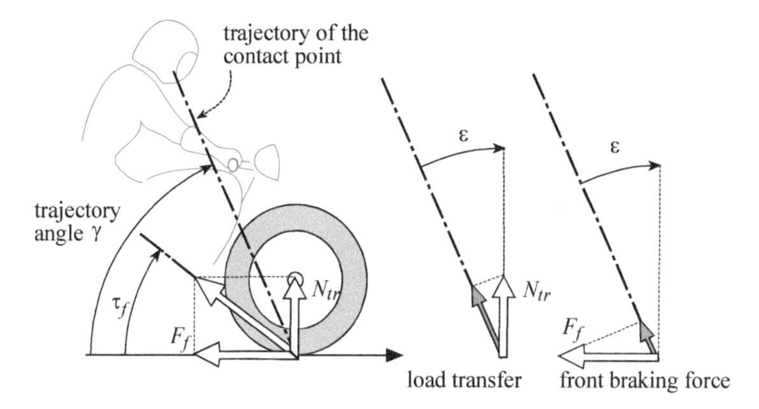

Fig. 6-30 Forces compressing the suspension.

Let us suppose that the braking force applied to the rear wheel is zero. In this case, the normalized compression force on the fork, expressed by the ratio between the sum of the compressing components on the suspension and the front braking force, depends only on the steering inclination angle and on the load transfer angle:

$$\frac{N_{tr}\cos\varepsilon + F_f\sin\varepsilon}{F_f} = \frac{F_f\dfrac{h}{p}\cos\varepsilon + F_f\sin\varepsilon}{F_f} = \frac{h}{p}\cos\varepsilon + \sin\varepsilon$$

The dive behavior of the suspension is maximum when the load transfer angle τ_f corresponds to the fork inclination angle γ.

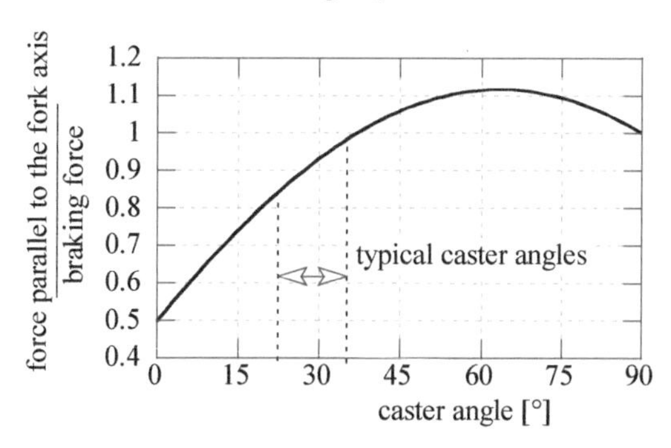

Fig. 6-31 Normalized compression force on the telescopic fork [h/p = 0.5].

Figure 6-31 shows that the dive displacement, proportional to the compression force, is at its maximum when the fork has an inclination angle of about 63°. We can observe that under normal usage, the dive behavior is more pronounced as the inclination angle of the fork becomes higher.

Another typical effect of the telescopic fork is the decrease in the trail as fork compression increases.

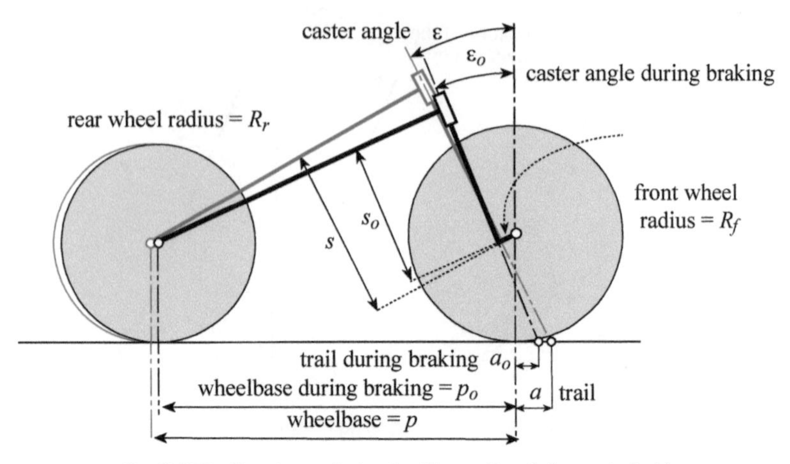

Fig. 6-32 Trail variance during braking action (telescopic fork).

Let's suppose that the wheels have the same radius and that we can ignore the rear suspension deformation. It is easy to determine a formula, which would indicate the trail variation, as a consequence of the front suspension compression. The steering inclination angle during braking is, therefore:

$$\varepsilon_o = \arctan\left(\frac{p\sin\varepsilon + \Delta s}{p\cos\varepsilon}\right)$$

where $\Delta s = s - s_o$ represents the front suspension compression.

The trail during braking is expressed by the following formula:

$$a_o = \frac{R_f \sin\varepsilon_o - d}{\cos\varepsilon_o}$$

Fig. 6-33 shows an example of the variation of trail and steering inclination directly related to the variation of the front suspension compression.

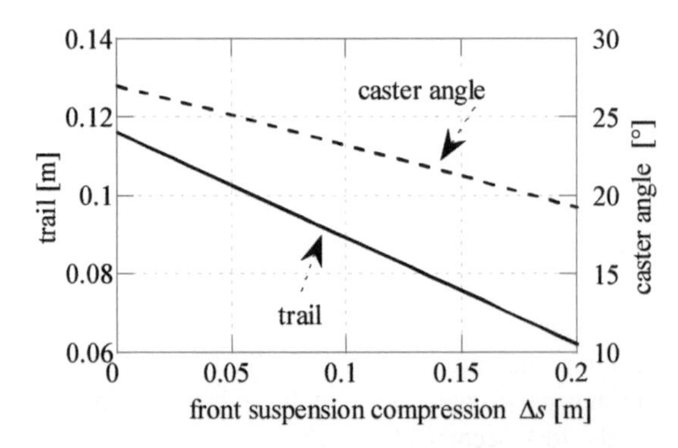

Fig. 6-33 Trail and caster angle versus the front suspension compression.
[p = 1.4 m, a = 0.116 m, ε = 27°, R_r = R_f = 0.36 m]

Neutral suspension and anti-dive suspension

If the braking behavior of the front suspension has to be neutral, that is, without any effect from the braking force, the trajectory described by the contact point should be a vertical line (see Fig. 6-34). In this case compression of the suspension is caused only by the load transfer.

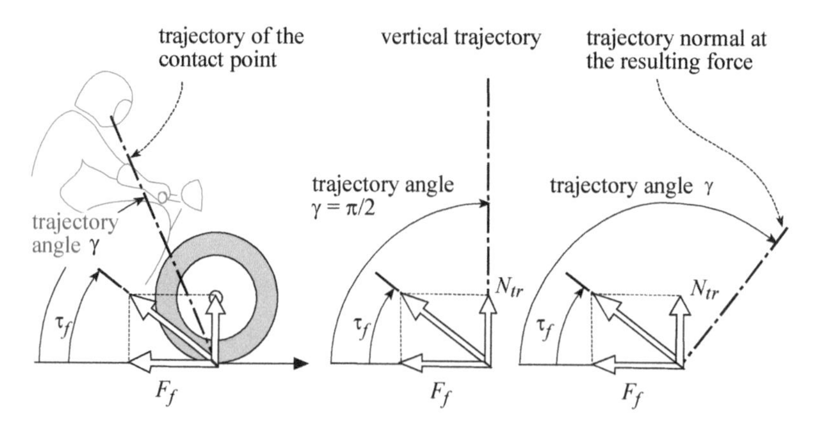

Fig. 6-34 Neutral front suspension and anti-dive suspension.

Vertical axle paths during braking are easily obtained by applying four-bar linkage suspensions with the wheel attached to the connecting link of the four-bar itself or with a simple leading or trailing arm.

If the trajectory of the wheel is normal in relation to the force generated by the sum of the load transfer and the braking force (Fig. 6-34), the front suspension spring will not be stressed during braking. This functional diagram is also obtained by applying a four-bar linkage suspension, or with the suspension push-arm.

Suspension systems, which diminish the suspension compression due to the load transfer, are called anti-dive suspensions.

Four-bar linkage suspension

The "Telelever' suspension (see Fig. 6-35), applied as we all know by BMW, is generated by the four-bar linkage. The mechanism is in fact a spatial mechanism with two degrees of freedom:

- the 1st degree of freedom, given by the rotation of the two components connected to the prismatic couple around the axis that passes through the center of the two spherical couples. It corresponds to the steering action.
- the 2nd degree of freedom, which implies the motion of all the components of the mechanism, corresponds to the front suspension movement.

From a kinematic point of view, the behavior shown in this diagram is quite different from that of a traditional suspension with telescopic fork. One of the advantages of this scheme is that it allows a certain degree of anti-dive suspension.

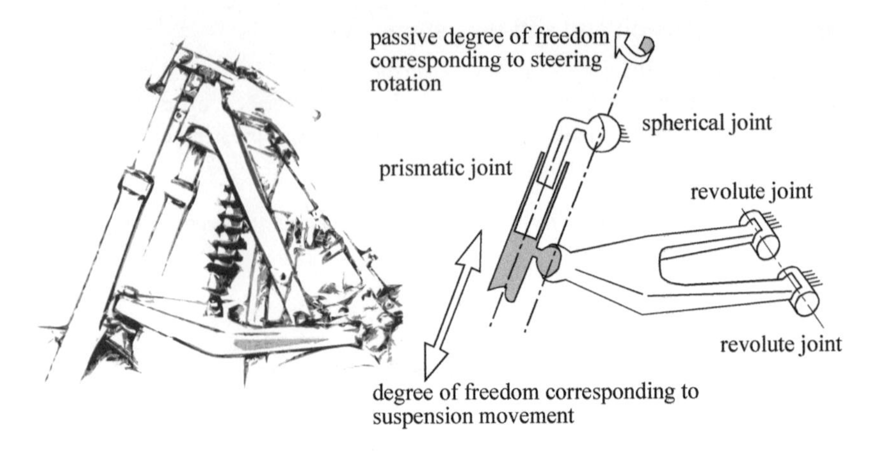

Fig. 6-35 Telelever front suspension based on a spatial mechanism.

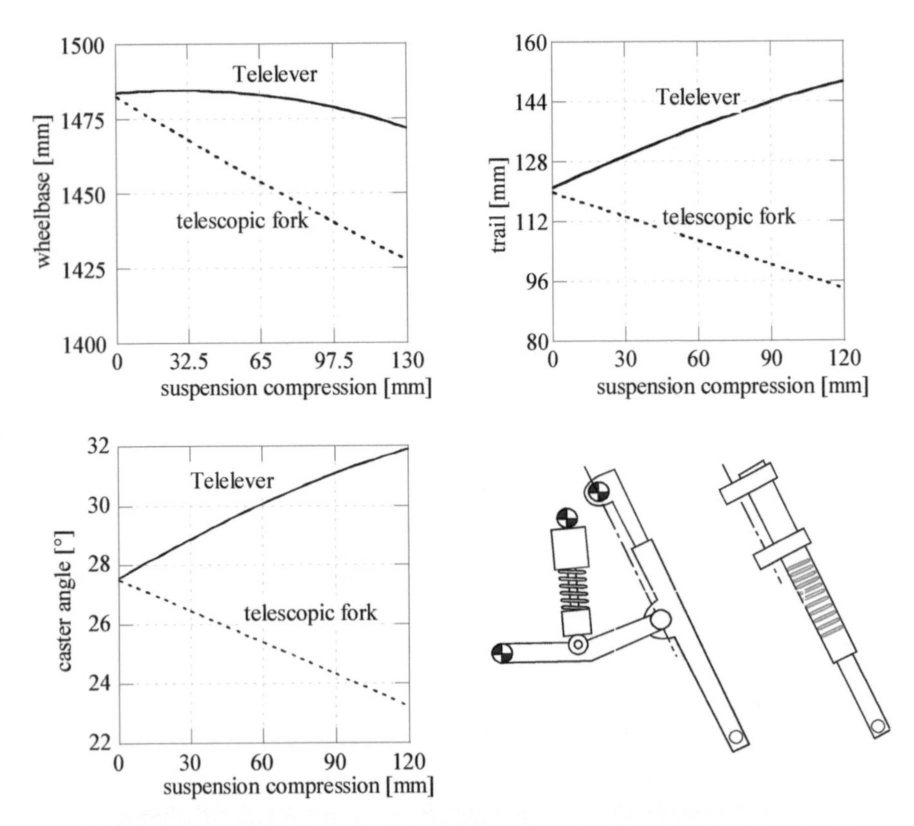

Fig. 6-36 Variance of wheelbase, trail and inclination angle versus the suspension movement.

The diagrams of Fig. 6-36 show the variance of the wheelbase, the trail and the steering head angle, for two motorcycles, geometrically similar in static equilibrium though equipped with different front suspensions. For the two suspensions we can observe a different, almost opposite behavior: with the traditional suspension the increase of the suspension compression causes a decrease in the wheelbase, the trail and the caster angle, whereas with the Telelever suspension, the wheelbase stays almost constant and the other two parameters augment.

Push-arm suspension

If the pivot trajectory of the front wheel is circular (as for example with the 'Earles' fork) the load transfer compresses the suspension, while the braking force extends it, as shown in Fig. 6-37.

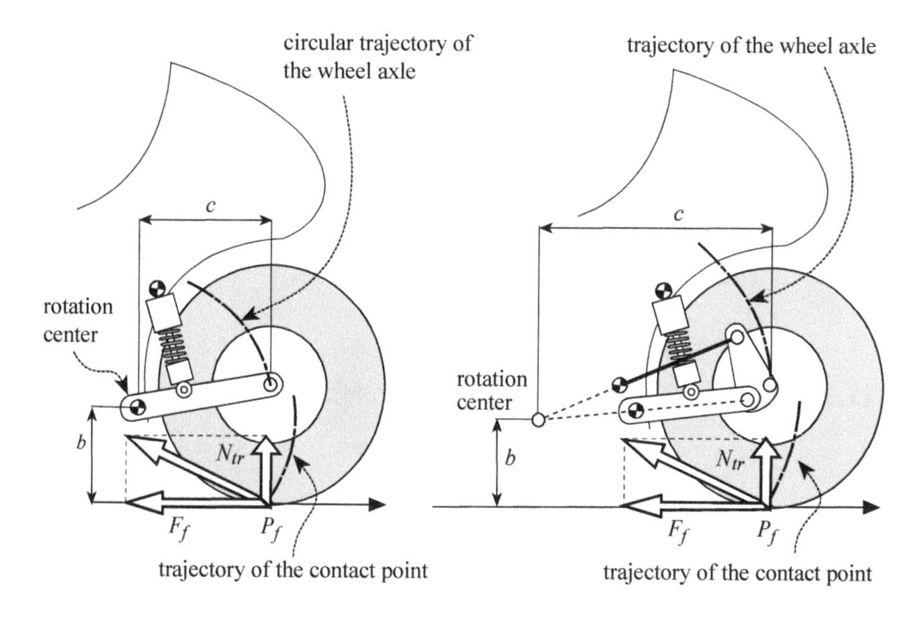

Fig. 6-37 Front push-arm suspension and four-bar suspension.

If we apply only the front brake (F_r = 0), the equation of the moments around the rotation center changes to:

$$M = N_{tr}\, c - F_f\, b = F_f(\frac{h}{p}c - b)$$

The suspension is therefore either compressed or extended, depending on its geometrical characteristics, and in particular depending on the distances c and b. If the moment is positive, the suspension is compressed. If the moment is negative, the suspension is extended.

If the brake caliper is not a part of the arm, but connected to the frame with a rod, the previous formula is still true. In this case the values c and b will represent the distances measured along the x and y axes between the wheel axle and the intersection point of the straight lines (see Fig. 6-37).

6.5.2 The rear suspension

Let us consider the rear suspension and suppose that the chain does not transfer any force (during braking the lower section of the chain is slack).

The direction of the sum of the rear braking force F_r, with the load transfer N_{tr}, may be more or less inclined in relation to the straight line which connects the contact point with the swinging arm pivot, depending on the values of the two corresponding forces.

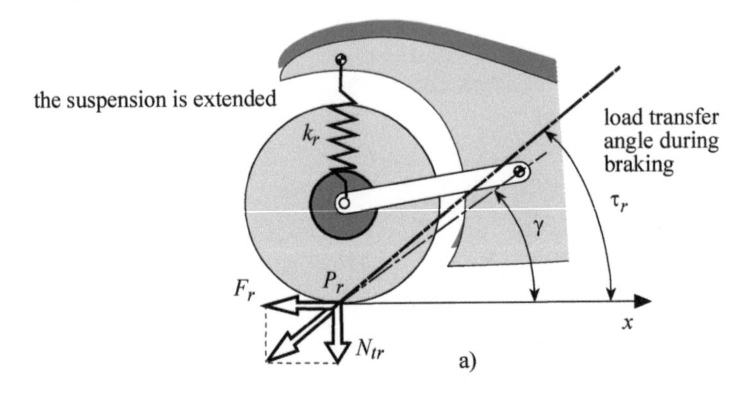

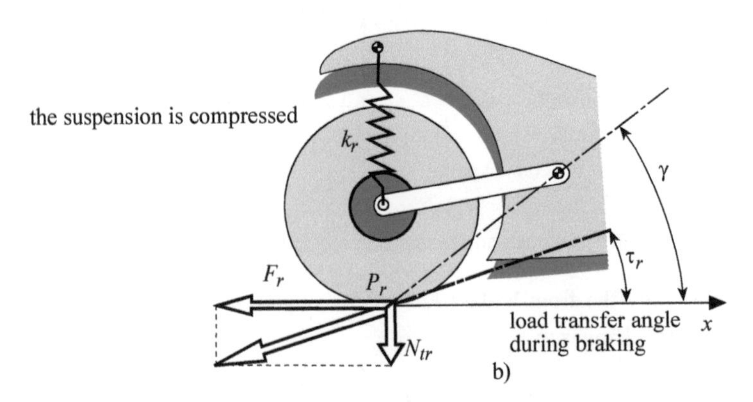

Fig. 6-38 Swinging arm suspension during braking.

The angle of the load transfer during rear braking τ_r corresponds to the inclination of the straight line of the force acting on the rear wheel in relation to the road plane (see Fig. 6-38).

$$\tau_r = \arctan \frac{N_{tr}}{F_r}$$

If the load transfer N_{tr} is high, due to the braking action of both the front and rear brakes ($F_f \neq 0$) and if the rear braking force F_r is low, the resulting force will generate a moment that will tend to extend the rear suspension (Fig. 6-38a). In this case, the load transfer angle is larger than the γ angle. This angle is formed by the straight line which connects the contact point with the swinging arm pivot and the x axis.

On the contrary, if we apply only the rear brake, and therefore the rear braking force F_r has a high value, the angle γ could be larger than the load transfer angle. In this case, the rear suspension will be compressed (Fig. 6-38b).

In the above diagram, we have supposed that the brake caliper is connected to the swinging arm. This means that the forces exchanged by the braking elements will react within the rear swinging arm-wheel assembly.

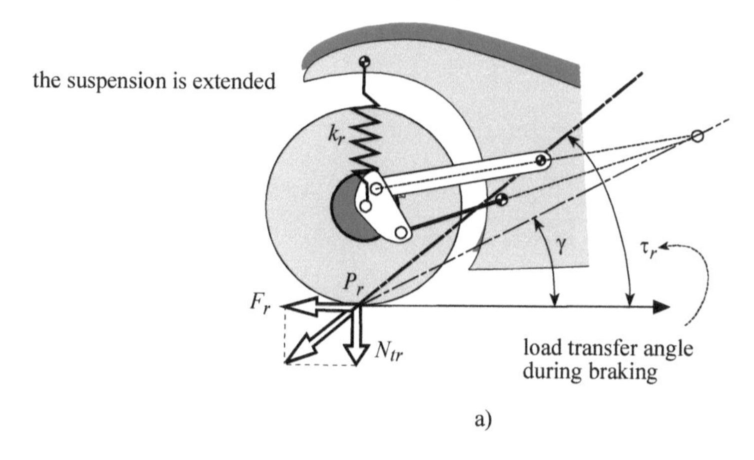

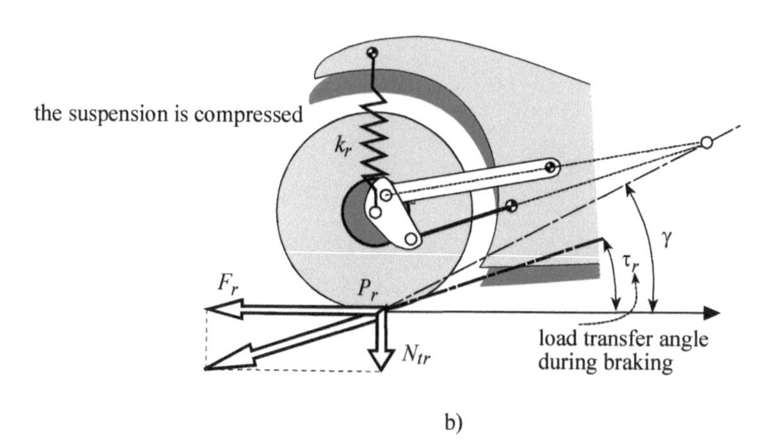

Fig. 6-39 Four-bar linkage rear suspension during braking.

The braking angle will change if the support of the brake caliper is free to rotate around the wheel axle and is fixed to the frame by means of a rod. The swinging arm, the connecting rod, the support and the frame form a four-bar linkage. The intersecting point of the swinging arm axis and the rod corresponds to the instantaneous center of rotation of the connecting rod in relation to the frame. The wheel-swinging arm equilibrium, in relation to this point, shows that the suspension is compressed or extended depending on its position, as shown in Fig. 6-39.

Motorcycle Ducati 125 cc. of 1956

7 Motorcycle Vibration Modes and Stability

As everyone knows, the front and/or rear end of a motorcycle in motion can start to oscillate around the steering axis, even if the wheels are well balanced. This phenomenon is easy to observe experimentally, for example by gradually slowing down the motorcycle from a fairly high speed. Oscillations can be observed at certain speeds especially if the front wheel is out of balance. They reach their maximum amplitude and then decrease as speed decreases until they disappear completely. Rear-end oscillations can be observed when traveling over a transverse bump or by exciting the rear frame with an impulsive movement of the rider's trunk. At a low speed it can also be easily observed that the motorcycle tends to fall over sideways, regardless of what the rider does. These experimental observations of motorcycle dynamics show that there are three major modes:

- capsize, a non-oscillating mode used and controlled by the rider;
- weave, an oscillation of the entire motorcycle, but mainly the rear end;
- wobble, an oscillation of the front end around the steering axis which does not involve the rear end in any significant way.

The rider's control task can be considered to involve either fixed control or free control, i.e. with or without their hands grasping the handlebar, respectively. With the steering rotation fixed the motorcycle-rider system is unstable in roll at all speeds, like a capsizing ship, whereas in the unconstrained condition the steering system is free to steer itself, potentially relieving the rider of the need to apply steering control action for stabilization. A viable motorcycle needs to self-steer effectively, thus contributing to automatic stabilization, without becoming too oscillatory under certain running conditions.

At very low speeds a motorcycle is unstable because of capsize, two non-oscillating modes involving respectively roll and steer motion of the vehicle. Around 1 m/s these real poles meet, coalesce and become the complex pole pair associated with the weave vibration mode. Weave mode is usually unstable up to 7-8 m/s. Over this speed motorcycles usually enter into a stable zone, such that the rider may remove his hands from the handlebar without falling. As speed increases, the weave, wobble, or capsize modes may become unstable, depending on the motorcycle characteristics, and the rider has to counteract these modes with a torque applied at the handlebar. Weave mode is usually poorly damped or unstable at high speeds, whereas wobble is poorly damped or unstable in the mid-range of speeds. Capsize instability at mid to high speeds is typically not very significant.

In this chapter, we will first study these modes using simplified models and later the in-plane and out-of-plane modes will be studied by means of an eleven degree of freedom model. Finally, the effect of frame compliance and rider mobility on motorcycle stability will be presented.

7.1 Simplified model

7.1.1 Capsize

This mode is deeply influenced by rider action on the handlebar, i.e. by the mechanical impedance (inertia, stiffness, damping) which the rider provides. Therefore the mode easily shifts from the unstable zone to the stable zone.

Capsize is a mode actually used by the rider to roll the motorcycle. This rolling action is achieved through the rider's effort to hold or move the steering head rotation to some non-equilibrium position (fixed-control). As mentioned, this mode is always unstable, since in essence motorcycle exists as an inverted pendulum. This mode is also present in free-control condition (i.e. without the rider's hands on the handlebar) as highlighted by eigenvalue analysis. In the latter case the mode is usually somewhat stable at low speeds because of the steering modal component of its eigenvector; whereas at higher speeds it may become slightly unstable.

The capsize mode consists mainly of a roll motion combined with a lateral displacement plus some less important steering and yaw movements (Fig. 7.1). It depends on a number of factors:

- speed of the motorcycle;
- wheel inertia (gyroscopic effect);
- position of the center of gravity;
- motorcycle mass;
- motorcycle roll inertia;
- caster angle;
- mechanical trail;
- properties of the tires, primarily cross sectional size of the tires, twisting torque and pneumatic trail of the front tire.

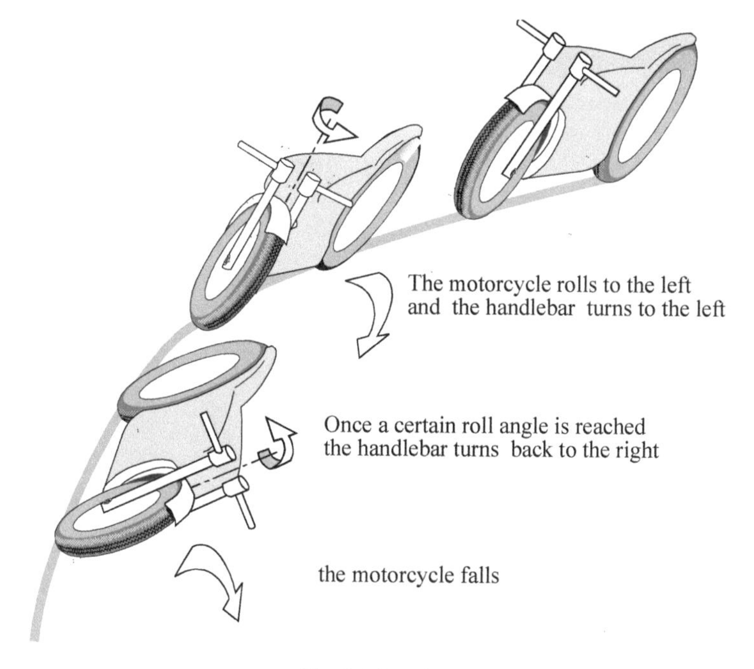

Fig. 7-1 Capsize.

To highlight the influence of some geometrical and inertial properties of the motorcycle on the capsize mode it is useful to analyze the fall motion of a motorcycle with the steering head locked.

In this hypothetical case, within the limits of linear approximation, capsize can be expressed as an exponential law:

$$\varphi = \varphi_o e^{\frac{t}{\tau}}$$

where τ is a positive time constant, therefore, the capsize is always unstable.

Basically the time constant is a measure of how easily the motorcycle will lean over. For example, racing motorcycles need a small time constant so they can corner and change trajectory quickly. Touring motorcycles need to roll more slowly, making them easier for the rider to control.

Capsize instability should not be viewed as a drawback, however, since it is precisely this phenomenon that, given proper control action, enables the motorcycle to lean into and execute curves correctly. The smaller the time constant (i.e., the greater the capsize instability), the less lead time is needed to start leaning the motorcycle into a curve.

The simplified models, with the steering head locked and negligible gyroscopic effects, yield smaller time constant values than the ones which can be obtained by studying the complete model of the motorcycle. The simplified models clearly show how geometric and inertial properties affect the capsize time constant.

Model using thin disk wheels

Mathematical modeling of motorcycle capsize is complicated by the presence of the steering head, gyroscopic effects, and tire contact forces arising from the slip and camber angles. Here we will concentrate on a very simple model to understand a specific aspect of capsize, the falling time.

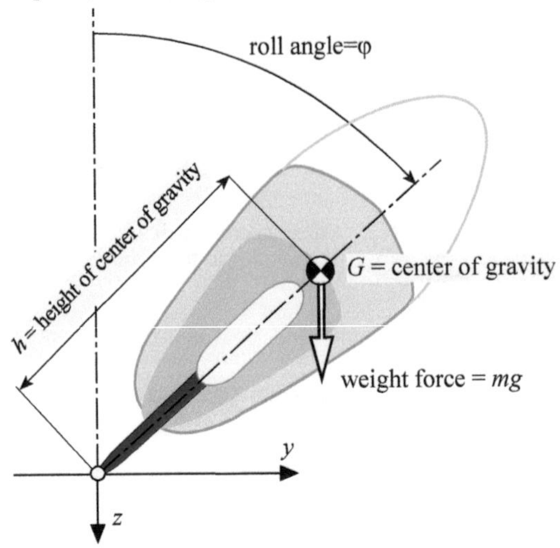

Fig. 7-2 Capsize for a motorcycle with thin disk wheels.

This simplified model makes the following assumptions:

- the motorcycle is moving in direction x at speed V;
- the thickness of the cross section of the tires is null;
- there is no slippage between the tires and the road;
- the steering head is locked in place;
- gyroscopic effects are negligible.

Based on these assumptions, capsize is a simple rotation of the motorcycle around the axis defined by the points in which the tires come into contact with the roadway (Fig. 7-2).

The equilibrium of moments with respect to the contact point gives the following equation:

$$(I_{x_G} + mh^2)\ddot{\varphi} = mgh\sin\varphi$$

Linearizing the equation around the vertical equilibrium position:

$$(I_{x_G} + mh^2)\ddot{\varphi} - mgh\varphi = 0$$

and introducing the solution:

$$\varphi = \varphi_o e^{st}$$

yields the following frequency equation:

$$(I_{x_G} + mh^2)s^2 - mgh = 0$$

Its solution is a real number and therefore corresponds to a non-oscillating motion:

$$s = \pm\sqrt{\frac{mgh}{I_{x_G} + mh^2}}$$

The time constant of interest τ is given by the inverse of the positive real eigenvalue:

$$\tau = \sqrt{\frac{I_{x_G} + mh^2}{mgh}}$$

Note that the time constant is determined by the height of the center of gravity, the mass of the motorcycle, and the motorcycle's moment of inertia about the x-axis through its center of mass.

Using the radius of gyration ρ ($I_{x_G} = m\rho^2$) to express the motorcycle's moment of inertia, the time constant τ takes on the following form:

$$\tau = \sqrt{\frac{h}{g}}\sqrt{1 + \frac{\rho^2}{h^2}}$$

Now let us assume that the height of the center of gravity and the mass of the motorcycle are constant, but that the mass can be distributed differently to vary the moment of inertia I_{x_G}, i.e. the radius of gyration ρ.

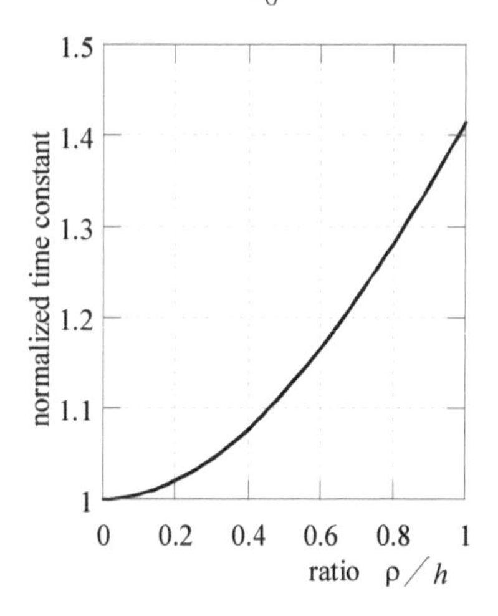

Fig. 7-3 Normalized time constant for capsize as a function of the ratio of gyration radius to center of gravity height.

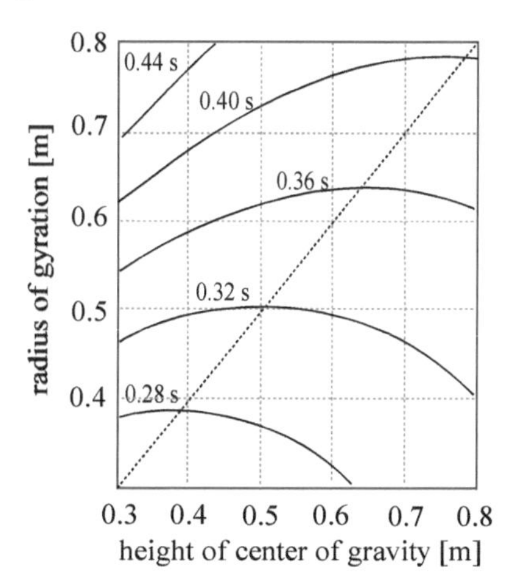

Fig. 7-4 Time constant for capsize as a function of radius of gyration and height of center of gravity.

The value of the time constant τ increases as the radius of gyration ρ increases (mass located further away from the center of gravity). Figure 7-3 shows the normalized time constant curve with respect to the minimum value for the ideal case of all the mass being concentrated at the center of gravity ($\tau = \sqrt{h/g}$).

Figure 7-4 is a contour plot that shows how the time constant varies as a function of the height of the center of gravity and the radius of gyration.

Note that, for a given radius of gyration value, the time constant decreases as the height of the center of gravity increases until it reaches a minimum value, and then increases. This means that once the radius of gyration ρ is set, the time constant is at its lowest value when the height of the center of gravity is equal to the radius of gyration, as can readily be shown analytically.

Example 1

Based on the following data, determine the capsize time constant.

- total mass of motorcycle: $m = 248$ kg
- height of mass center: $h = 0.648$ m
- moment of inertia about the x-axis through its center of mass: $I_{x_G} = 20$ kgm^2 (radius of gyration $\rho = 0.284$ meters)

The resulting time constant is: $\tau = 0.281$ s.

In the ideal case of all the mass being concentrated at the center of gravity ($I_{x_G} = 0$ kgm^2) the time constant decreases to 0.257 s.

Model using tires with circular cross section

A second simplified model can be built from the first, by removing the assumption that the wheels are thin disks, and instead assuming a motorcycle with circular tire cross sections which do not slip laterally on the roadway during capsize (Fig. 7-5). Once again, the system has only one degree of freedom.

The resulting equations are as follows:

$$m\,\ddot{z}_G = -N + m\,g$$

$$m\,\ddot{y}_G = F_s$$

$$I_{xG}\ddot{\varphi} = N(y_G - y) + F_s\,z_G$$

where the third equation is the equilibrium of moments with respect to the mass center.

Under pure forward rolling conditions the system still has just one degree of freedom, and therefore y, y_G and z_G can be expressed as a function of φ:

$$y = \varphi t$$
$$y_G = \varphi t + h_o \sin\varphi$$
$$z_G = -t - h_o \cos\varphi$$

Simple mathematical substitution yields the following frequency equation:

$$\left[I_{xG}+m(h_o+t)^2\right]s^2-mgh_o=0$$

The resulting time constant is given by:

$$\tau=\sqrt{\frac{h_o}{g}}\sqrt{\left(1+\frac{t}{h_o}\right)^2+\frac{\rho^2}{h_o^2}}$$

This equation shows that the time constant increases with the radius of the tire cross section. Therefore, entering a curve, a motorcycle with large tires takes longer to lean than one with small tires.

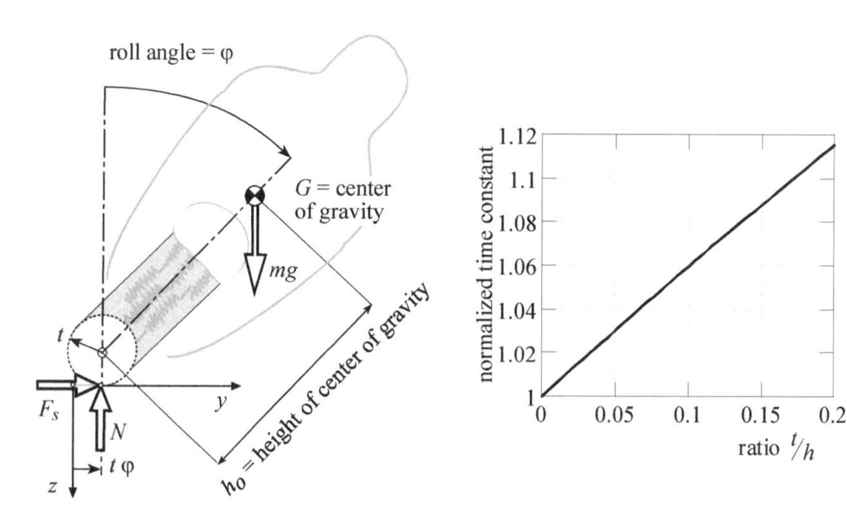

Fig. 7-5 Capsize with pure lateral rolling on the tire.

Fig. 7-6 Capsize time constant as a function of the ratio between the tire cross section radius and the height of center of gravity.

Figure 7-6 shows the ratio between the time constant for a motorcycle with tires of non-zero thickness and the time constant for the same motorcycle with thin disk tires as a function of the ratio between the tire cross section radius and the height of the center of gravity.

Example 2

Using the motorcycle in Example 1 but replacing each thin disk tire ($t = 0$) with a tire having cross section radius ($t = 0.10$ m). Determine the time change in the capsize time constant.

The time constant increases from $\tau = 0.281$s to $\tau = 0.305$s.

Model using tires with lateral sideslip

This model is the same as the previous one, except that the pure lateral rolling condition on the tire-to-road has been removed. In other words, the tires can slip sideways on the roadway when the motorcycle leans over. This simplified model has two degrees of freedom:

- rotation φ around the x-axis (leaning the motorcycle);
- displacement in direction y (the contact point sideslips).

The motorcycle equilibrium equations that assume pure forward rolling motion are valid even when the tire shows sideslip.

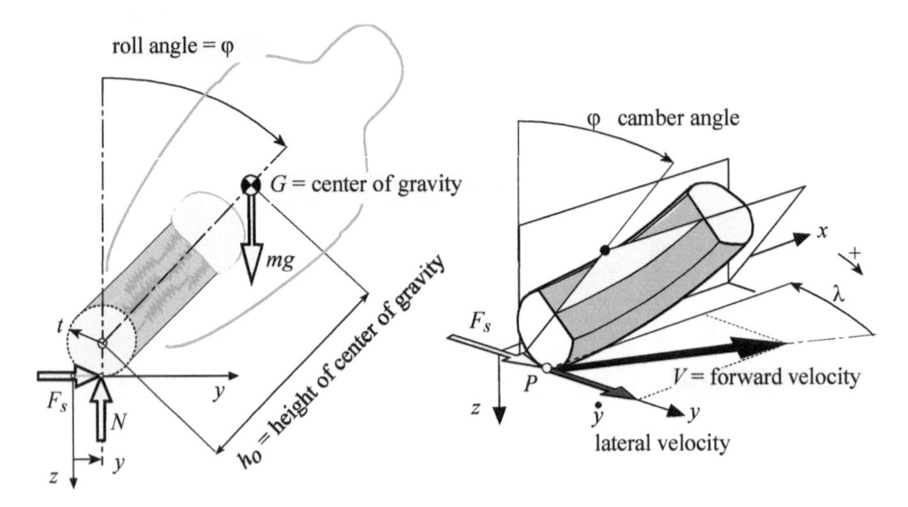

Fig. 7-7 Capsize with tire slippage.

In this case, however, the lateral force exerted on the tires by the roadway is defined by both the sideslip angle and the camber angle. This lateral force can be described by linear law as a function of the tires sideslip and camber angles:

$$F = (k_\lambda \lambda + k_\varphi \varphi) mg$$

where the sideslip angle is given by $\lambda = -\dot{y}/V$

The linearization around the vertical equilibrium position yields:

$$m h_o \ddot{\varphi} + m\ddot{y} = \left(-k_\lambda \frac{\dot{y}}{V} + k_\varphi \varphi\right) mg$$

$$I_{xG} \ddot{\varphi} + m h_o \ddot{y} = m \ g \ h_o \varphi - h(-k_\lambda \frac{\dot{y}}{V} + k_\varphi \varphi) mg$$

Manipulation of the equations to eliminate time dependence yields the following characteristic equation:

$$V\frac{I_{xG}}{mg}s^3 + k_\lambda(\frac{I_{xG}}{m} + (h_o + t)h_o)k_\lambda s^2 + V\,k_\varphi(h_o + t)\,s\ - k_\lambda g h_o = 0$$

It can be shown that just one of the four roots (eigenvalues) of the characteristic polynomial has a positive real part. A second root is zero, and the other two have negative real parts.

The positive root relates to capsize instability, whereas the two negative ones represent two stable motions. Remember that the time constant for capsize is given by the inverse of the real eigenvalue. In this model the time constant depends on the speed of forward motion V.

Example 3

Assume that the motorcycle in Example 2 has tires with the following stiffness values:

- cornering stiffness coefficient: $k_\lambda = 11.0\ \text{rad}^{-1}$;
- camber stiffness coefficient : $k_\varphi = 0.93\ \text{rad}^{-1}$;

Determine the effect on the capsize time constant as a function of speed.

Figure 7-8 shows the time constant for capsize as a function of speed. The time constant increases with speed. The camber stiffness k_φ has a towing effect on capsize (increasing the time constant) by generating a sort of elastic resisting torque which acts against roll $k_\varphi \varphi N t$. Without the camber component there would be more sideslip and the capsize time would be shorter as a result.

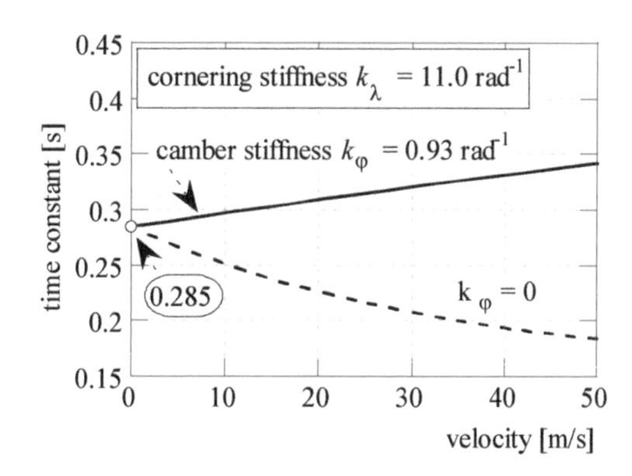

Fig. 7-8 Time constant for capsize as a function of speed.

Figure 7-8 also shows the borderline case of zero camber stiffness in which the time constant decreases as speed increases. This is due to the fact that to maintain the sideslip angle or rather the sideslip force constant, the lateral velocity increases as speed V increases. Thus, higher speed contributes to the lateral displacement of the contact point, thereby decreasing the time constant.

Example 4

In the borderline case of zero stiffness values for both k_λ and k_φ (that is, as if the road surface were a sheet of ice) determine the time constant for the motorcycle described in Example 3.

In this case there is no force counteracting tire sideslip and the time constant decreases to its limit value of:

$$\tau = \frac{\rho}{\sqrt{g h_o}}$$

which is obtained by putting $k_\lambda = 0$ and $k_\varphi = 0$ into the characteristic equation.

Substituting numerical values yields τ = 0.122 s, which is the minimum value that can be obtained with the steering head locked in place without taking gyroscopic effects into consideration.

7.1.2 Wobble

Wobble is an oscillation of the front assembly around the steering axis that can become unstable at fairly low to middle speeds (Fig. 7-9).

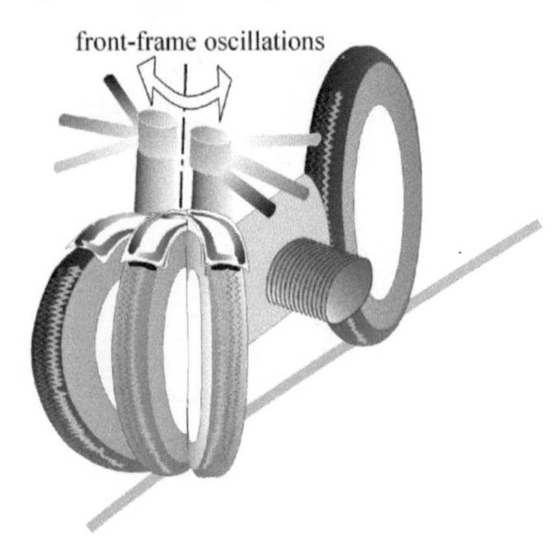

Fig.7-9 Wobble mode.

Wobble oscillations resemble shimmy of a car's front wheels or airplane landing gear. Typical frequency values range from 4 Hz for heavy motorcycles to 10 Hz for lightweight motorcycles.

Wobble frequency goes up as trail increases and front-frame inertia decreases, and is determined mainly by the stiffness and damping of the front tire, although the lateral flexibility of the front fork also plays a part.

In the forward speed range from 10 to 20 m/s (40 to 80 km/h), wobble is only slightly damped and can therefore become unstable. Adding a steering damper increases the damping effect and, consequently, the stability.

Model of wobble with one degree of freedom

Wobble can first be thought of in complete isolation from rear-assembly motion and roll. Thus, the front-end is a rigid body that can rotate around the steering axis while the rear frame is fixed (Fig. 7-10).

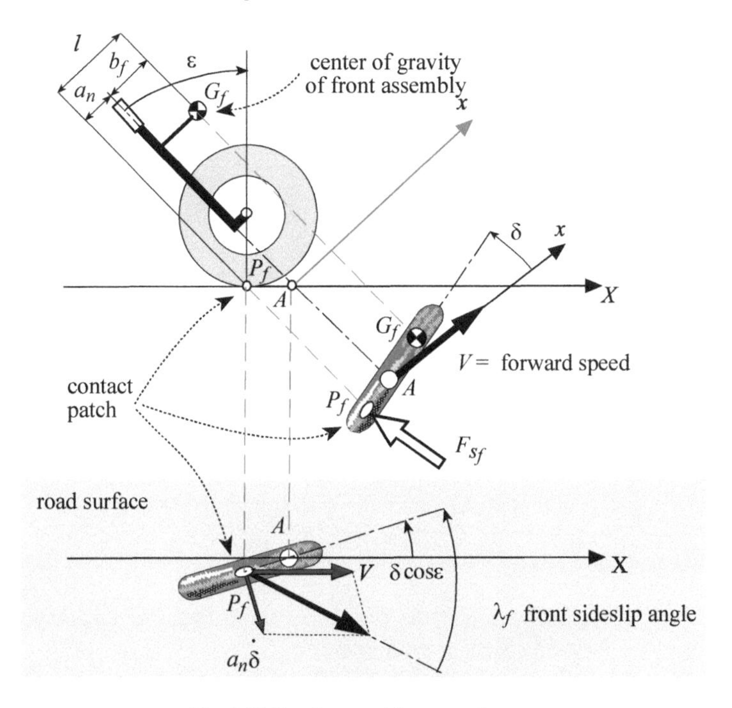

Fig. 7-10 Front assembly geometry.

The equilibrium equation around the steering axis leads to the following relationship:

$$I_{A_f}\ddot{\delta} = -c\,\dot{\delta} - F_{s_f}a_n + M_f g\, b_f \delta \sin\varepsilon + N_f a_n \delta \sin\varepsilon$$

where:

- $I_{A_f} = I_{G_f} + M_f b_f^2$ is the front assembly moment of inertia (including the front wheel) around the steering axis;
- c is the damping coefficient of the steering damper;
- F_{s_f} is the lateral force acting on the tire.

This last term is assumed proportional to the sideslip angle λ according to the following equation:

$$F_{s_f} = K_{\lambda_f}\lambda_f$$

with null value of the relaxation length. For small displacements the following equation can be used to calculate the sideslip angle:

$$\lambda_f = \frac{\dot{\delta} a_n}{V} + \delta \cos\varepsilon$$

The sideslip angle is therefore the sum of two components:

- the first depends on the lateral speed of the contact point due to steering velocity $\dot{\delta}$;
- the second on the steering angle measured at the road surface.

The effects due to the front tire normal load and front frame weight force are significantly smaller than that due to tire lateral force. So, making the proper substitutions the motion equation for small oscillations becomes:

$$I_{A_f}\ddot{\delta} + (c + \frac{K_{\lambda_f} a_n^{\,2}}{V})\dot{\delta} + K_{\lambda_f} a_n \delta \cos\varepsilon = 0$$

Introducing an oscillating solution into the equation and eliminating time-dependence gives the following characteristic equation:

$$I_{A_f} s^2 + (c + \frac{K_{\lambda_f} a_n^{\,2}}{V})s + K_{\lambda_f} a_n \cos\varepsilon = 0$$

which yields the following roots:

$$s_{1,2} = -\frac{cV + K_{\lambda_f} a_n^2}{2 I_{A_f} V} \pm \sqrt{\left(\frac{cV + K_{\lambda_f} a_n^2}{2 I_{A_f} V}\right)^2 - \frac{K_{\lambda_f} a_n}{I_{A_f}} \cos\varepsilon}$$

The system is oscillating when the discriminant is negative, i.e., neglecting the damping c , for forward speeds greater than :

$$V > \frac{1}{2}\sqrt{\frac{K_{\lambda_f} a_n^3}{I_{A_f} \cos\varepsilon}}$$

The frequency for the damped system ν is:

$$\nu = \frac{1}{2\pi}\sqrt{\frac{K_{\lambda_f} a_n}{I_{A_f}} \cos\varepsilon}\sqrt{\left(1-\zeta^2\right)}$$

The damping ratio ζ is given by:

$$\zeta = \frac{cV + K_{\lambda_f} a_n^{\,2}}{2V\sqrt{I_{A_f} K_{\lambda_f} a_n \cos\varepsilon}}$$

Note that ζ decreases as the forward speed of the motorcycle increases.

Figures 7-11 and 7-12 show how the natural frequency and damping ratio, respectively, of the wobble mode vary with speed. Specifically, Fig. 7-12 shows how a steering damper affects the damping ratio of the wobble mode, especially at high speeds.

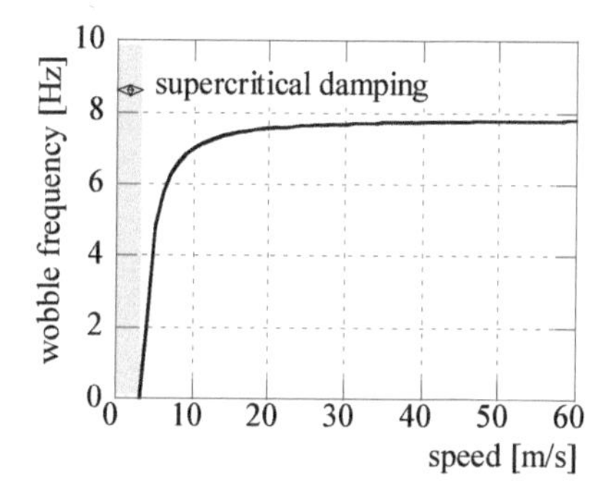

Fig. 7-11 Natural frequency of wobble as a function of speed.

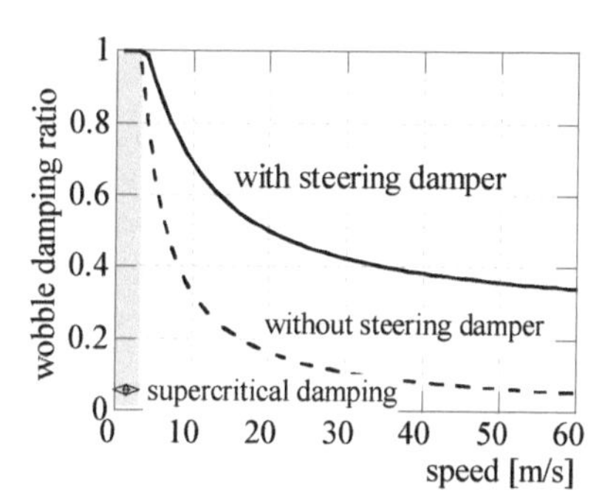

Fig. 7-12 Damping ratio for wobble as a function of speed.

7.1.3 Weave

Weave is an oscillation of the entire motorcycle, but mainly the rear end, as shown in Fig. 7-13.

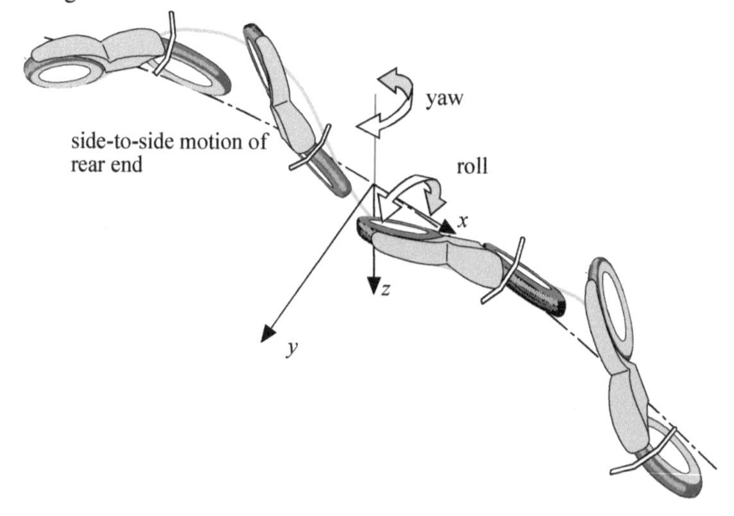

Fig.7-13 Weave.

The natural frequency of this side-to-side motion is zero when the forward speed is also zero and ranges from 0 to 4 Hz at high speed. Weave is determined by many factors:

- position of the center of gravity of the rear assembly (and secondarily that of the front assembly);
- wheel inertia;
- caster angle;
- trail;

- cornering stiffness of the rear tire.

The damping of weave can be shown to decrease as speed increases.

Weave is usually unstable at a low speed (up to 7-8 m/s). It is generally stable in the middle speed range, but it may be uncontrollable from the practical standpoint at high speed since its damping may decrease substantially and its natural frequency may be too high for the rider to control..

The weave mode is generated by the coalescence, at very low speeds, of two unstable non-oscillating modes: body-capsize and steering-capsize (Fig. 7-14).

Body capsize

Body-capsize indicates–capsize of the entire motorcycle, and can be estimated with one of the three simple models presented in section 7.1.1. The time constant with the steering free decreases slightly. In fact, suppose the machine begins to fall to the rider's right. In body capsize mode, steering geometry causes the machine to steer left thereby moving the front wheel contact point towards the rider's right. Consequently, the ground contact line that joins the front and rear wheel ground contact points rotates to the rider's right. This means the gravitational torque increases and so the vehicle capsizes less quickly.

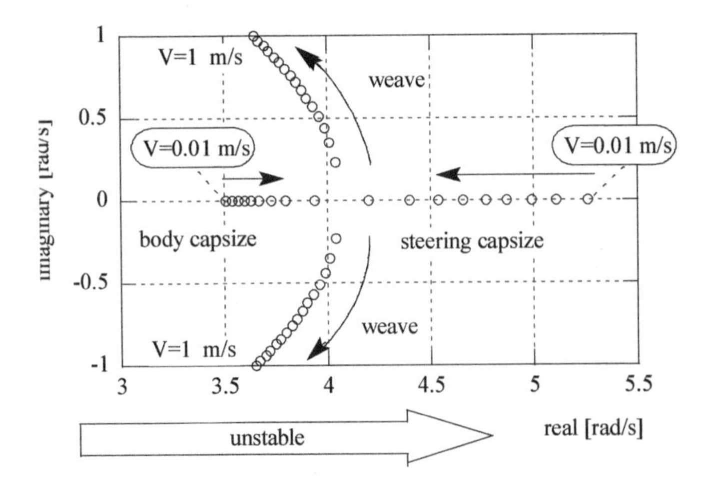

Fig.7-14 Weave coalescence.

Steering capsize

Steering-capsize is a capsize of the steering head, due to the disaligning effect of both the front tire normal load and front frame weight force. Once again consider the simplified situation in which the rear frame is fixed whereas the front frame is free to steer. The equilibrium equation, neglecting the steering damper and the tire lateral force, which does not act at the beginning of motion, because of its lag, yields the following relationship:

$$I_{A_f}\ddot{\delta} - M_f g\, b_f \sin(\varepsilon)\delta - N_f a_n \delta \sin(\varepsilon) = 0$$

The roots are:

$$s_{1,2} = \sqrt{\frac{(M_f g b_f + N_f a_n)\sin(\varepsilon)}{I_{A_f}}}$$

The time constant of the steering capsize has values in the range 0.1-0.2s for speed less then 1 m/s.

Model of weave with one degree of freedom

Weave can first be seen as an oscillation of the rear end around the steering head axis almost independent of front-assembly motion and roll motion. This model of weave with one degree of freedom is based on the following assumptions:

- the roll value for the motorcycle is null;
- the steering head is locked in place.

This assumption arises from the observation of real weave oscillations, in which lateral displacement of the steering head axis is substantially lower than the lateral displacement of the rear tire.

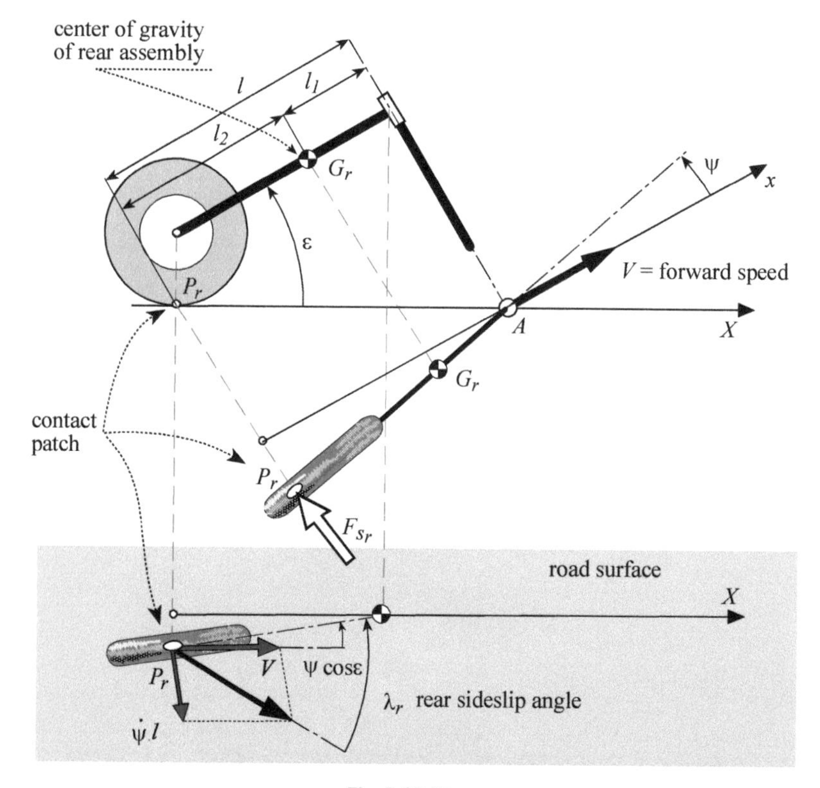

Fig. 7-15 Weave.

The motion equation is obtained by imposing equilibrium on rotation around the steering head axis. The relationship obtained is once again the same relationship presented in the previous steering-capsize section, where normal rear trail l substitutes normal trail a_n. The effects due to the front tire normal load and front frame weight force are significantly smaller than that due to tire lateral force, so we assume the following simplified expression of steering equilibrium:

$$I_{A_r}\ddot{\psi} = -c\dot{\psi} - F_{s_r} l$$

where (Fig. 7-15):

- $I_{A_r} = I_{G_r} + M_r l^2$ is the rear assembly moment of inertia (including the rear wheel) around the steering axis;
- c is the damping constant for the steering head rotation;
- F_{s_r} is the lateral force acting on the tire.

This last term is assumed proportional to the rear sideslip angle λ according to the following equation:

$$F_{s_r} = K_{\lambda_r}\lambda_r$$

with null value of the relaxation length. For small displacements the following equation can be used to calculate the sideslip angle:

$$\lambda_r = \frac{\dot{\psi} l}{V} + \psi \cos\varepsilon$$

The sideslip angle is therefore the sum of two components:

- the first results from the lateral speed of the contact point due to yaw velocity $\dot{\psi}$;
- the second from the rear frame yaw angle measured at the road surface.

Making the proper substitutions the motion equation for small oscillations becomes:

$$I_{Ar}\ddot{\psi} + \left(c + \frac{K_{\lambda r} l^2}{V}\right)\dot{\psi} + K_{\lambda r} l \cos\varepsilon\, \psi = 0$$

Introducing an oscillating solution into the equation and eliminating time-dependence gives the following characteristic equation:

$$I_{A_r} s^2 + \left(c + \frac{K_{\lambda_r} l^2}{V}\right)s + K_{\lambda_r} l \cos\varepsilon = 0$$

which yields the following roots:

$$s_{1,2} = -\frac{cV + K_{\lambda_r} l^2}{2 I_{A_r} V} \pm \sqrt{\left(\frac{cV + K_{\lambda_r} l^2}{2 I_{A_r} V}\right)^2 - \frac{K_{\lambda_r} l}{I_{Ar}} \cos\varepsilon}$$

The system is oscillating when the discriminant is negative, i.e., for forward speeds greater than:

$$V > \frac{1}{2}\sqrt{\frac{K_{\lambda_r} l^3}{I_{A_r} \cos\varepsilon}}$$

In this case, the frequency for the damped system ν and the damping ratio ζ are given by:

$$\nu = \frac{1}{2\pi}\sqrt{\frac{K_{\lambda_r} l}{I_{A_r}}\cos\varepsilon}\sqrt{\left(1-\zeta^2\right)} \qquad \zeta = \frac{cV + K_{\lambda_r} l^2}{2V\sqrt{I_{A_r} K_{\lambda_r} l \cos\varepsilon}}$$

Thus, we can draw the following conclusions:

- the damping ratio ζ falls rapidly and as speed increases it approaches a limit value:

$$\zeta_O = \frac{c}{2\sqrt{I_{A_r} K_{\lambda_r} l \cos\varepsilon}}$$

- the damped natural frequency ν decreases as inertia increases and increases with speed and tire stiffness; with negligible damping c, increasing the speed it approaches the value:

$$\nu_O = \frac{1}{2\pi}\sqrt{\frac{K_{\lambda_r} l}{I_{A_r}}\cos\varepsilon}$$

- this last equation confirms that the weave mode frequency increases with tire stiffness and the length of the wheelbase, but decreases as the caster angle and inertia of the motorcycle increase.

Example 5

Let us consider a motorcycle with the following data:

- caster angle $\varepsilon = 27°$
- rear frame inertia about steering axis $I_{A_r} = 119$ kgm^2
- distance between rear contact point and steering axis $l = 1.38$ m
- distance between mass center and steering axis $l_1 = 0.67$ m
- steering damper c =6.8 Nm/rad/s
- tire cornering stiffness: $K_{\lambda_r} = 15.80$ kN/rad

We will investigate how the characteristics of the weave mode change with speed. Figures 7-16 and 7-17 show how the natural frequency and damping ratio, respectively, of the weave mode vary with speed.

The weave frequency tends to the undamped natural frequency value as the velocity increases. The simplified model shows that the steering damper increases the weave damping ratio. In this case the simplified mathematical results are not true because in reality the steering damping has a negative effect on weave stability.

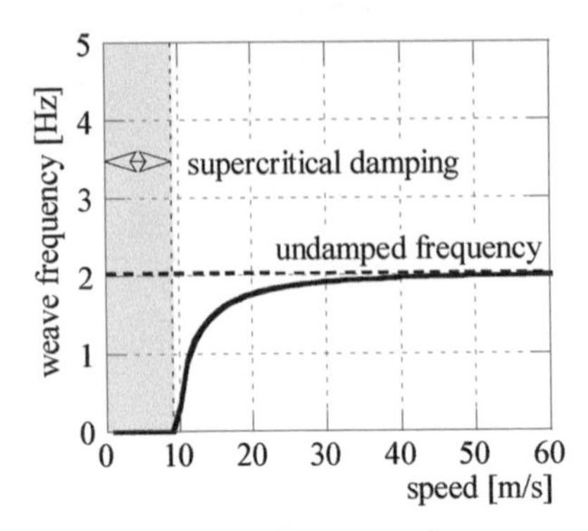

Fig. 7-16 Natural frequency of weave as a function of forward speed.

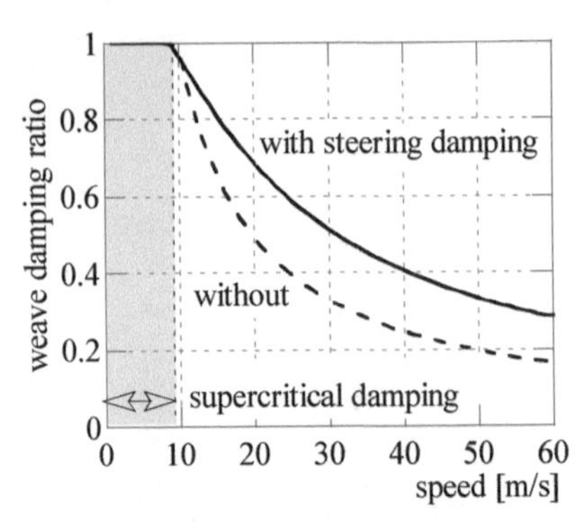

Fig. 7-17 Damping ratio for weave as a function of forward speed.

7.1.4 Combined model for weave and wobble

Up to this point, we have looked at weave and wobble in isolation assuming that the steering head does not displace laterally in either vibration mode. Now we will use a system with three degrees of freedom to represent the motorcycle. Looking at the motorcycle in the direction of the steering axis, it is easy to see that the three degrees of freedom are the absolute rotations of the front assembly θ_f, and rear assembly θ_r around the steering axis, and the lateral displacement y of the steering axis (Fig. 7-18).

The resulting equations of motion are as follows:

$$[M]\begin{Bmatrix}\ddot{y}\\ \ddot{\theta}_r\\ \ddot{\theta}_f\end{Bmatrix}+[C]\begin{Bmatrix}\dot{y}\\ \dot{\theta}_r\\ \dot{\theta}_f\end{Bmatrix}+[K]\begin{Bmatrix}y\\ \theta_r\\ \theta_f\end{Bmatrix}=0$$

where $[M]$ is mass and $[K]$ is stiffness matrix:

$$[M]=\begin{bmatrix}M_r+M_f & -M_r l_1 & M_f b_f\\ -M_r l_1 & M_r l_1^2+I_r & 0\\ M_f b_f & 0 & M_f b_f^2+I_f\end{bmatrix}$$

$$[K]=\begin{bmatrix}0 & -K_{\lambda_r}\cos\varepsilon & -K_{\lambda_f}\cos\varepsilon\\ 0 & K_{\lambda_r} l\cos\varepsilon & 0\\ 0 & 0 & K_{\lambda_f} a_n\cos\varepsilon\end{bmatrix}$$

$[C]$ is the damping matrix:

$$[C]=\frac{1}{V}\begin{bmatrix} K_{\lambda_f}+K_{\lambda_r} & -K_{\lambda_r}l & -K_{\lambda_f}a_n \\ -K_{\lambda_r}l & K_{\lambda_r}l^2+cV & -cV \\ -K_{\lambda_f}a_n & -cV & K_{\lambda_f}a_n^2+cV \end{bmatrix}$$

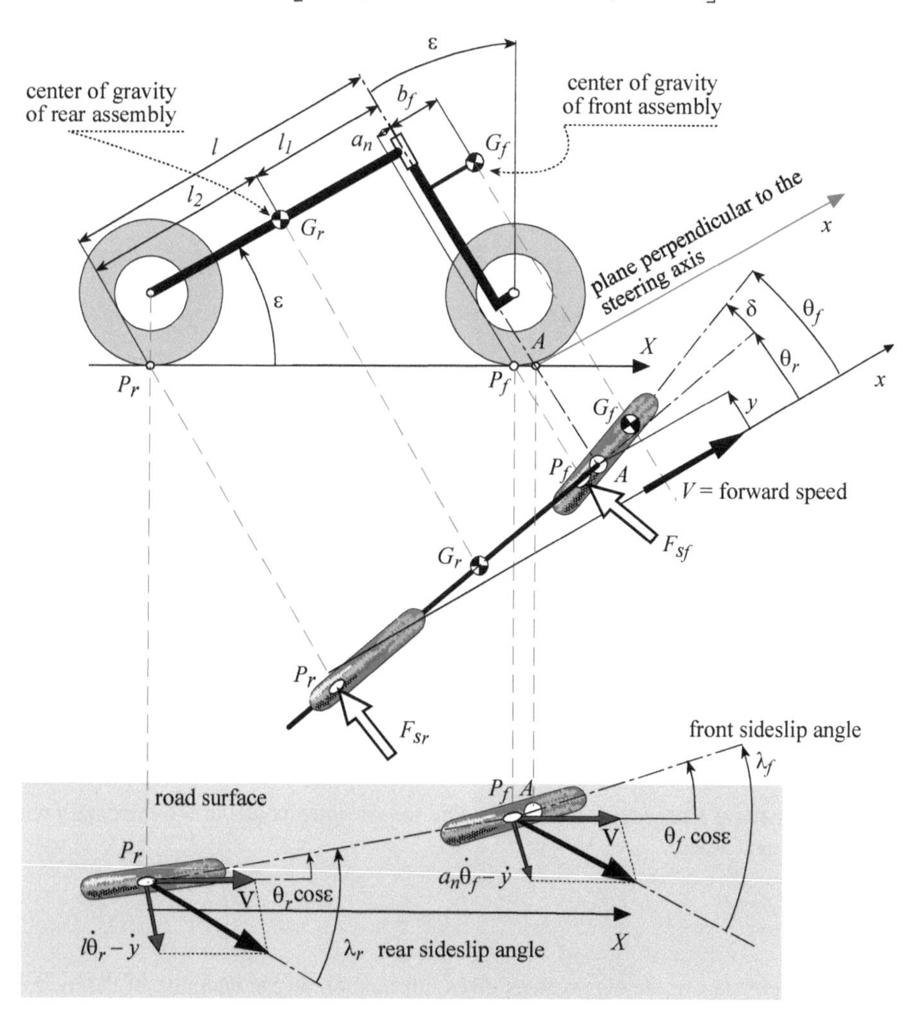

Fig. 7-18 Front-end and rear-end geometry.

Figures 7-19 and 7-20 confirm that:

- the two modes of vibration are substantially independent of each other;
- there are no significant differences between the model with one degree of freedom and the one with three degrees of freedom with respect to the frequency values or damping ratios.

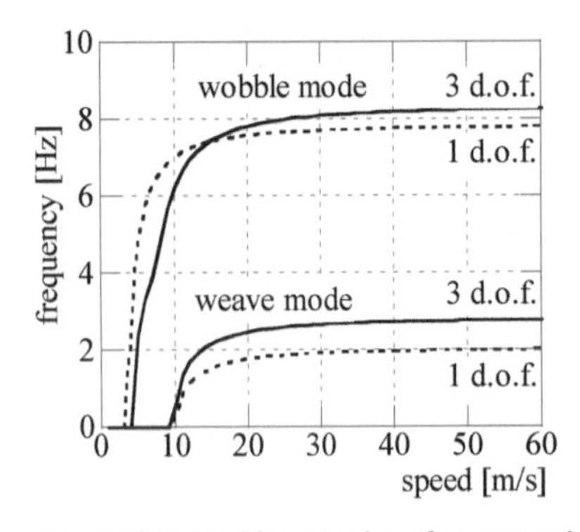

Fig. 7-19 Natural frequencies of weave and wobble as a function of speed.

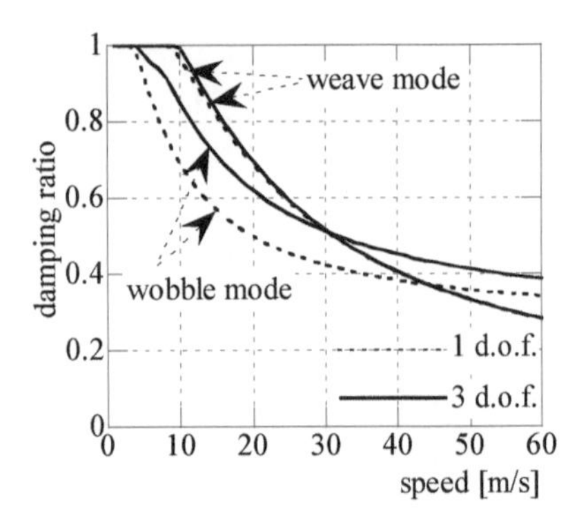

Fig. 7-20 Damping ratios for weave and wobble as a function of speed.

7.2 Multi-body Model

7.2.1 Introduction

The motorcycle has two kinds of modes:

- the *in-plane* modes, which involve frame, suspension and wheel motion in the vertical plane,
- the *out-of-plane* modes, which involve roll, yaw, steering angle and steering head lateral displacement.

The in-plane modes are related to ride comfort and road-holding, whereas the out-of-plane modes are related to vehicle stability and handling. In straight running, in-plane and out-of-plane modes are uncoupled and they can be examined separately.

The eigenvalues of the in-plane and out-of-plane modes are complex:

$$s = s_r + is_i$$

The natural frequency of the oscillating modes corresponds to the imaginary part of the eigenvalue:

$$\nu = \frac{s_i}{2\pi}$$

The real part of the eigenvalues gives information on the damping of the modes. In the case of the out-of-plane modes, the real part of the eigenvalues provides information on the stability of the motorcycle. The motion is unstable if the real part is positive, and damped if the real part is negative.

The damping ratio is given by:

$$\zeta = \frac{s_r}{\sqrt{s_i^2 + s_r^2}}$$

If the imaginary part is zero the mode is a non-oscillating one and the time evolution of the generic modal component can be represented by a decreasing or increasing exponential law:

$$\alpha = \alpha_o e^{s_r t} = \alpha_o e^{\frac{t}{\tau}}$$

where τ is the time constant, positive for unstable modes and negative for the stable ones.

The properties of the eigenvalues are clearly represented in the root locus graph, as shown in Fig. 7-21. The position of the eigenvalues in the complex plane represents different cases. The eigenvalues located on the right, in relation to the vertical axis, represent unstable modes while the eigenvalues located on the left are stable modes. To clarify the meaning of the eigenvalues, the time evolution laws corresponding to different eigenvalues are shown in the same figure.

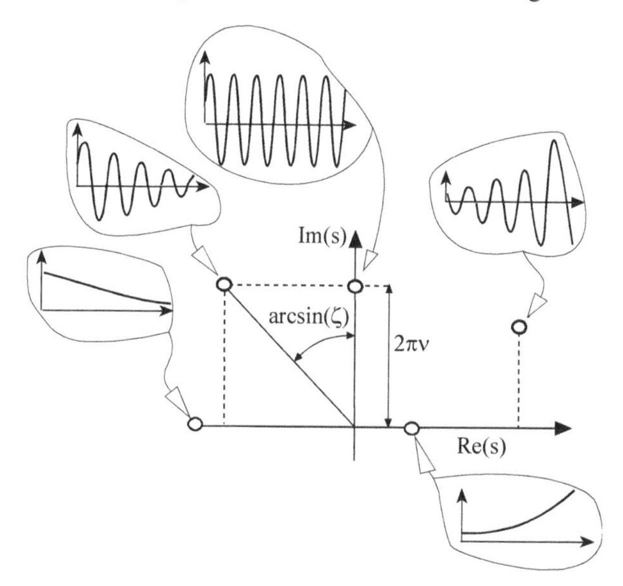

Fig. 7-21 Root locus and examples of time evolution laws for different eigenvalues.

The first out-of-plane motorcycle equations, linearized around the vertical equilibrium position, were developed by R.S. Sharp in an influential article [Sharp, 1971].

Sharp's model of the motorcycle has four degrees of freedom (Fig. 7-22):

- rear-frame roll;
- rear-frame yaw;
- rotation of the front frame around the steering axis;
- lateral displacement of the rear frame.

The tire-to-road contact forces are taken as linear functions of the sideslip and camber angles. The lag in the forces with regards to the sideslip angles is introduced in the model using two first-order differential equations.

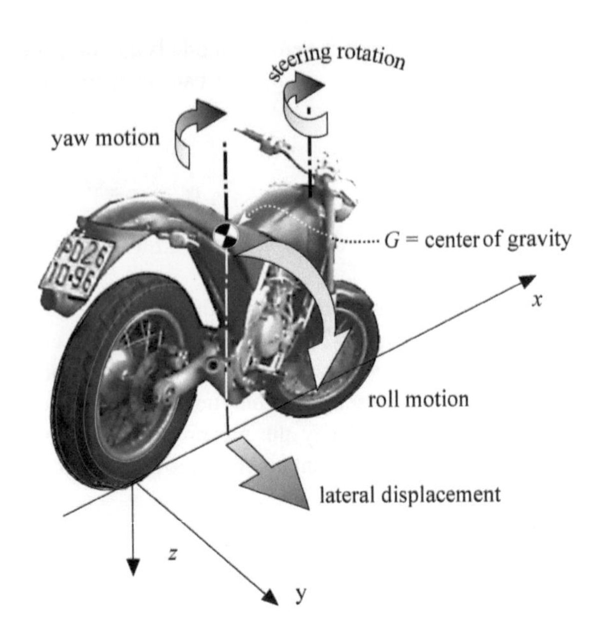

Fig. 7-22 Motorcycle model with four degrees of freedom.

Geometry		Cornering stiffness:	K_{λ_f} =11.2 kN/rad
Wheelbase	p = 1.414 m	Camber stiffness:	K_{φ_f} = 0.94 kN/rad
Trial	a =0.116 m	***Rear assembly***	
caster angle:	ε = 27°	Mass:	M_r = 217.5 kg
Front assembly		Longitudinal position of G_r:	b_r = 0.480 m
Mass:	M_f = 30.7 kg	Height of G_r:	h_r = 0.616 m
x - position of G_f:	b_f = 0.024 m	*moments of inertia about c. of g.*	
z - position of G_f:	h_f = 0.461 m	- around x -axis:	I_{x_r} = 31.20 kgm^2
moments of inertia about c. of g.		-around z -axis:	I_{z_r} = 21.08 kgm^2
- around x -axis	I_{x_f} =1.23kgm^2	Product of inertia about x - z	I_{xz_r} = 1.74 kgm^2
-around z -axis:	I_{z_f} = 0.44 kgm^2	***Rear Wheel***	
Steering damper:	c = 6.8 Nm/rad/s	Wheel radius:	R_r = 0.305 m
Front Wheel		Axial inertia:	I_{w_r} = 1.05 kgm^2
Wheel radius:	R_f = 0.305 m	Cornering stiffness:	K_{λ_r} = 15.8 kN/rad
Axial inertia:	I_{w_f} = 0.72 kgm^2	Camber stiffness:	K_{φ_r} = 1.32 kN/rad

Table 7-1.

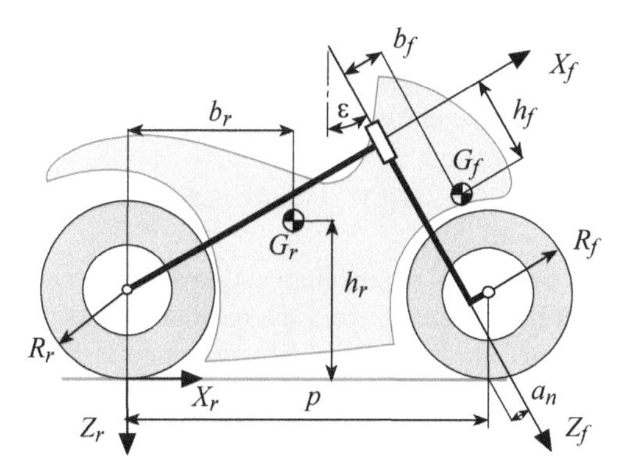

Fig. 7.23 Motorcycle diagram.

Figure 7-24 shows the frequencies and real parts of vibration modes as a function of the forward velocity, derived with Sharp's model.

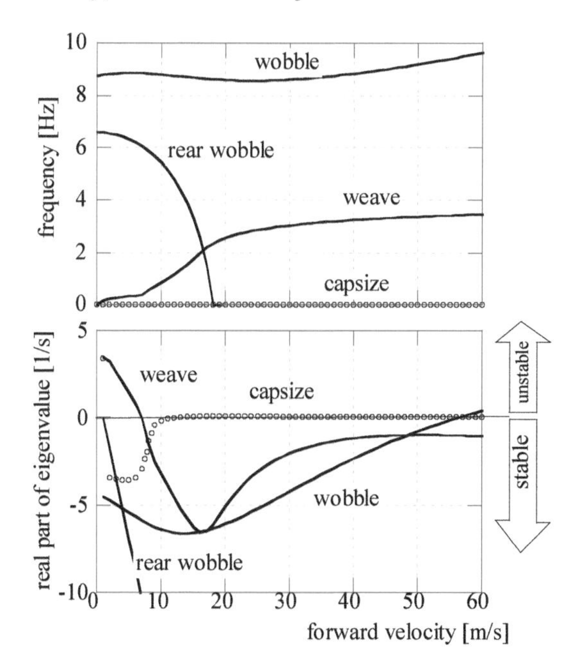

Fig. 7-24 Frequencies and real parts of vibration modes as a function of speed.

The wobble covers the frequency range 8.5 to 9.6 Hz; the speed of the motorcycle has little effect on wobble frequency.

The maximum rear wobble frequency is approximately 6.5 Hz at the minimum velocity; the rear wobble frequency drops off markedly at speeds of 18 m/s and above and ceases to vibrate.

The maximum weave frequency is approximately 3.6 Hz at the maximum velocity. In fact the weave mode frequency increases with speed.

The Fig. 7-24 also shows how the real part of the eigenvalues varies with speed. Note that:

- weave mode is unstable at low speed, damped in the medium speed range, and weakly damped at high speed;
- rear wobble mode by contrast is strongly damped and becomes overcritical above a certain speed;
- front wobble is damped in the medium and low speed ranges, but becomes a little unstable at high speed. In fact, the oscillations associated with wobble are the most dangerous since their high frequency makes them hard for the rider to control.

Above approximately 18 m/s the rear wobble becomes supercritical and splits into two modes. The first one is very stable and characterized by nearly constant damping, while the second one is characterized by a decreasing real part which gradually becomes more stable.

Figure 7-24 shows that the real part of the capsize eigenvalue is always negative, it is very stable at low velocity and tends to become borderline unstable at high velocity. Stability at low velocity is in contrast with the experimental evidence. However, it is worth remembering that the first Sharp model is based on wheels with zero thickness; since the capsize mode depends on the location of the front contact point, the evaluation of the capsize mode needs a more accurate tire model.

The frequencies and dampings of the out-of-plane modes in straight running are quite well validated by experiment results. In cornering, the frequencies and the damping ratios, of both in-plane and out-of-plane modes, vary with respect to the straight running condition. Furthermore, the two types of modes are coupled to each other.

7.2.2 Motorcycle multi-body model

A more complex mathematical model is needed to give a more accurate description of motorcycle behavior than the current one. For this reason the first Sharp model can be developed further, by adding the suspension's degrees of freedom and describing the tire properties accurately.

The multi-body model of the motorcycle from which the remarks included in this chapter arise is composed of six rigid bodies, respectively: (for more information see [Cossalter and Lot, 2002]):

- the rear frame (which includes chassis, engine, tank, rider, part of the rear suspension and part of swinging arm),
- the front frame (which includes handlebars, sprung fork components and the steering head),
- the rear unsprung mass (which includes part of the swinging arm and the rear brake caliper),
- the front unsprung mass (which mainly includes part of the forks and the front brake calipers),
- the rear wheel and the front wheel.

This set of bodies has eleven degrees of freedom in all, as can be seen in Fig. 7-25, they are:

- the position of the rear frame center of mass (three coordinates);
- the orientation of the rear frame given by pitch, roll and yaw angles;
- the travel of both suspensions (the front one is telescopic, the rear one is a swinging-type);
- the spin angles of the wheels;
- the steering angle.

The dynamics of this multi-body system is very complex because of the large number of bodies and forces. The aerodynamic effects are reduced to only three forces (drag, lift, and lateral) which act on the center of pressure of the rear frame. The brake torques are applied on the respective wheel axes, whereas the propulsive force is transmitted from the engine to the rear wheel by a chain transmission. The steering torque acts between the front and rear frame along the steering axis.

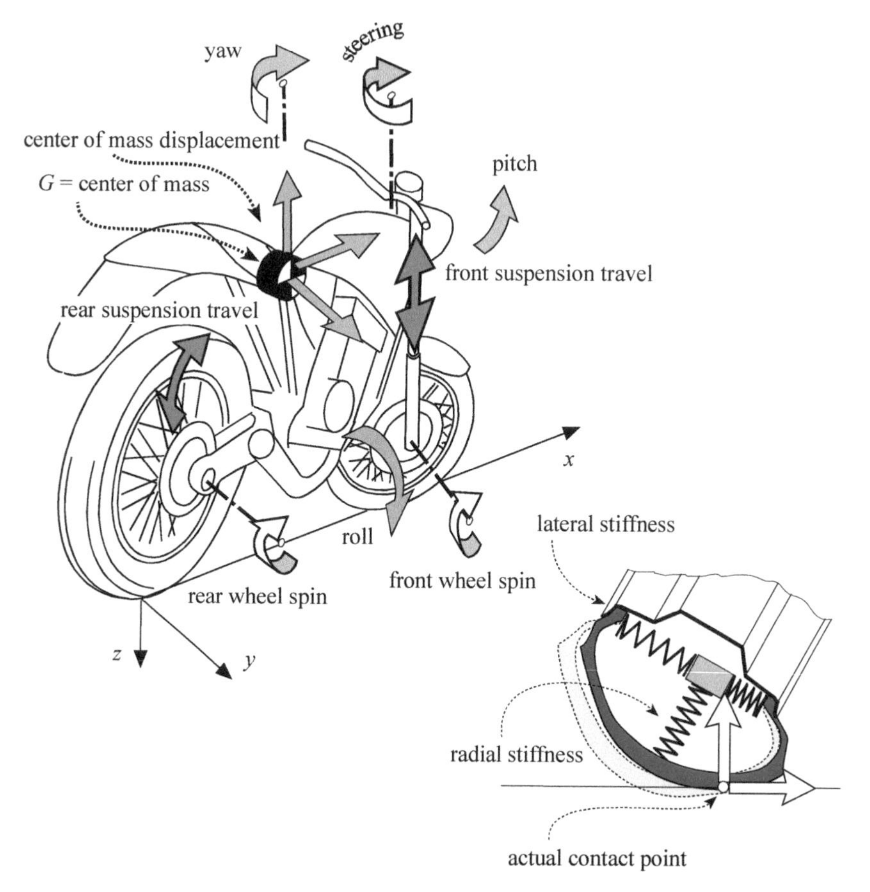

Fig. 7-25 Motorcycle model with eleven degrees of freedom.

The tire model accurately describes the geometry of the tread which is fundamental in evaluating the tire behavior at large camber angles. The carcass is thought of as elastically deformable along radial, lateral and tangential directions. The contact forces (see also chapter 2) are applied on a point whose position is defined by the equilibrium between the external forces (due to the camber angle, the longitudinal slip and the sideslip) and the internal elastic reactions. This point represents the center of the contact patch. The contact yaw torque is applied on the same point and acts around the vertical axis, whereas the rolling resistance acts in the wheel plane and tends to slow the wheel's rotation.

7.2.3 Modes of vibration in straight running

Fig. 7-26 shows the root locus plot of both in-plane and out-of-plane modes for a sport motorcycle in straight running. The direction of the arrows shows the increase of the velocity from 3 to 60 m/s. The grey lines starting from the origin are constant damping ratio loci.

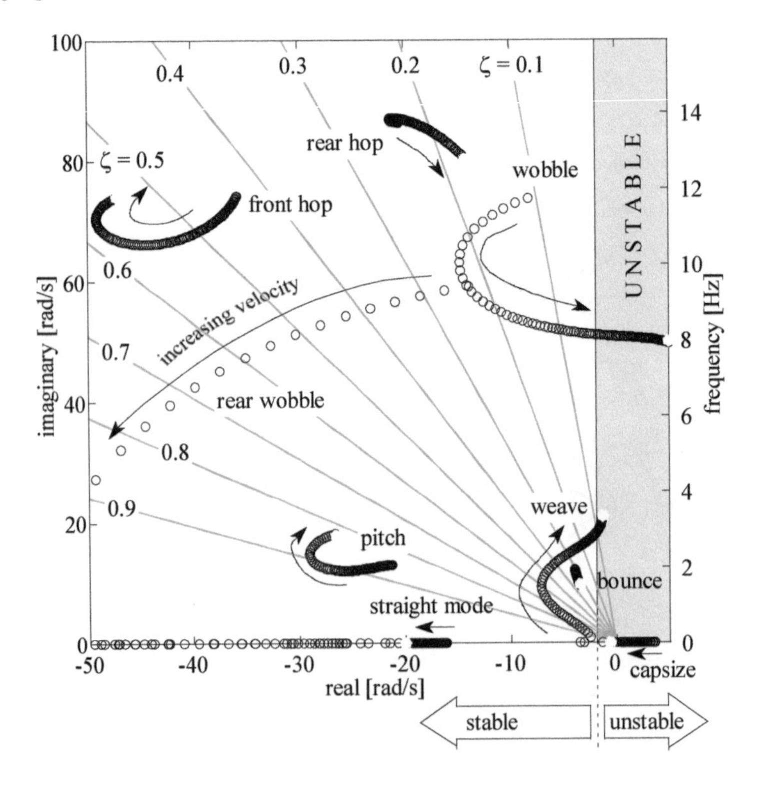

Fig. 7-26 Root-locus plot in straight running at different speeds, (speed from 3 to 60 m/s).

The in-plane modes, pitch and bounce, are also discussed in chapter 5. In the case plotted here the pitch and bounce are coupled together, the pitch mode involves a not negligible vertical displacement of the motorcycle mass center while the bounce mode involves a not negligible pitch rotation. Generally the pitch mode is more damped than the bounce mode. In fact, due to the damping selected for the shock absorbers, the bounce mode has a damping ratio of about 0.3-0.5 whereas the pitch mode generally has a damping ratio of about 0.9 and can also be overdamped. The frequency of the bounce mode is in the range 1.4-2.0 Hz; the pitch frequency increases from 2 Hz to 2.9 Hz at 60 m/s.

The front hop mode (around 10-11 Hz) and the rear hop mode (around 13-14 Hz), respectively, of the front and rear unsprung mass are characterized by negligible motion of the sprung mass.

Moreover, Fig. 7-26 highlights the values and the dependence of the real and imaginary parts of the main out of plain modes on the velocity:

- capsize;
- wobble;
- weave;
- rear wobble.

The wobble and weave modes are vibrating modes in the entire range of velocity considered here while the rear wobble becomes overcritical when velocity increases. The capsize mode is a non-vibrating mode that is quite unstable in the example considered. The wobble mode becomes unstable when increasing the velocity due to the fact that in this model the steering damper was set to zero. Figure 7-27 shows the frequencies and the damping properties of the main out-of-plane modes as a function of the forward velocity.

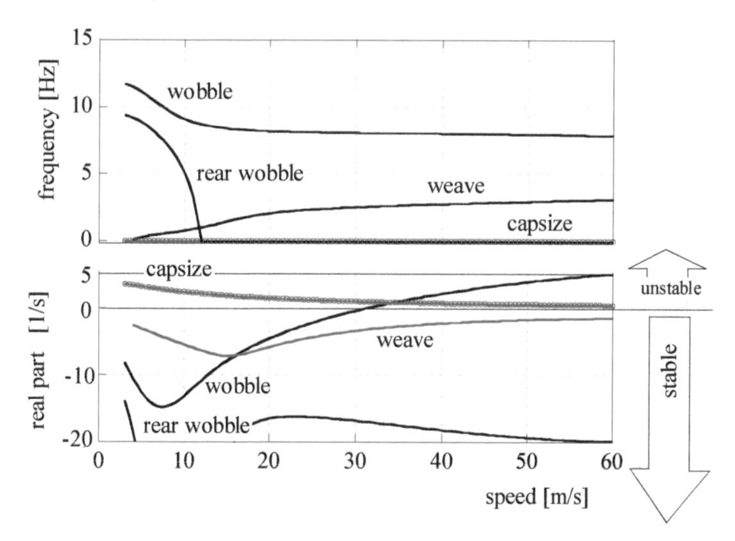

Fig. 7-27 Frequencies and real parts of vibration modes as a function of speed.

The speed of the motorcycle has little effect on wobble frequency; the maximum wobble frequency is approximately 11 Hz at the minimum velocity and after 15 m/s remains constant and equal to 7.8 Hz. The wobble mode is damped in the low and medium speed ranges but becomes unstable at high speed.

The rear wobble ceases to vibrate at a velocity of about 14 m/s; it is a strongly damped mode that becomes overcritical.

The weave mode frequency increases with speed; its maximum value is approximately 3.0 Hz at the maximum velocity. The weave mode is well damped in the medium speed range, and weakly damped at high speed.

The capsize eigenvalue is always positive. It is unstable at low velocity and tends to become less unstable at high velocity due to the gyroscopic effects.

In the following sections details of the out-of-plane modes in straight running are presented.

Capsize

Simplified models with locked steering show that capsize is always unstable. Actually the capsize mode may be unstable or moderately stable depending on velocity and on motorcycle and tire properties.

A detailed analysis shows that capsize becomes more unstable by reducing the mechanical trail of the motorcycle and increasing the caster angle. As far as tires are concerned, we can summarize that only the front one has appreciable influence due to the presence of the steering mechanism. An increase of the cross section size (section radius) reduces the stability, but does not make the mode unstable. On the contrary, the yaw torque parameters can alter the sign of the real part of the eigenvalue: twisting torque tends to stabilize whereas the self-aligning torque tends to destabilize the capsize. Figure 7-28 shows the time evolution of a stable capsize at low velocity, whereas Fig. 7-29 shows the case of an unstable capsize at the same velocity. We can observe that in the stable case, roll and steering motion have opposite phases.

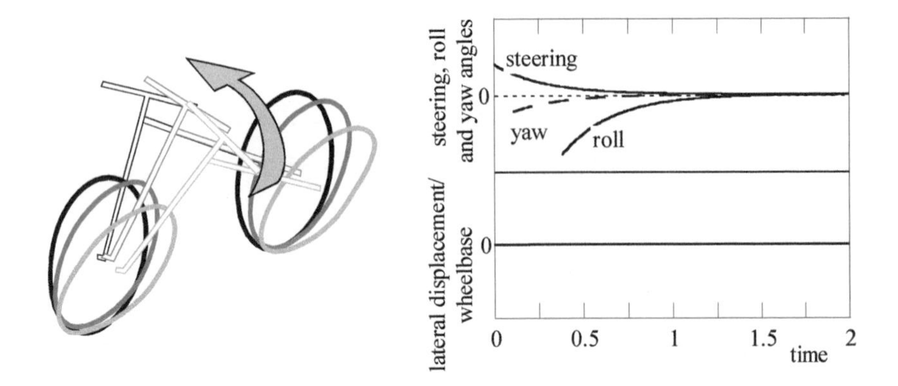

Fig. 7-28 Example of a motorcycle with stable capsize (speed 4 m/s).

Figure 7-30 highlights the time evolution of a stable capsize at high velocity. From these three figures one can observe that at the lower velocity of 4 m/s the steering, roll, and yaw variations are much more pronounced than any variation in lateral displacement. Conversely at 30 m/s the lateral displacement is the dominate modal content with much smaller variations in the other signals. This is due to the wheel gyroscopic effects that become important at high velocity. Therefore, the cornering maneuver at high speed involves other out-of-plane modes to a greater extent than the capsize mode.

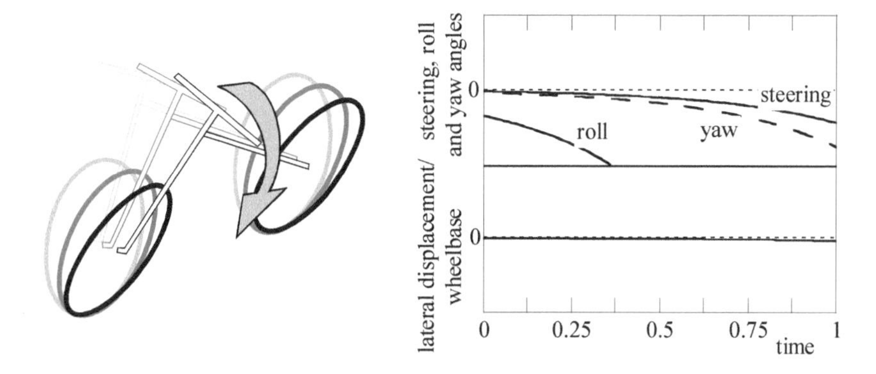

Fig. 7-29 Example of a motorcycle with unstable capsize (speed 4 m/s).

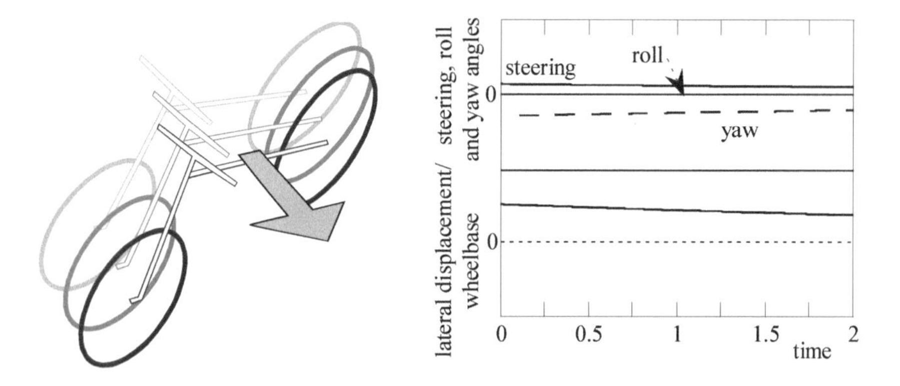

Fig. 7-30 Capsize at a speed of 30 m/s.

Figure 7-31 shows how the real part of the eigenvalue of the capsize mode changes when the values of several geometric and inertial parameters of the motorcycle and several tire properties are changed. In each case only one parameter was changed by 10% and all the other parameters were maintained constant. In reality it would be difficult, in several cases, to vary just one parameter at a time. For example, the increase in the radius of the front wheel, causes variations in the trail, the caster angle, and other parameters at the same time.

The figure shows that the stability of the capsize mode can be improved in the following ways:

- decreasing:

- the caster angle;
- the cross section radius of the front tire;
- the height of the motorcycle mass center;
- the front wheel spin inertia;
- the trail of the front tire;
- the camber stiffness of the front tire.

- increasing:

- the twisting torque of the front tire;
- the mechanical trail;
- the distance between the motorcycle mass center and the rear wheel axis;
- the cross section radius of the rear tire;
- the motorcycle roll inertia;
- the front wheel radius;
- the camber stiffness of the rear tire.

The figure also shows that the influence of all the parameters decreases when the speed increases; in fact as the speed increases the gyroscopic effects become dominant.

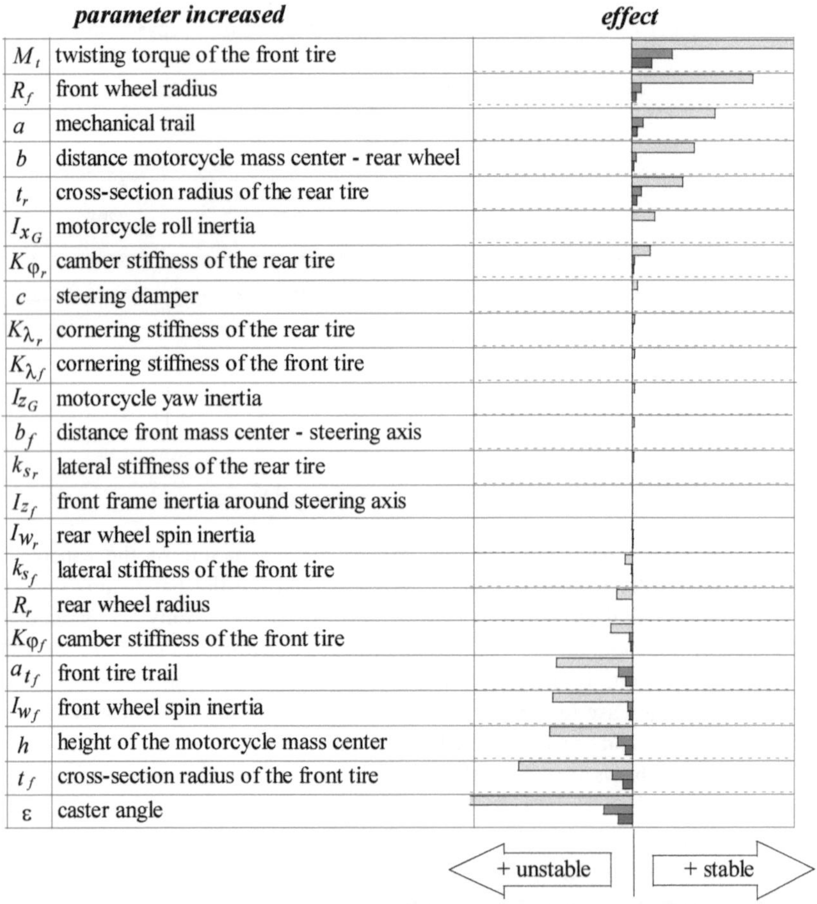

	parameter increased	*effect*
M_t	twisting torque of the front tire	
R_f	front wheel radius	
a	mechanical trail	
b	distance motorcycle mass center - rear wheel	
t_r	cross-section radius of the rear tire	
I_{x_G}	motorcycle roll inertia	
K_{φ_r}	camber stiffness of the rear tire	
c	steering damper	
K_{λ_r}	cornering stiffness of the rear tire	
K_{λ_f}	cornering stiffness of the front tire	
I_{z_G}	motorcycle yaw inertia	
b_f	distance front mass center - steering axis	
k_{s_r}	lateral stiffness of the rear tire	
I_{z_f}	front frame inertia around steering axis	
I_{w_r}	rear wheel spin inertia	
k_{s_f}	lateral stiffness of the front tire	
R_r	rear wheel radius	
K_{φ_f}	camber stiffness of the front tire	
a_{t_f}	front tire trail	
I_{w_f}	front wheel spin inertia	
h	height of the motorcycle mass center	
t_f	cross-section radius of the front tire	
ε	caster angle	

Fig. 7-31 Predicted change in capsize mode at 10 m/s (light gray) at 30 m/s s (gray), at 60 m/s (dark gray) for parameter increased 10% with respect to the reference case.

Wobble

As shown in Fig. 7-32, wobble is characterized by rotations of the front-frame while the rear frame is only slightly affected. The lateral displacement of the motorcycle and the yaw and roll oscillations are substantially smaller than the steering oscillations.

Figure 7-33 shows sensitivities of the wobble mode damping to the variation in the motorcycle and tire parameters. The figure shows that the stability of the wobble mode can be improved in the following ways:

- increasing:
  - the lateral stiffness of the carcass of the front tire;
  - the value of the steering damper;
  - the front wheel radius;
- decreasing:
  - the distance between the motorcycle mass center and the rear wheel axis;
  - the cornering stiffness of the front tire;
  - the front wheel spin inertia.

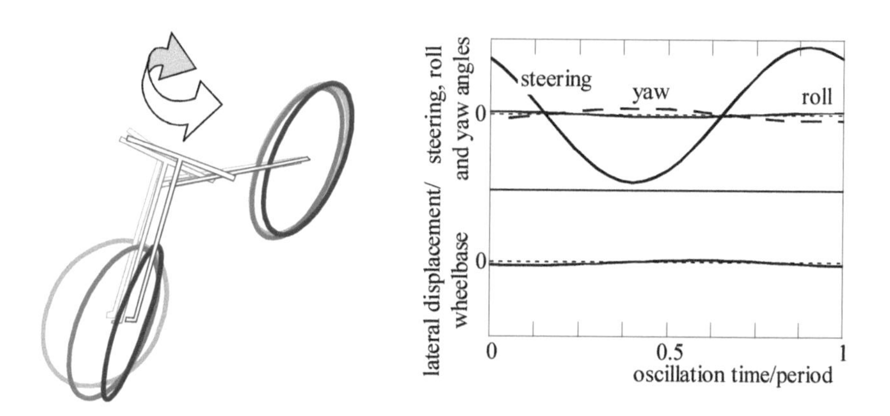

Fig. 7-32 Wobble at a speed of 30 m/s.

The need to increase the lateral stiffness and, at the same time, to decrease the cornering stiffness of the front tire means that the relaxation length of the front tire sideslip force should be shorter to obtain strong improvements in stability.

Note that an increase in:

- the height of the motorcycle mass center;
- the caster angle;
- the rear wheel radius;

gives advantages at low velocity and disadvantages at high velocity.

The increase of the roll inertia of the motorcycle and of the mechanical trail gives an opposite behavior, i.e., gives advantages at high velocity and disadvantages at low velocity.

Note that increasing the damping constant of the steering damper is a very easy way to increase the damping ratio for the wobble mode, although it has the undesired effect of slightly reducing the weave mode damping.

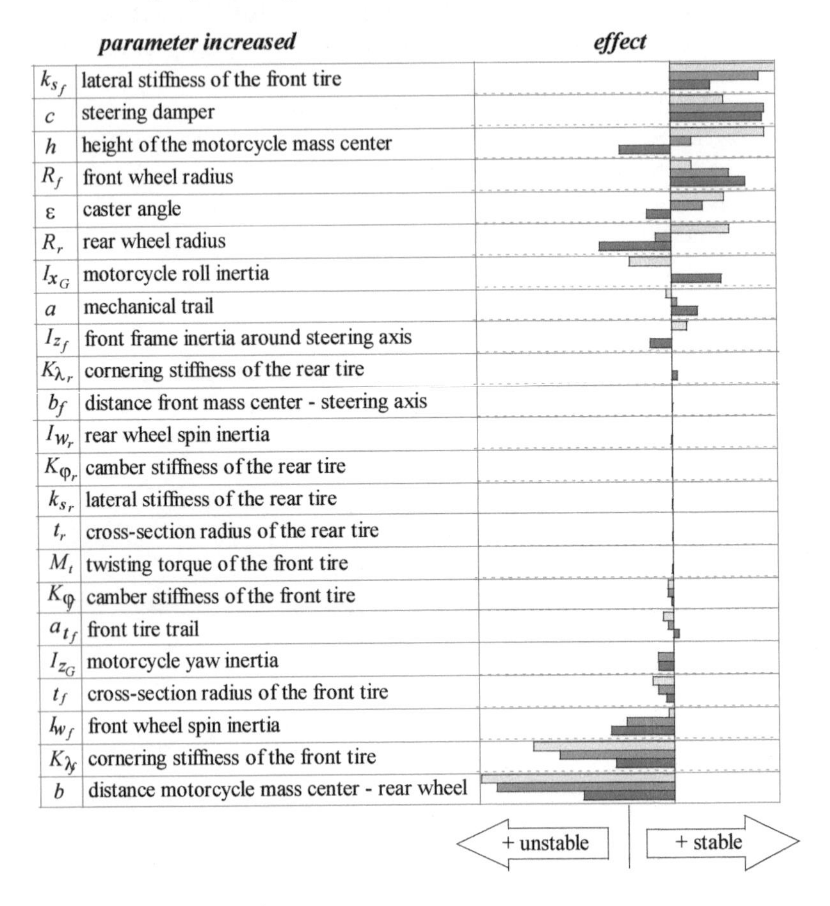

Fig. 7-33 Predicted change in wobble damping at 10 m/s (light gray) at 30 m/s s (gray) , at 60 m/s (dark gray) for parameter increased 10% with respect to the reference case.

Weave

Figure 7-34 shows the weave mode with the motorcycle moving at a forward speed of 10 m/s.

Note that this mode is characterized by sizable oscillations in roll, yaw and steering angle. The rotation of the steering head is opposite in phase with respect to the yaw oscillation, which is in turn 90° out of phase with the roll oscillation. The lateral displacement of the motorcycle lags behind roll.

As speed increases, the lateral displacement becomes more pronounced, as does the rotation of the steering head with respect to the roll and yaw oscillations (see Fig. 7-35).

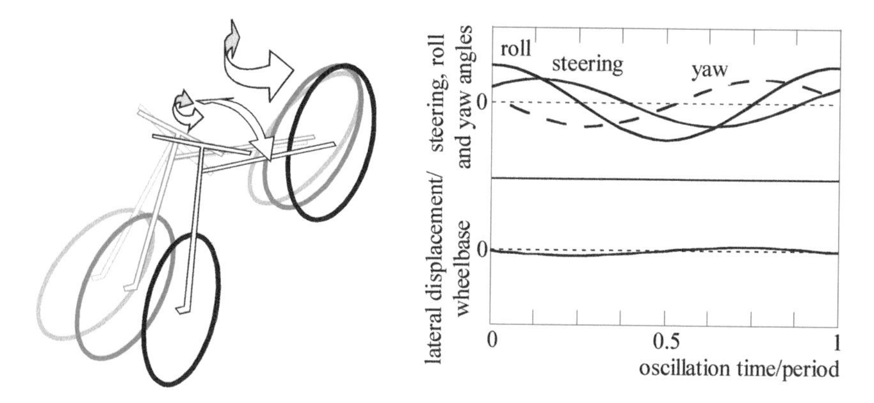

Fig. 7-34 Weave mode (V = 10 m/s).

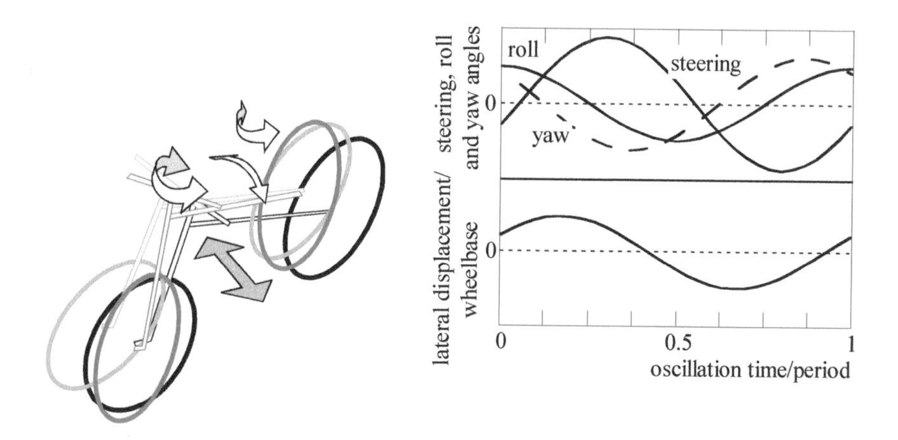

Fig. 7-35 Weave mode (V = 30 m/s).

Figure 7-36 shows how the damping ratio for the weave mode varies when the values of several parameters of the motorcycle are changed. Note that the increase of the damping constant of the steering damper has the undesired effect of slightly reducing damping of the weave mode.

The figure shows that for the motorcycle used in this example the weave mode stability at high velocity can be improved in the following ways:

- increasing:
  - the distance between the motorcycle mass center and the rear wheel axis;
  - the caster angle;
  - the front wheel spin inertia;
  - the lateral stiffness of the carcass of the rear tire;
- decreasing:
  - the front wheel radius;
  - the motorcycle yaw inertia.

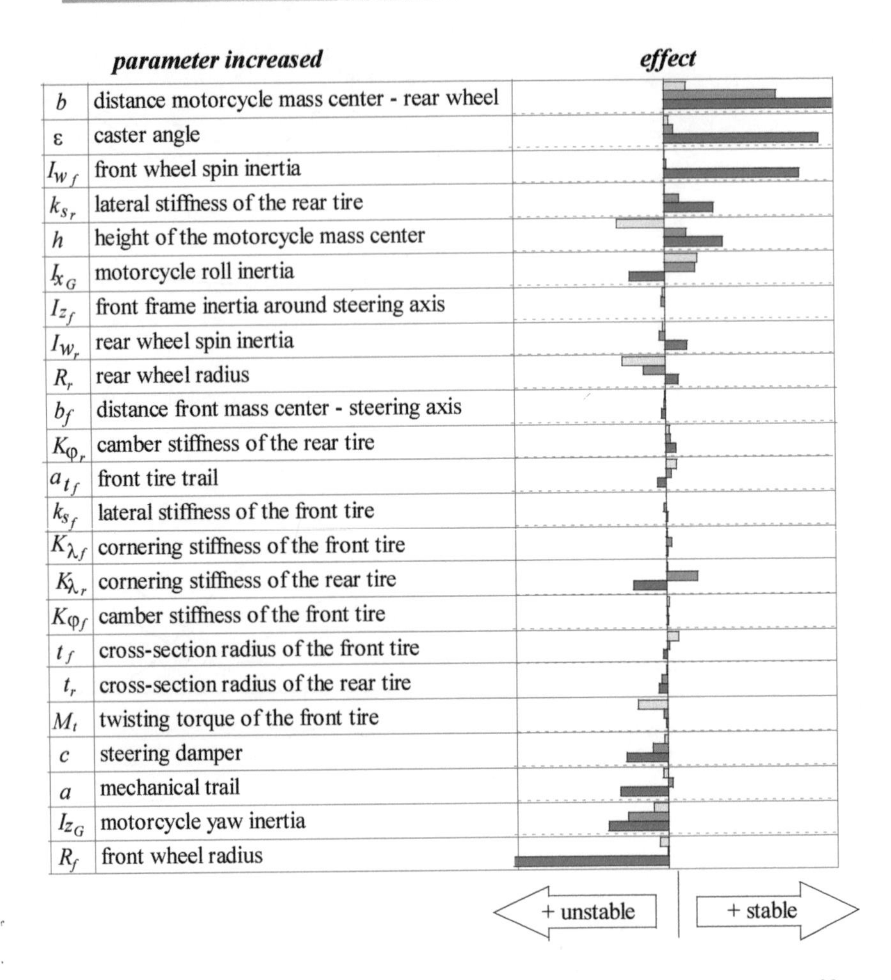

Fig. 7-36 Predicted change in weave damping at 10 m/s (light gray) at 30 m/s s (gray) , at 60 m/s (dark gray) for parameter increased 10% with respect to the reference case.

The increases in the height of the motorcycle mass center and of the rear wheel radius give a disadvantage at low velocity and an advantage at high velocity. On the contrary, the increase of roll inertia is advantageous at low velocity and proves a disadvantage at high velocity.

The increase of the mechanical trail gives a slight advantage at medium velocities but a disadvantage at high velocity. An increase in the cornering stiffness of the rear tire produces similar behavior.

It is worth pointing out that the increase in front wheel spin inertia is stabilizing due to the increase of the gyroscopic phenomenon. On the contrary, the increase in the front wheel radius (keeping front wheel spin inertia constant) is destabilizing because it causes a reduction in front wheel spin angular velocity and, hence, a reduction in gyroscopic effects.

In the simulation the rider is assumed to be rigid and attached to the main frame.

Effectively the rider acts like a damper that improves both wobble and weave stability. For example, a pillion passenger improves stability whereas the same rigid mass attached in the same position causes a decrease of the stability.

Rear wobble

Rear wobble is characterized by roll and yaw fluctuations that are almost in phase with each other, and in phase opposition to the lateral displacement of the motorcycle and the steering angle (Fig. 7-37).

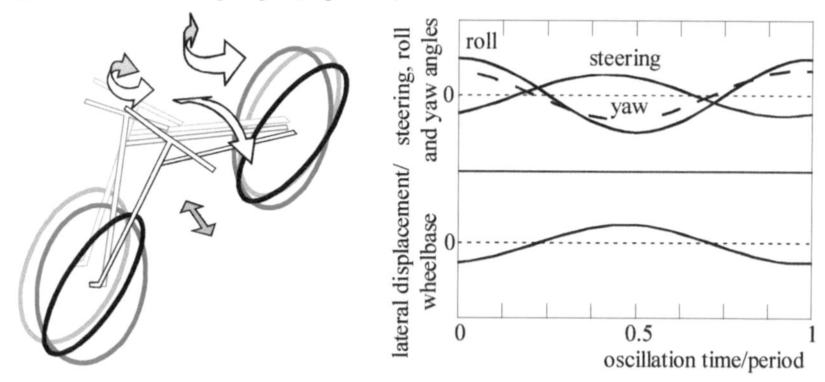

Fig. 7-37 Rear wobble (V = 10 m/s).

7.2.4 Modes of vibration in cornering

In straight running the in-plane and the out-of-plane modes are uncoupled. Since the roll angle is null, the tire vertical loads lie in the motorcycle symmetry plane to which the tire lateral forces are orthogonal. For this reason the normal loads only excite the in-plane modes, whereas the lateral forces only act on the out-of-plane vibration.

When the motorcycle corners, it is inclined by some roll angle and each of the previous forces has components in both directions. It should now be clear that in some modes both in-plane and out-of-plane degrees of freedom are involved. This phenomenon is called “modal coupling”.

In Fig. 7-38 the root locus plot in cornering at constant speed is presented. In the same figure, for comparison, the root locus in straight running is plotted in grey.

For the motorcycle considered here the main differences between the loci are:

- the rear hop moves a little towards a more stable region;
- the front hop does-not present significant modifications;
- there is an interaction between the bounce and the weave modes at medium speed;
- there are two pitch modes that differ in phase between the pitch motion and the yaw and steering motions;
- the capsize mode becomes more unstable.
- the frequency of the wobble mode increases slightly.

In the following sections the time evolution of some modes in cornering are presented.

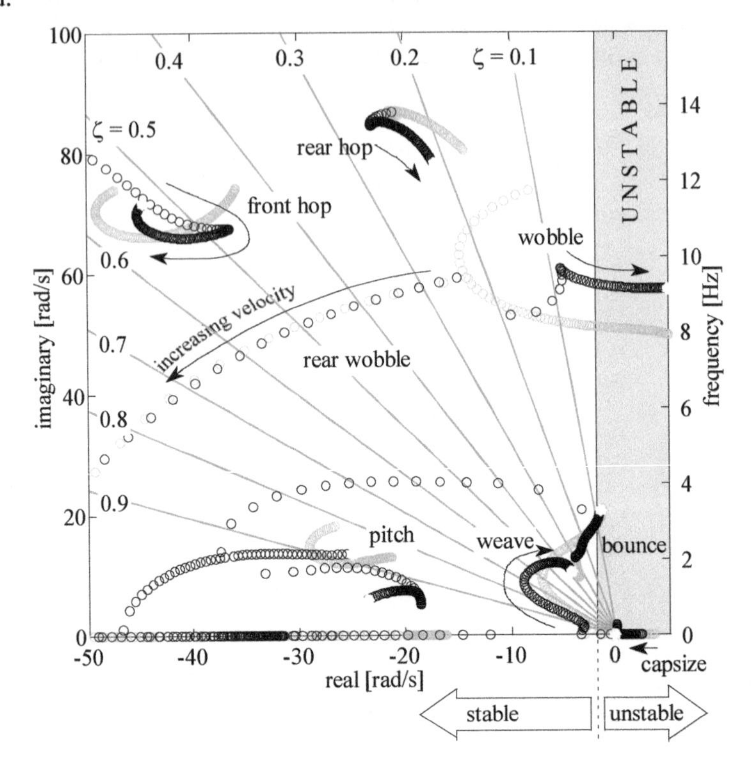

Fig. 7-38 Root-locus plot in cornering at different speeds, (speed from 3 to 60 m/s, centripetal acceleration = 0.5 g).

Capsize

When the motorcycle is rolled the capsize mode involves both in-plane and out-of-plane degrees of freedom, as Fig 7-39 shows.
In this configuration (that is with a speed of 4 m/s and a camber angle of 30 degrees) the mode shows a pronounced instability; in fact, the motion components depart from the steady state value as the time increases.

Wobble

The plot of the wobble (Fig 7-40) shows only slight coupling, because only the out-of-plane components are involved in the oscillation, whereas the in-plane quantities keep their constant trim values.

Weave

The weave is the mode showing the main coupling effect: all components of the eigenvector are involved. Figure 7-41 shows that the mode is stable and the oscillation vanishes after 1 second.

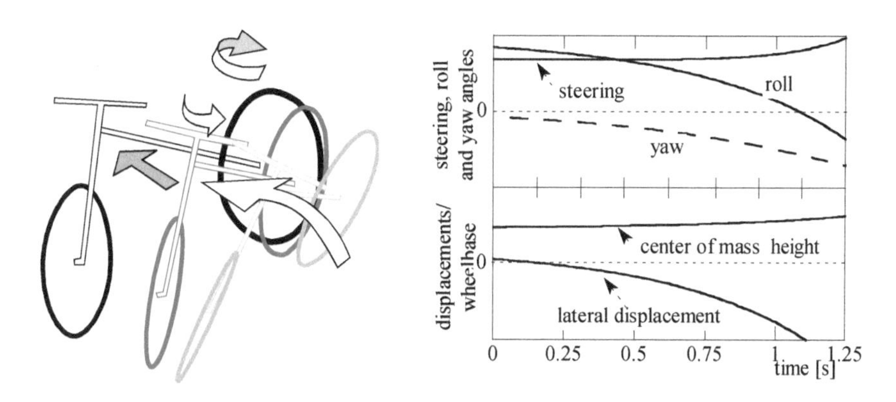

Fig. 7-39 Unstable capsize in cornering at a speed of 4 m/s.

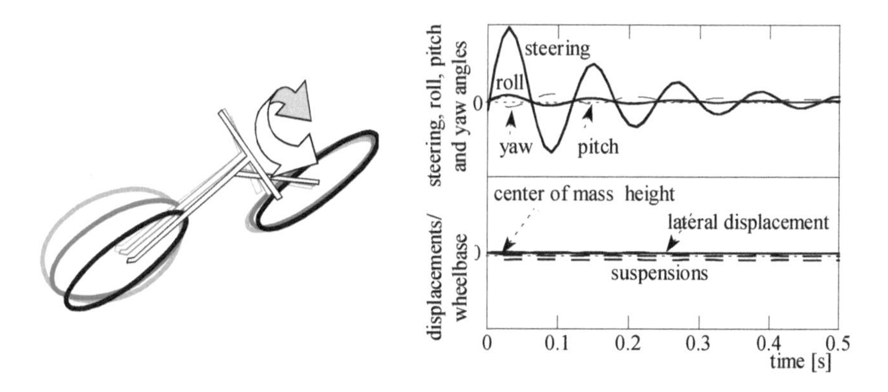

Fig. 7-40 Wobble in cornering at a speed of 25 m/s; roll angle=30°.

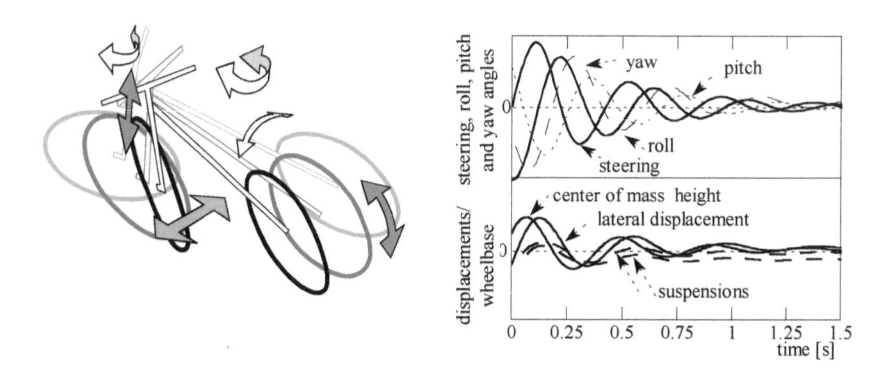

Fig. 7-41 Weave in cornering at a speed of 25 m/s; roll angle=30°.

Bounce

The time evolution of the motion components for the bounce mode shows the strong interaction between in-plane and out-of-plane oscillations.

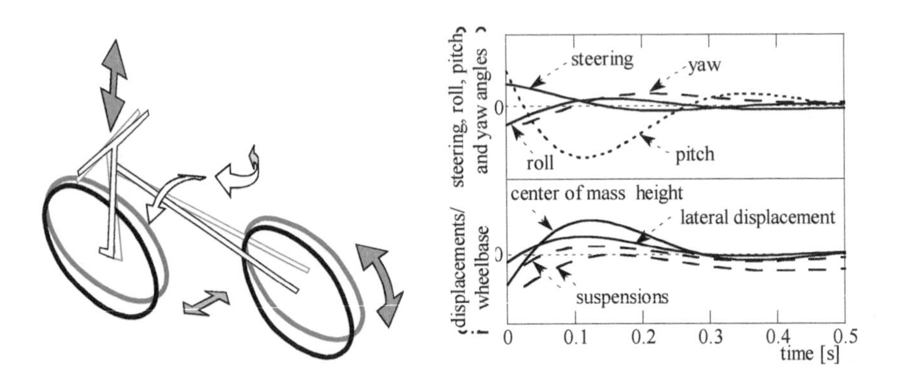

Fig. 7-42 Bounce in cornering at a speed of 25 m/s; roll angle=30°.

7.2.5 Effect of frame flexibility on modes of vibration

The assumptions of rigid bodies and a fixed rider in motorcycle modeling are not strictly true, particularly for a less rigid chassis as in the case of a scooter. Therefore, in order to highlight the effect of flexibility, a scooter was considered. As is common in scooters low-price front forks have a non-negligible bending and torsion compliance; the engine is elastically connected to the main chassis and the low-slung, cradle shape of the frame does not make it possible to design a high, stiff chassis. Inadequate structural stiffness may notably reduce stability and handling of these vehicles. In order to investigate these phenomena, a mathematical model of the scooter which also includes vehicle compliances and rider mobility is used (Fig. 7-43).

The vehicle compliances are taken into account by means of a fork bending stiffness (25-75 kN-m/rad), a fork torsion stiffness (4-10 kN-m/rad), a swingarm bending stiffness (30-70 kN-m/rad) and a swingarm torsion stiffness (10-20 kN-m/rad). Rider-vehicle mobility is taken into account according to the suggestion of the reference, [Katayama et al., 1997].

Simulation results are presented in terms of the root-loci in Fig. 7-44. The well-known weave, wobble and capsize modes are clearly visible, and differences between the rigid and lumped stiffness model estimation are also evident. Moreover the lumped stiffness model shows two additional modes that correspond to the rider lean and shake.

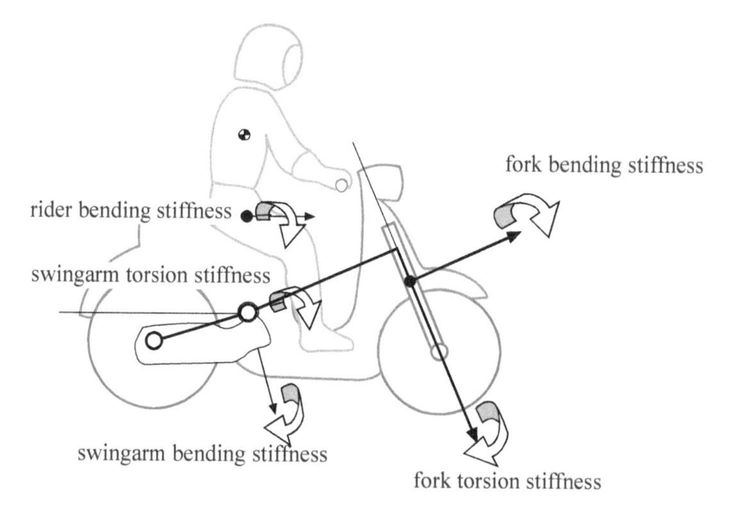

Fig. 7-43 Flexible bodies model.

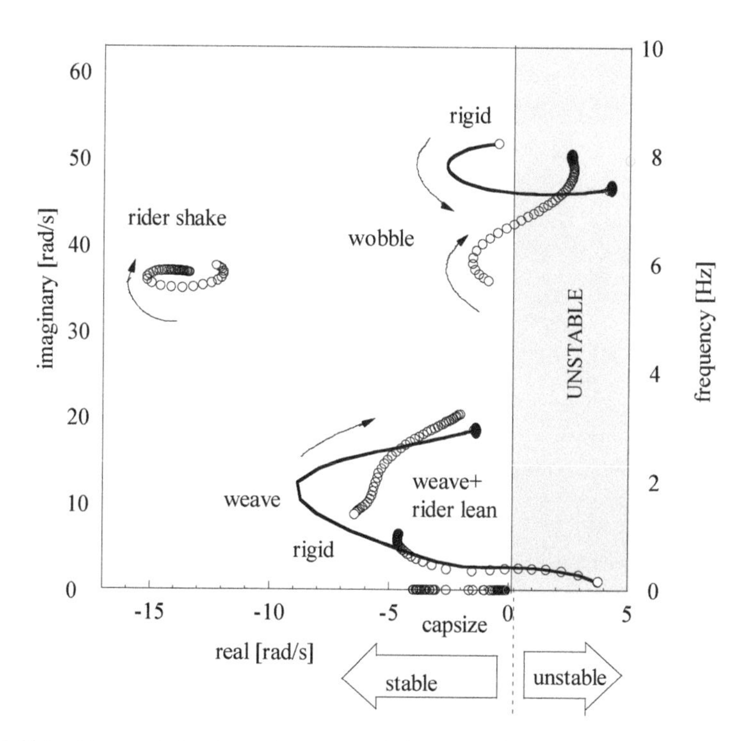

Fig. 7-44 Root-locus plot in straight running at different speeds: rigid model vs. flexible model (speed from 1 to 40 m/s).

In more detail, the rigid model estimates a wobble frequency that decreases from 8 to 7.5 Hz as the speed increases; on the contrary the lumped stiffness model predicts a frequency which rises from 6 to 8 Hz. Moreover, at low speeds the wobble stability of the rigid model is greater than that of the lumped stiffness model, whereas at high speeds the wobble instability of the rigid model is greater than that of the lumped stiffness model.

The main differences in the weave mode are due to the mode branching in the lumped stiffness model, which gives rise to two modes coupled with the rider lean motion. Both models show that the weave mode has a very low frequency at low speeds; this frequency climbs to 3 Hz as the speed increases. Weave mode is unstable at very low speeds, becomes very stable in the medium speed range and its damping ratio decreases at high speeds.

The capsize mode is non oscillatory in the whole speed range and there are no differences between the two models. The rider shake mode is very stable and does not appear to influence other modes.

Figure 7-45 presents the simulation results, obtained taking into account only the front fork bending compliance. The wobble mode is deeply influenced by fork compliance. In particular a flexible fork decreases low speed stability, but increases stability in the upper speed range. This behaviour can be attributed to a pair of opposing effects: the increment of fork flexibility tends to reduce stability, at the same time the combination of wheel spin and fork bending generate a gyroscopic torque around the steering axis which tends to stabilize the wobble. At a low speed the first negative effect is predominant. At high speed the second positive effect dominates.

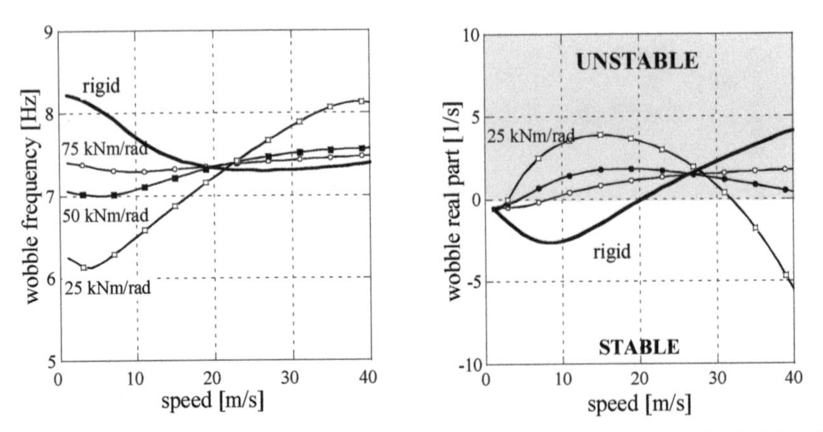

Fig. 7-45 Frequency and stability of the wobble mode taking into account only the fork bending compliance.

The weave mode is only slightly affected by fork compliance. Figure 7-46 shows the effect of the swingarm bending compliance on the weave mode: the flexibility may increase weave stability slightly at high speeds.

Finally, Fig. 7-47 shows that the torsional compliance of the swingarm always worsens the weave stability at medium-high speeds.

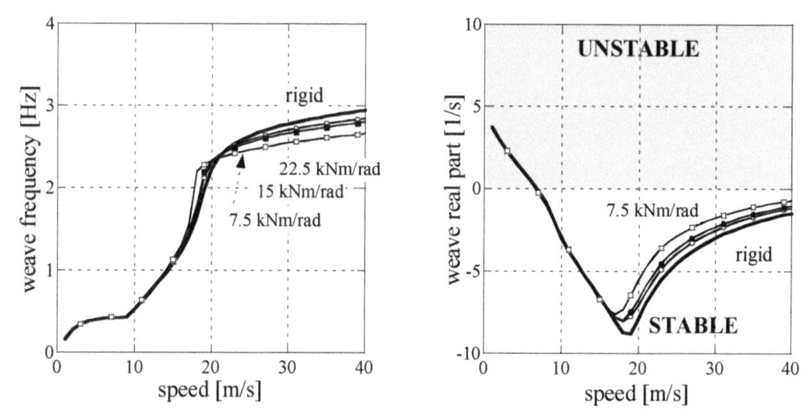

Fig. 7-46 Frequency and stability of weave mode taking into account the swingarm bending compliance only.

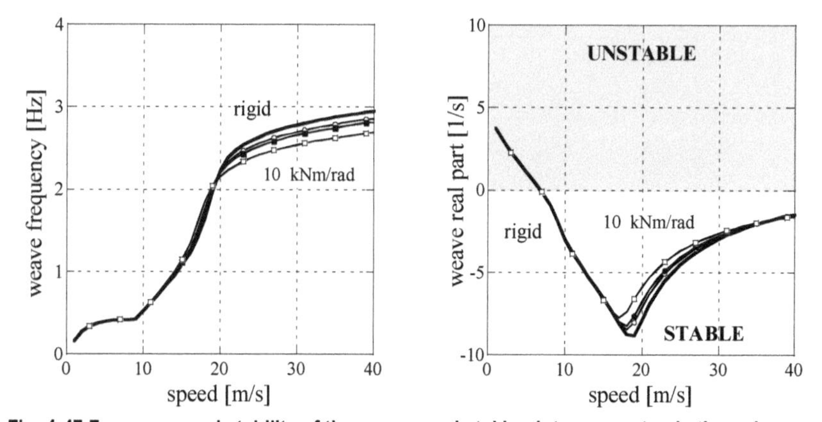

Fig. 4-47 Frequency and stability of the weave mode taking into account only the swingarm torsional compliance.

Moto Guzzi Falcone touring motorcycle, 1957 (owned by Vittore Cossalter)

8 Motorcycle Maneuverability and Handling

A motorcycle's dynamic properties are described using terms like maneuverability, handling and stability. Maneuverability and handling describe the motorcycle's ability to execute complicated maneuvers, and how difficult it is for the rider to perform them. Stability, on the other hand, means a motorcycle's ability to maintain equilibrium in response to outside disturbances like an uneven road surface or gusts of wind.

8.1 Directional stability of motorcycles

Motorcycles in motion need to be controlled by the rider at all times. Rider input affects the motorcycle's equilibrium and direction of forward motion.

In rectilinear motion, a motorcycle is called "directionally stable" if it is easy to control or naturally tends to maintain its equilibrium and follow a rectilinear path.
It is easy to see, however, that a large tendency towards directional stability makes a motorcycle hard to handle, i.e., cumbersome to turn and control through twists and turns.

This section discusses the directional stability of motorcycles, which is determined by a number of factors:

- inertial properties of the motorcycle;
- forward speed;

- geometric properties of the steering head (which collectively determine the aligning effect of the trail);
- gyroscopic effects;
- tire properties.

Obviously, the greater the motorcycle's quantity of motion (mV), the less it will deviate from its rectilinear trajectory as a result of outside disturbances.

Taking a gust of wind as an example (Fig. 8-1). Suppose the aerodynamic pressure generated by the gust acts on the motorcycle for short time interval Δt that tends to zero. The disturbance causes an angular deviation of the motorcycle from the rectilinear trajectory equal to:

$$\alpha = \arctan\frac{\Delta V}{V} = \arctan\frac{F\,\Delta t}{mV}$$

The angle of deviation is inversely proportional to the mass of the motorcycle and its forward speed, and directly proportional to the lateral aerodynamic force. The length of the motorcycle's wheelbase also plays a rather important role in determining directional stability. Figure 8-2 shows how a motorcycle with a short wheelbase behaves differently from one with a long wheelbase. If a disturbance causes a displacement of the front wheel, the angle of deviation from the rectilinear trajectory is inversely proportional to the length of the wheelbase.

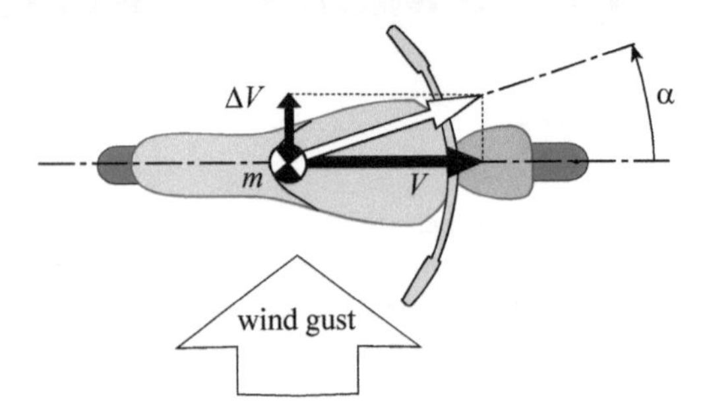

Fig. 8-1 Directional behavior of a motorcycle struck by a gust of wind.

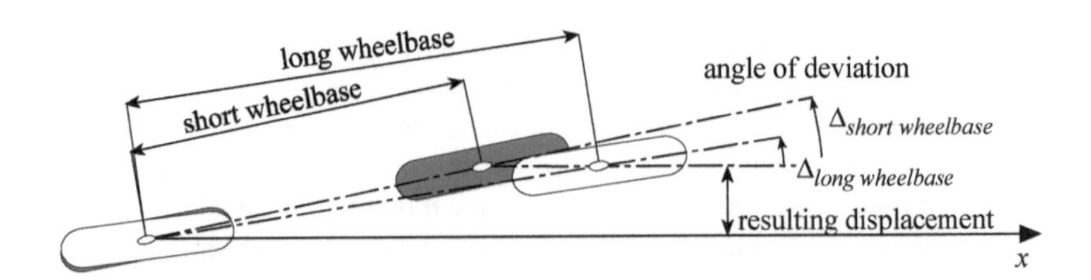

Fig. 8-2 Directional behavior of a motorcycle as a function of length of wheelbase.

In terms of the effect of motorcycle geometry, we have already seen that the moment exerted by resistance on the front tire has an aligning effect that increases with forward speed and the length of the trail.

We can calculate this aligning effect using a simplified model: a motorcycle in rectilinear motion traveling at constant speed. Suppose that an outside disturbance causes the front end to rotate to the right, and therefore the motorcycle begins to follow a curved trajectory to the right with a large radius. Let us also assume that the roll angle is negligible. Based on these simplifying assumptions, we can calculate the moment exerted around the steering head axis.

The following forces are acting on the vehicle as a whole:

- thrust S, exerted at the contact point of the rear wheel;
- drag F_D, which is assumed to be exerted at the center of gravity;
- rolling resistance $F_{w_f} = \mu N_f$ exerted at the contact point of the front wheel;
- lateral forces F_{s_f}, F_{s_r} exerted at the contact points of the wheels;
- vertical loads N_f, N_s exerted at the contact points of the wheels.

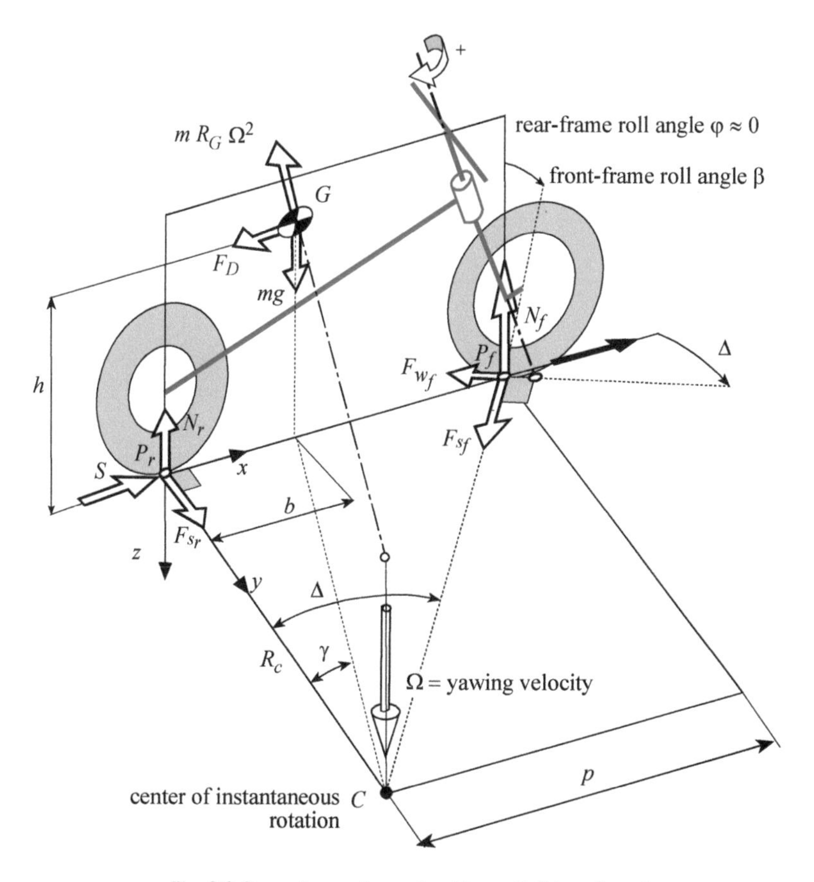

Fig. 8-3 Cornering motorcycle with negligible roll angle.

The equilibrium equations for forces and moments (Fig. 8-3):

$$(\Rightarrow) \qquad F_{s_r} + F_{s_f} \cos\Delta - F_{w_f} \sin\Delta - m\frac{V^2}{R_c}\cos\gamma = 0$$

$$(\Downarrow) \qquad mg - N_r - N_f = 0$$

$$(\Rightarrow) \qquad S + m\frac{V^2}{R_c}\sin\gamma - F_D + F_{s_f}\sin\Delta - F_{w_f}\cos\Delta = 0$$

$$(\subset) \qquad N_f\, p + F_D h - mg\, b = 0$$

$$(\cap) \qquad F_{s_r} p - m\frac{V^2}{R_c}(p-b)\cos\gamma = 0$$

give the dynamic vertical loads:

$$N_f = mg\frac{b}{p} - F_D\frac{h}{p} \qquad\qquad N_r = mg\frac{(p-b)}{p} + F_D\frac{h}{p}$$

the front lateral force:

$$F_{s_f} = \frac{1}{\cos\Delta}\left(\frac{mV^2}{R_c}b\cos\gamma + \mu N_f \sin\Delta\right) \approx \frac{mV^2}{R_c}b + \mu N_f \frac{p}{R_c}$$

and the thrust needed to make the motorcycle travel through the turn at constant speed:

$$S = \frac{1}{2}\rho C_D A V^2 - m\frac{V^2}{R_c}\sin\gamma + F_{s_f}\sin\Delta - F_{w_f}\cos\Delta \approx$$

$$\approx \frac{1}{2}\rho C_D A V^2 - m\frac{V^2}{R_c^2}b + F_{s_f}\frac{p}{R_c} - F_{w_f}$$

A moment is exerted around the steering head axis (Fig. 8-4) through:

- the component of the vertical load N_f perpendicular to the steering head axis which has the effect of increasing the steering angle: $N_f \sin\beta$,
- the component of lateral force F_{s_f} perpendicular to the steering head axis, which has the effect of aligning the wheel: $F_{s_f}\cos\beta$.

The resulting moment exerted by the two forces, neglecting the rolling resistance, is given by:

$$M = -\left(F_{s_f}\cos\beta - N_f \sin\beta\right)\cdot a_n \approx$$

$$\approx -\left[\left(m\frac{b}{R_c p}\cos\beta + \frac{1}{2}\rho C_D A\frac{h}{p}\sin\beta\right)V^2 - mg\frac{b}{p}\sin\beta\right]\cdot a_n$$

Since the positive term is proportional to the square of the speed, the aligning moment increases with speed. The longer the normal trail is, the more marked the effect is. As expected, motorcycle stability is strongly influenced by the length of the trail.

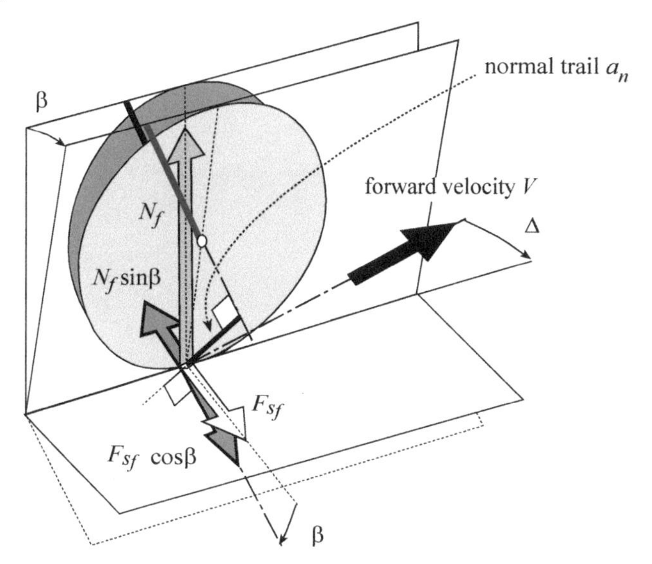

Fig. 8-4 Forces acting on the front wheel.

We have suggested that gyroscopic effects play an especially important role in directional stability and maneuverability. Many gyroscopic effects are experienced while cornering, and entering or exiting turns. The rotation of the steering head, wheels and rotary engine parts generate gyroscopic moments as a result of motorcycle roll and/or yaw motions.

8.2 Gyroscopic effects on the motorcycle

A gyroscopic effect is generated by a rigid body rotating around an axis $a-a$, which in turn is rotating around a second axis $b-b$ askew (not parallel) to the first axis $a-a$. The gyroscopic effect takes the form of a couple exerted around an axis perpendicular to both $a-a$ and $b-b$. The value of the gyroscopic moment is equal to the vector product of the angular momentum $I\omega$ of the body around axis $a-a$ and the speed of rotation Ω around the second axis $b-b$. Angular momentum is equal to the polar moment of inertia of the body I around axis $a-a$ multiplied with the speed of rotation ω around the same axis.

Motorcycle dynamics incorporate a variety of gyroscopic effects, which may be broken down according to the second axis of rotation $b-b$:

- **yaw gyroscopic effects:** where axis $b-b$ passes through the turn center of the path and is perpendicular to the roadway;
- **roll gyroscopic effects:** where axis $b-b$ is the straight line lying in the plane of the roadway which passes through the tire contact points;
- **steering gyroscopic effects:** where axis $b-b$ is the steering head axis.

8.2.1 Gyroscopic effects generated by yaw motion

Gyroscopic effect generated by the wheels during cornering (wheel rotation - yaw motion)

Let us consider the front wheel alone, rotating at a constant speed ω_f as the motorcycle travels through a turn of radius R_c at a constant yaw velocity Ω (Fig. 8-5).

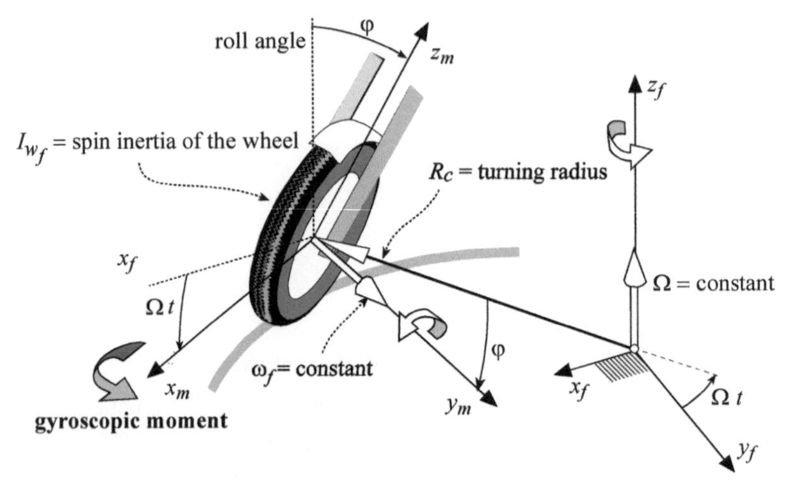

Fig. 8-5 Gyroscopic effect generated by the front wheel during cornering (the coordinate system with subscript m is attached to the fork of the motorcycle).

The motion of the wheel as it corners generates a gyroscopic moment around the horizontal axis, which has the effect of straightening the wheel:

$$M_g = I_{w_f}\left(\omega_f \Omega - \frac{\Omega^2 \sin\varphi}{2}\right)\cos\varphi \qquad\qquad M_g \cong I_{w_f}\,\omega_f \Omega \cos\varphi$$

The second approximate expression is valid if the yaw velocity Ω can be considered small with respect to the speed of rotation ω_f. This assumption is verified in practice because the turning radius is much greater than the wheel radius. Axis x_m is fixed to the fork, and therefore, it is a mobile axis.

Looking now at the effect of both wheels and setting aside the fact that the wheels have slightly different roll angles and directions during cornering, their gyroscopic effects can be added together:

$$M_g \approx (I_{w_f}\omega_f + I_{w_r}\omega_r)\Omega\cos\varphi = I_w\,\omega\,\Omega\cos\varphi$$

Motorcycle equilibrium occurs when the resultant of the weight force and the centrifugal force intersects the line joining the contact points of the two wheels. Disregarding the gyroscopic effect and assuming zero thickness wheels, the ideal roll angle for a motorcycle in steady state cornering is given by the following simple equation:

$$\varphi = \arctan\left(\frac{R_c\Omega^2}{g}\right)$$

As we have seen, the gyroscopic effect of the wheels during cornering is manifested by a righting moment. To counteract the gyroscopic effect of the two wheels and thereby maintain equilibrium, the rider can lean into the turn in such a way that the resultant of the weight force and the centrifugal force generates a moment equal and opposite to the gyroscopic moment of the two wheels, as shown in Fig. 8-6.

$$M = -\sqrt{(mg)^2 + \left(mR_c\Omega^2\right)^2}\cdot d = -M_g$$

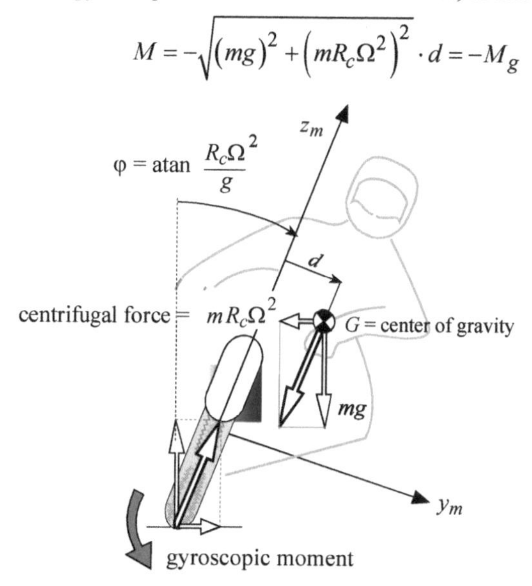

Fig. 8-6 Influence on equilibrium of gyroscopic effect generated by wheels during cornering.

Of course, the rider can achieve equilibrium without displacing his trunk in order to produce a displacement of the mass center towards inside of the curve, but the lean angle of the motorcycle will be greater than the ideal roll angle calculated on the assumption that the gyroscopic effect is zero (Fig. 8-7).

In this case, the righting moment generated by the centrifugal force and the moment generated by the gyroscopic effect (which also has a righting effect) are both offset by the overturning moment of the weight force. The gyroscopic effect makes the actual roll angle greater than the ideal roll angle that would be achieved if the gyroscopic effect were absent.

The increase in the roll angle $\Delta\varphi$ needed to counterbalance the gyroscopic effect is given by:

$$\Delta\varphi = \arcsin\frac{d}{h} = \arcsin\frac{I_w\ \omega\cdot\Omega\cos(\varphi+\Delta\varphi)}{h\sqrt{(mg)^2+\left(mR_c\Omega^2\right)^2}}$$

Since $\Delta\varphi$ is small with respect to φ, it can be disregarded in the numerator on the right hand side, the following simpler equation holds:

$$\Delta\varphi \cong \frac{I_w \ \omega \cdot \Omega \cos\varphi}{h\sqrt{(mg)^2 + \left(mR_c\Omega^2\right)^2}}$$

Here, the numerator represents the gyroscopic moment generated by the two wheels of the motorcycle. The moment M which counterbalances the gyroscopic moment is generated by the resultant of the weight force and the centrifugal force. The increase $\Delta\varphi$ makes the motorcycle less maneuverable, since the motorcycle takes more time to reach the incrementally larger equilibrium roll angle (which is greater).

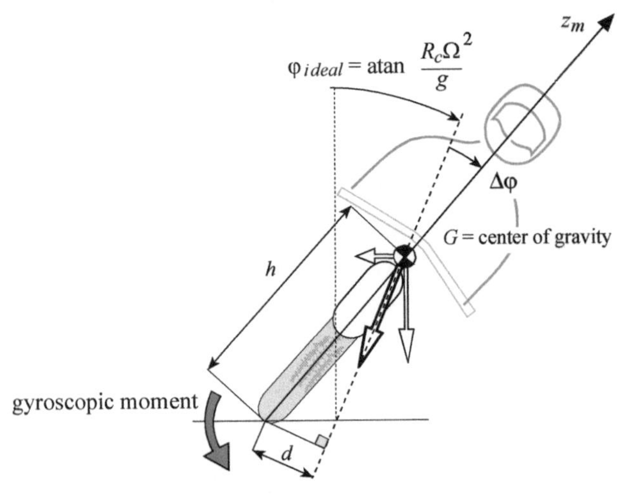

Fig. 8-7 Increase in roll angle caused by yaw gyroscopic effect.

Example 1

Assume a motorcycle in stationary motion during cornering:

- turning radius: $R_c = 200\text{m}$;
- forward speed: $V = 40$ m/s;

The properties of the motorcycle are as follows:

- mass: $m = 200$ kg;
- height of center of gravity: $h = 0.6$ m;
- wheel radius: $R_f = R_r = 0.32$ m;
- spin inertia of front and rear wheels: $I_{w_f} = I_{w_r} = 0.6\ \text{kgm}^2$.

Now determine the gyroscopic moment generated by the motion of the two wheels, and the resulting increase in the roll angle.

Since we know the turning radius and forward speed, we can calculate the angular velocity values:

- yaw velocity of the motorcycle: $\Omega = 0.2$ rad/s;
- angular velocity of the wheel: $\omega = 125$ rad/s;

and therefore:

- ideal roll angle: $\varphi_{ideal} = 39.20°$;
- gyroscopic moment generated by the motion of the two wheels: $M_g = 23.25$ Nm.

The rider can achieve equilibrium in one of two ways:

- either displacing the mass center toward the inside of the turn by: $d = 9.2$ mm;
- or increasing the roll angle by: $\Delta\varphi = 0.88°$.

Gyroscopic effect generated by transversally mounted engine (engine rotation - yaw motion)

The gyroscopic effect generated by the engine is determined by the engine's speed of rotation, which depends on what gear the motorcycle is in.

Let us assume a motorcycle in steady-state cornering motion and disregard the inertia of the wheels. In other words, we are going to look at the gyroscopic effect generated by the rotation of the engine only. The main shaft of the engine generally rotates in the same direction as the wheels, as shown in Fig. 8-8.

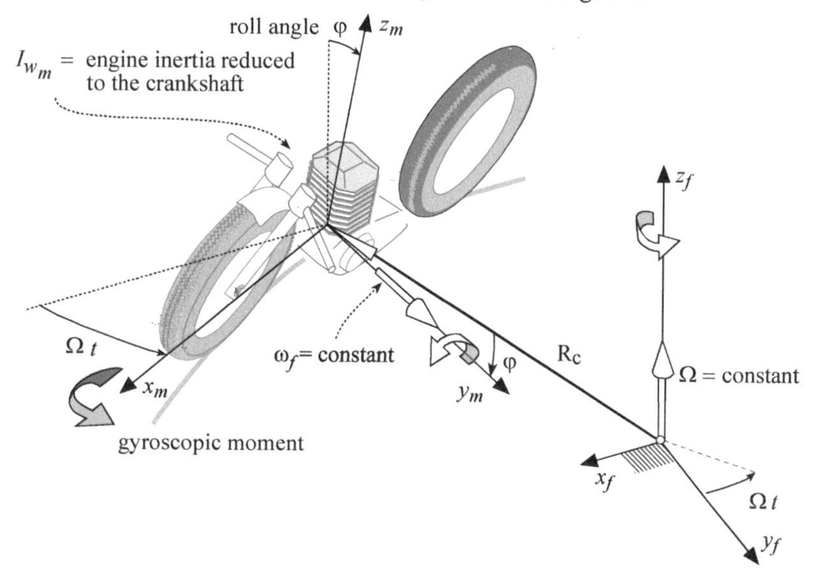

Fig. 8-8 Gyroscopic effect generated by a transverse engine during cornering.

As before, the gyroscopic effect generated by the engine causes equilibrium to be achieved by leaning the motorcycle over at an actual roll angle greater than the ideal angle that would be necessary if the gyroscopic effect were absent.

The resulting increase in the roll angle is equal to:

$$\Delta\varphi \approx \frac{I_{w_m}^{*} \omega_m \cdot \Omega \cos\varphi}{h\sqrt{(mg)^2 + \left(mR\,\Omega^2\right)^2}}$$

The sign is positive when the engine rotates in the same direction as the tires and negative for a counter-rotating engine.

The term $I^*_{w_m}\,\omega_m$ expresses the engine's total angular momentum, incorporating the angular momentum of the drive shaft, transmission shafts and any other rotating shaft parallel to rear wheel axis and rotating with same sense:

$$I^*_{w_m}\,\omega_m = \sum I_{w_i}\,\omega_i$$

In calculating the resulting angular momentum, the sign of the angular velocity is assumed to be positive if the direction of shaft rotation agrees with the direction of rotation of the wheels. Otherwise, it is assumed to be negative:

For example, if the engine's rotation is in the same sense as concordant with the wheel spin the main gear shaft rotates with an opposing velocity, and its contribution is therefore negative.

Of course, the engine's contribution must be added to, or subtracted from, the contribution of the wheels, depending on the direction of rotation following the convention established above (added if the direction of rotation is concordant with the wheel spin, and subtracted if not).

To reduce the gyroscopic effect, the momentum of the rotating bodies must be reduced. Since lightweight materials can only be used to reduce the moment of inertia, an attractive option is to reduce the angular momentum of the engine, or even to give it a negative sign, by choosing a rotation in the direction opposite to the wheel spin.

For example, in some two-cylinder racing engines the two drive shafts rotate in opposite directions. Thus, the gyroscopic effects of the two crankshafts cancel each other out, leaving only the effect of the transmission shafts.

However, the main gear shaft and transmission contribute less to the gyroscopic effect than the drive shaft does, since they have less inertia and rotate at slower angular velocities; the velocity ratio between the drive shaft and the main transmission shaft is of the order of 2 to 2.5.

Example 2

Using the motorcycle in Example 1 in steady-state cornering motion, determine the gyroscopic effect of an engine with the following properties:

- moment of inertia of crankshaft: $I_{w_m} = 0.015$ kgm^2;
- primary shaft inertia moment (including clutch): $I_{w_p} = 0.008$ kgm^2;
- secondary shaft inertia moment: $I_{w_s} = 0.008$ kgm^2;
- engine - primary shaft transmission ratio: $\tau_{m,p} = 3$;
- primary - secondary gear shaft transmission ratio: $\tau_{p,s} = 2$;
- engine rpm: $n = 12{,}000$ rpm.

Adding up the various components, the engine's total angular momentum is equal to:

$$I^*_{w_m}\omega_m = I_{w_m}\omega_m + I^*_{w_p}\omega_p + I^*_{w_s}\omega_s = (I_{w_m} - I_{w_p}\frac{1}{\tau_{m,p}} + I_{w_s}\frac{1}{\tau_{p,s}\tau_{m,p}})\omega_m$$

$$I^*_{w_m}\omega_m = 18.85_m - 3.35_p + 1.67_s = 17.17 \text{ kgm}^2/\text{s}$$

The gyroscopic moment is: M_g= 2.66 Nm; the increase in roll angle solely due to the gyroscopic effect generated by the engine is $\Delta\varphi = 0.1°$.

Note that the gyroscopic effect generated only by the engine is less than that generated by the wheels (M_{engine} is about 8% of M_{wheels}); the engine's contribution generally falls within the range of 5% to 15% of the gyroscopic effect generated by the wheels. If the contribution by the wheels is taken into account as well, the increase in the roll angle goes up to $\Delta\varphi = 0.97°$.

Gyroscopic effect generated by longitudinally mounted engine (engine rotation - yaw motion)

Now consider a motorcycle equipped with a longitudinal drive shaft cornering at constant speed, as shown in Fig. 8-9.

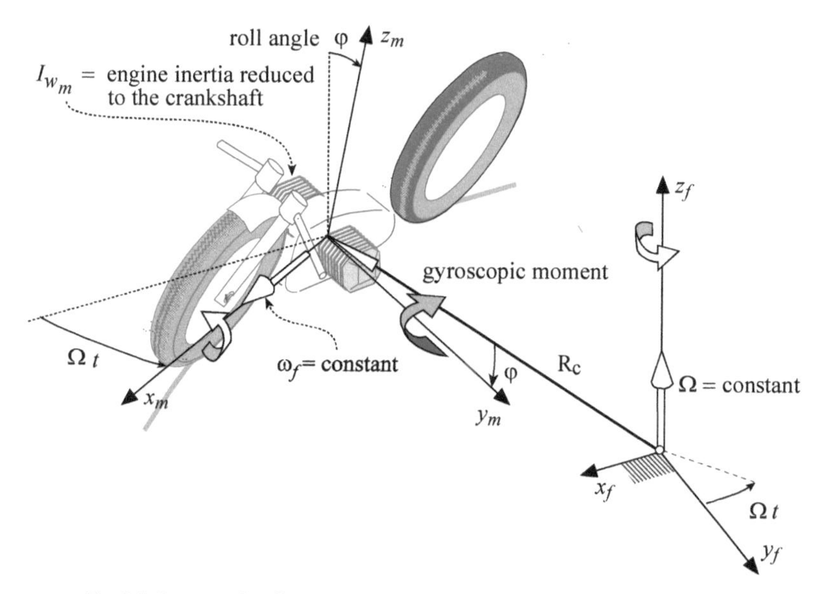

Fig. 8-9 Gyroscopic effect generated by an engine with a longitudinal axis.

If the turn is to the left with respect to the direction of forward motion the motorcycle leans over to the left. Assuming that the longitudinal drive shaft is rotating toward the outside of the turn, the gyroscopic moment acting around the y_m axis is equal to:

$$M_g = -I^*_{w_m}\omega_m \cdot \Omega$$

The gyroscopic moment has the effect of extending the front suspension and compressing the rear suspension to a greater degree making the motorcycle pitch backwards.

The gyroscopic moment has the opposite effects when cornering to the right, i.e., the front suspension compresses and the rear suspension extends.

Now consider the motorcycle cornering to the left again, but this time assume that the drive shaft rotates toward the inside of the turn.
The gyroscopic moment acting around the y_m axis reverses sign:

$$M_g = I^*_{w_m} \omega_m \cdot \Omega$$

In this instance, the gyroscopic moment has the effect of reducing the load on the rear suspension and increasing the load on the front suspension. The moment generated by the suspensions' forces balances out the gyroscopic moment, with the end result of the motorcycle characteristically pitching forward slightly.

Example 3

Now assume that the engine in example 2 is mounted longitudinally.
The stiffness values for the front and rear suspensions are: k_f = 9 kN/m and k_r = 2 kN/m, respectively. We want to calculate the change in trim generated by the gyroscopic effect of the engine.

The gyroscopic moment of the engine is equal to: M_g = 2.43 Nm.

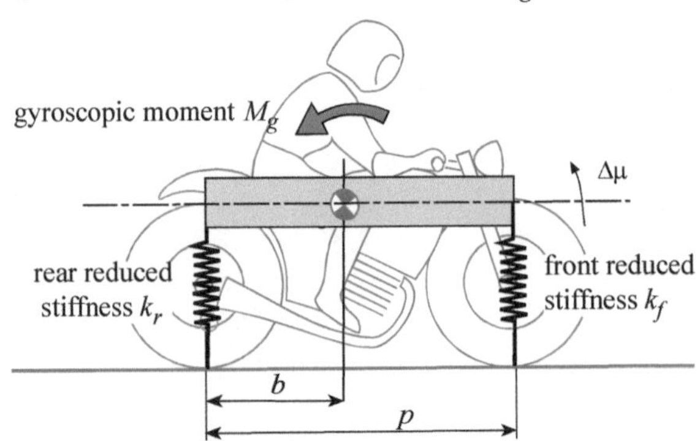

Fig. 8-10 Motorcycle riding trim.

To calculate the pitch angle, we need to take into account the equilibrium of the motorcycle on which the gyroscopic moment is acting (Fig. 8-10). The static equilibrium equations easily yield the vertical displacement of the center of gravity and the pitch angle:

$$\Delta h = \frac{M_g[b\ k_r - (p-b)\ k_f]}{k_f \cdot k_r p^2} \qquad \Delta\mu = \frac{(k_f + k_r)}{k_f \cdot k_r p^2} M_g$$

Both the pitch angle of 0.01° and the increase in height of the center of gravity are entirely negligible.

8.2.2 Gyroscopic effects generated by roll motion

Gyroscopic effect generated by the front wheel (front wheel rotation – roll motion)

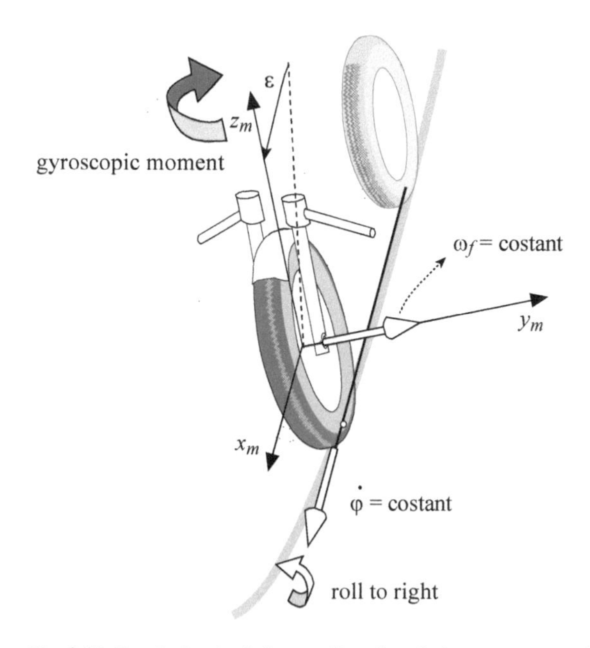

Fig. 8-11 Front wheel rotation - roll motion induce a gyroscopic moment acting on the front end.

Now we will look at the front wheel while the motorcycle is rolling to the right. The front-wheel spin, coupled with the roll to the right, generates a gyroscopic moment M_g that acts on the front frame around an axis lying in the plane of the motorcycle and perpendicular to the longitudinal roll axis, as shown in Fig. 8-11:

$$M_g = -I_{w_f} \omega_f \dot{\varphi}$$

The projection along the steering axis provides the beneficial moment around the steering axis:

$$M_{g_u} = -I_{w_f} \omega_f \dot{\varphi} \cos\varepsilon$$

Thus, the gyroscopic moment has the effect of turning the steering head to the right, thereby helping the motorcycle enter the turn (increasing the steering angle reduces the turning radius). Analogously, when the roll velocity changes sign as the motorcycle returns to the vertical position the gyroscopic moment has the effect of reducing the steering angle, thereby helping the motorcycle exit the turn and return to rectilinear motion.

Example 4

Now consider a motorcycle rolling from left to right. To calculate the gyroscopic moment acting on the steering head, assume that the motorcycle is rolling from the left to the right at a velocity of 0.5 rad/s.
The properties of the motorcycle are as follows:

- moment of polar inertia of front wheel: $I_{w_f} = 0.6$ kgm^2;
- motorcycle roll velocity: $\dot{\varphi} = 0.5$ rad/s;
- spin velocity of wheel: $\omega = 100$ rad/s;
- rake angle: $\varepsilon = 25°$;

The gyroscopic moment around the steering axis is equal to $M_g = 27$ Nm.

It can be demonstrated that in transient maneuvers such as the lane change most of the steering torque applied by the rider is used to overcome this gyroscopic moment.

Gyroscopic effect generated by wheels (wheels rotation - roll motion)

If the motorcycle is assumed to be a rigid body (i.e., with the steering head locked in place), the gyroscopic effect of the wheel spin during roll can easily be shown to generate a yawing moment, as shown in Fig. 8-12.

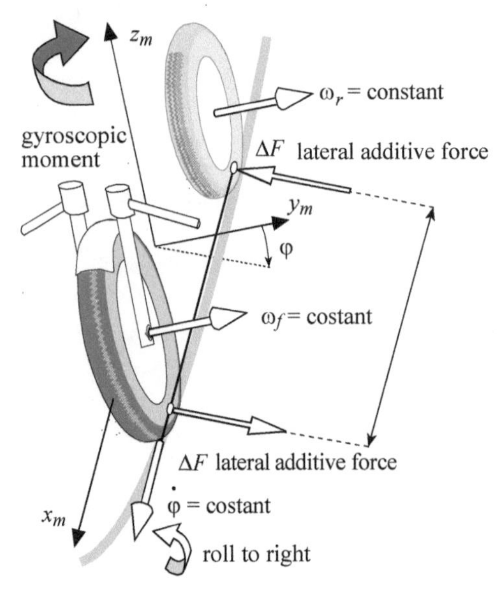

Fig. 8-12 Gyroscopic effect generated by motorcycle roll motion-wheels rotation.

Again, consider a motorcycle rolling from left to right. The gyroscopic moment acting on the motorcycle is equal to:

$$M_g = -(I_{w_f}\omega_f + I_{w_r}\omega_r)\dot{\varphi}$$

The gyroscopic moment tends to make the motorcycle yaw to the right, and is balanced by the lateral resistance exerted on the wheels by the ground. Thus, the front lateral force increases slightly ΔF, while the rear lateral force decreases by the same amount:

$$M_g = -(I_{w_f}\omega_f + I_{w_r}\omega_r)\dot{\varphi} = \Delta F \cdot p\cos\varphi$$

When exiting the turn the motorcycle rolls from right to left. The gyroscopic moment reverses sign, and, hence, also the variation in tire lateral forces changes sign.

Example 5

Now consider a motorcycle rolling from left to right at a roll velocity of 0.5 rad/s. The properties of the motorcycle are as follows:

- moment of polar inertia of wheels: $I_{w_f} = I_{w_r} = 0.6$ kgm^2;
- spin velocity of wheels: $\omega = 100$ rad/s;
- length of wheelbase: $p = 1.37$ m;
- motorcycle roll velocity: $\dot{\varphi} = 0.5$ rad/s.

Determine the gyroscopic effect generated by the wheels and the change in lateral force.

The gyroscopic moment acting on the motorcycle is equal to 60 Nm.

The change ΔF in lateral force needed to counterbalance the gyroscopic moment is 44 N when the motorcycle is in the vertical position.

This is a fairly high value compared to the values for the lateral force needed to maintain equilibrium under steady-state cornering. For example with speed $V = 30$ m/s, turning radius $R_c = 200$ m, and mass $m = 180$ kg the sum of the two lateral forces must be equal to 872 N.

If the total lateral force is distributed evenly between the two wheels, thereby exerting a transverse force of 436 N on each wheel, the variation due to the gyroscopic effect is on the order of 10 %.

8.2.3 Gyroscopic effects generated by steering

Since the wheel's direction of spin is perpendicular to the steering head axis, turning the handlebars from right to left generates a gyroscopic moment around an axis perpendicular to both the steering head axis and the axis of the front wheel, as shown in Fig. 8-13:

$$M_g = I_{w_f}\omega_f\dot{\delta}$$

This has the effect of leaning the motorcycle over towards the right. The projection of the gyroscopic moment on the roll axis (the line connecting the contact points of the two wheels) is as follows:

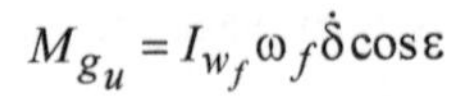

$$M_{g_u} = I_{w_f} \omega_f \dot{\delta} \cos\varepsilon$$

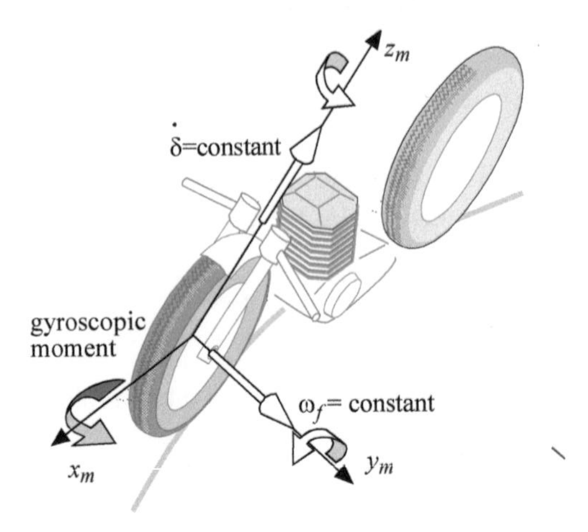

Fig. 8-13 Gyroscopic effect generated by the front wheel and steering head rotations.

Based on these gyroscopic effects, one might conclude that a motorcycle with zero wheel inertia is ideal. It is important to point out, however, that the gyroscopic effect generated by the front wheel and steering motion plays an important part in motorcycle stability during rectilinear motion.

8.3 Motorcycle equilibrium in rectilinear motion at low speed

Since the same principles govern control of any two-wheeled vehicle at low speed, a motorcyclist also knows how to ride a bicycle.

A child learning to ride a bicycle begins by rolling down a gentle hill, and quickly learns that, if the bicycle begins to lean to the right and he turns the handlebars in the same direction, the bicycle easily returns to the vertical position after turning right. If the bicycle leans to the left, the equilibrium is achieved by a similar maneuver.

The path the bicycle follows is influenced by the control actions continually made by the cyclist to keep the bicycle vertical. Therefore, as a result the bicycle follows a weaving path, and how much it weaves is determined by the skill of the rider.

The same principles are used to control the motorcycle's vertical equilibrium at low speed.

As a motorcycle leans to the right its fall is counteracted by turning the handlebars to the right. The motorcycle begins to turn right, creating a centrifugal force that straightens the motorcycle back up, as shown in Fig. 8-14.

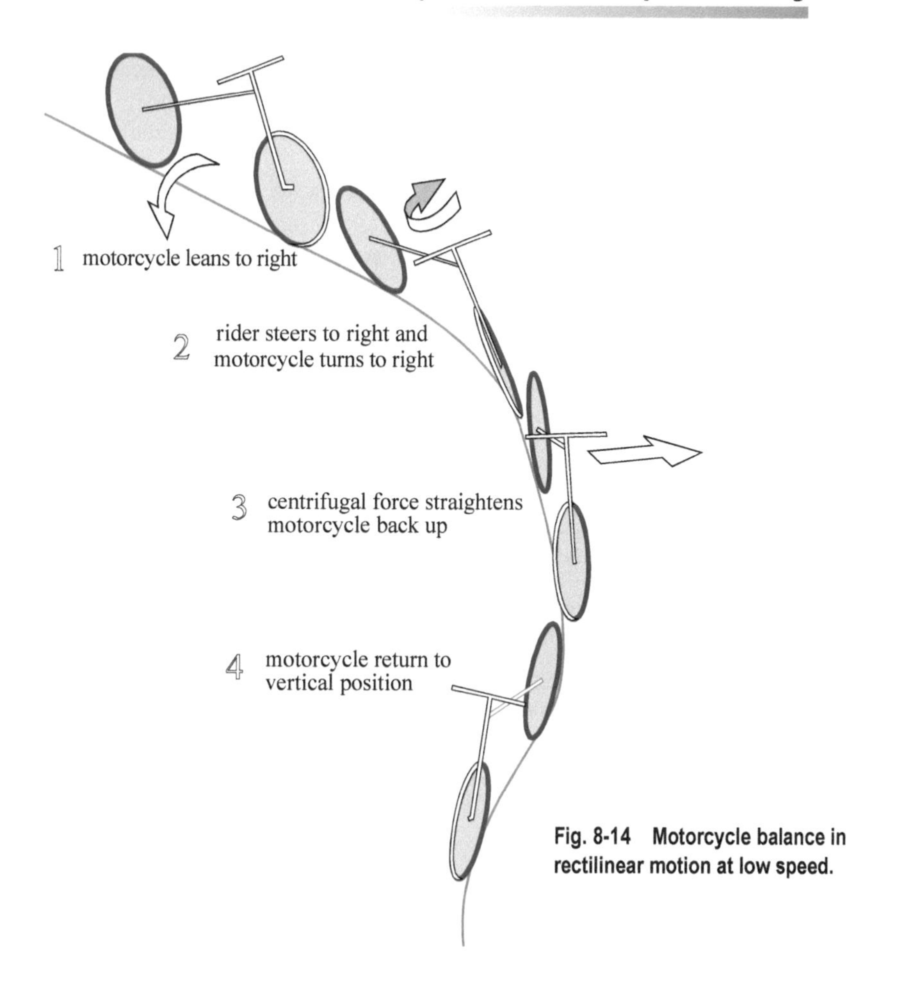

Fig. 8-14 Motorcycle balance in rectilinear motion at low speed.

8.4 Motorcycle equilibrium in rectilinear motion at high speed

Let us suppose an external disturbance that causes the motorcycle to roll to the right, as shown in phase 1 of Fig. 8-15 and that the rider does not apply any torque to the handlebars; in other words, he/she remains passive. Let ω be the angular velocity of the front wheel, $\dot{\varphi}$ the roll velocity, and $\dot{\delta}$ the steering velocity.

Due to the front wheel-spin and roll velocities a gyroscopic moment is generated (phases 2) that has the effect of turning the handlebars to the right (phases 3).
The handlebars steer to the right thereby reducing the turning radius (phase 3).
As the turning radius decreases, the centrifugal force increases, thereby straightening up the motorcycle (phases 3-4).

At the same time, due to the front wheel spin and steering and yaw velocities an overturning gyroscopic moment is generated that counteracts the roll motion towards the right (phase 3-4).

The motorcycle stops rolling to the right (phase 4). The centrifugal force has the effect of straightening the motorcycle up and reversing the roll motion (phase 5-6).

The roll-induced gyroscopic moment has the effect of turning the handlebars to the left, thereby steering the motorcycle to the left.

Finally, the motorcycle returns to the vertical position shown in phase 6, but it continues to roll to the left, and the resulting gyroscopic moment turns the steering head to the left. When rolling to the left the motorcycle goes through a similar sequence.

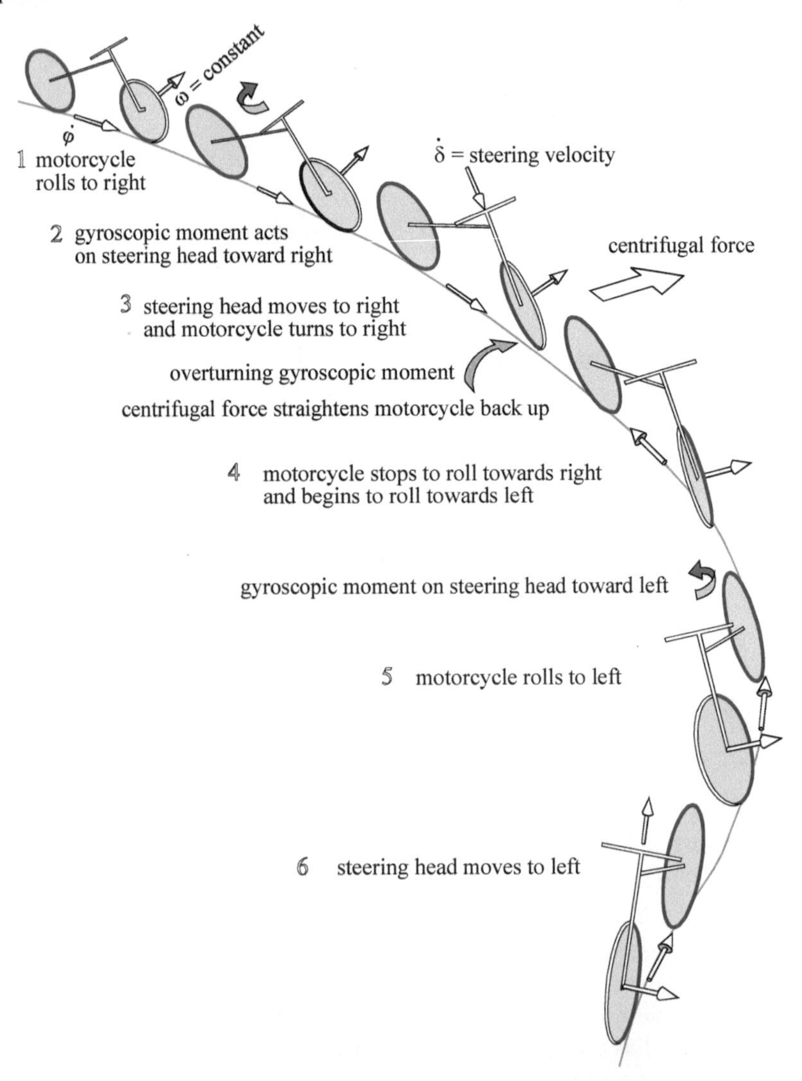

Fig. 8-15 Motorcycle balance in rectilinear motion at high speed.

8.5 Slow entering in a turn

After learning how to keep a two-wheeled vehicle in vertical equilibrium in rectilinear motion one must learn how to turn.

The beginner rider of a bicycle finds a gentle hill and sets off, worrying only about keeping the bicycle upright. At a certain point he spies an obstacle e.g., a pothole to his left, and decides to change direction, so he turns the handlebars to the right. The bicycle begins to follow a curved trajectory to the right. The centrifugal force, generated as the bicycle rounds the curve, rapidly leans the bicycle and rider toward the left, making a fall inevitable.

After several unsuccessful attempts, the rider comes up with a strategy that is neither simple nor intuitive. If he wants to change his or her direction toward the right he must first apply leftward torque to the handlebars ("you turn left to go right"). That makes the steering head turn to the left, and the bicycle begins to turn left, creating a centrifugal force that leans the bicycle to the right. Once the bicycle has begun rolling to the right the rider can turn the handlebars to the right to continue his entrance into a rightward turn.

Let us now consider a motorcycle traveling at 20 m/s entering a turn with a slow maneuver. The rider begins the maneuver well before (approx. 14 m) the turn in order to reach the steady state condition slowly (Fig. 8-16).

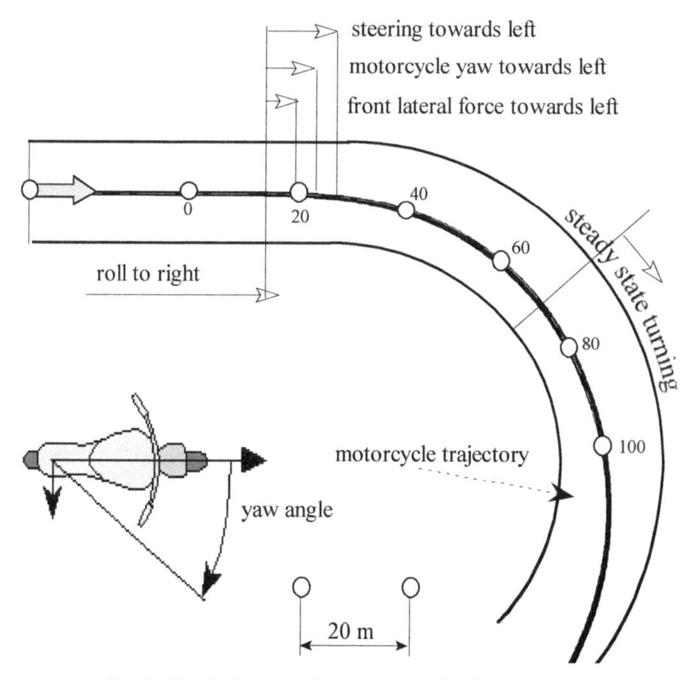

Fig. 8-16 Trajectory when entering slowly in a turn.

Initially the rider applies a torque towards left causing the front wheel to steer towards the same side (Fig. 8-17). The lateral force generated at the front tire contact point, causes a yaw motion to the left and the beginning of a roll motion to the right.

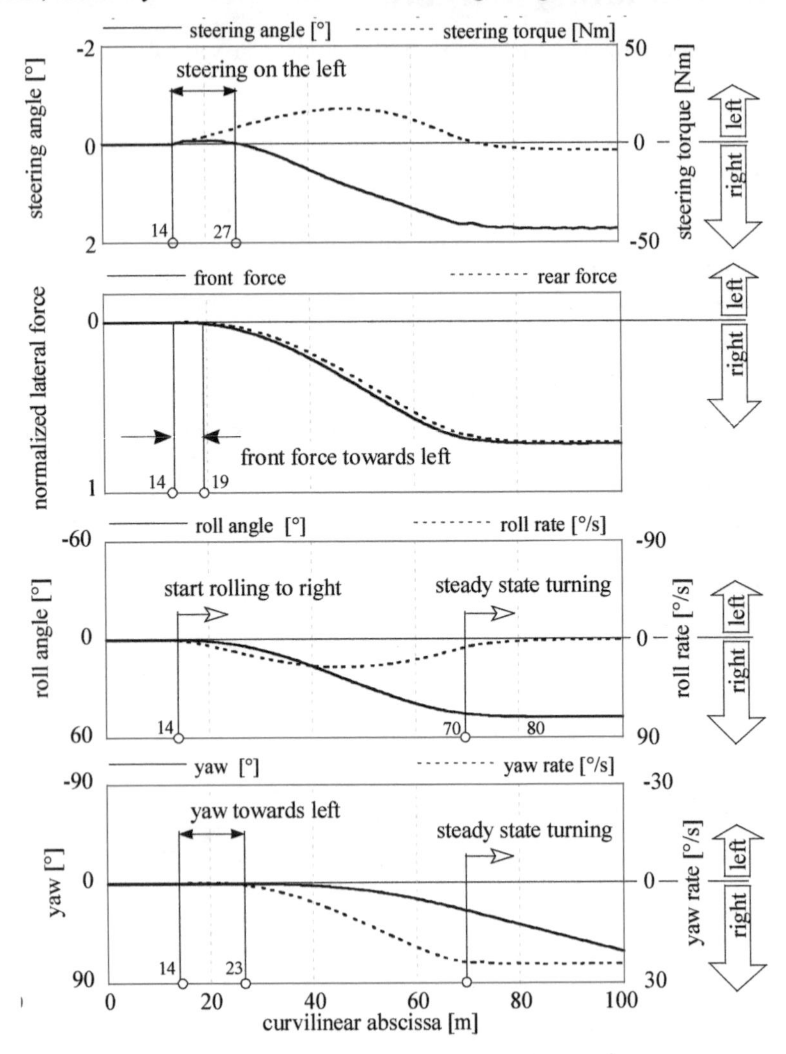

Fig. 8-17 Steering angle, steering torque, lateral force, roll and yaw when entering slowly in a turn (computed with FastBike code).

The vehicle follows a heading toward the left, opposite the direction of the desired trajectory.

After about 5 m from the initiation of the maneuver (at approx. 19 m), the lateral force changes direction from the outside direction toward the inside of the curve. After an additional 4 m (at approx. 23 m), with the motorcycle already rolled toward the right, the leftwards yaw motion ends and the motorcycle starts to yaw in the desired direction. Now the path is towards the right.

Now the rider follows the desired path controlling the vehicle with the steering torque. After about 13 m from the initiation of the maneuver (at approx. 27 m) the steering angle changes from left to right. The motorcycle reaches the steady-state turning condition about 70 m after the beginning of the maneuver. In the steady state condition the centripetal acceleration is equal to 0.9 g. It is worth noting that the steering torque value in the steady condition is small with respect to the maximum value of the torque applied during the entrance phase.

8.6 Fast entering in a turn

To enter fast in a right-hand turn, the rider applies a quick torque on the handlebars to the left. The movement of the front wheel around the steering head generates a front tire lateral force that has the effect of leaning the motorcycle to the right. The gyroscopic effect generated by the front wheel and the steering rotation also has an important effect on fast entering in a turn. This gyroscopic moment has the effect of leaning the motorcycle to the right. Once the motorcycle has begun to roll to the right the rider can slowly turn the handlebars to the right, and the motorcycle enters the turn.

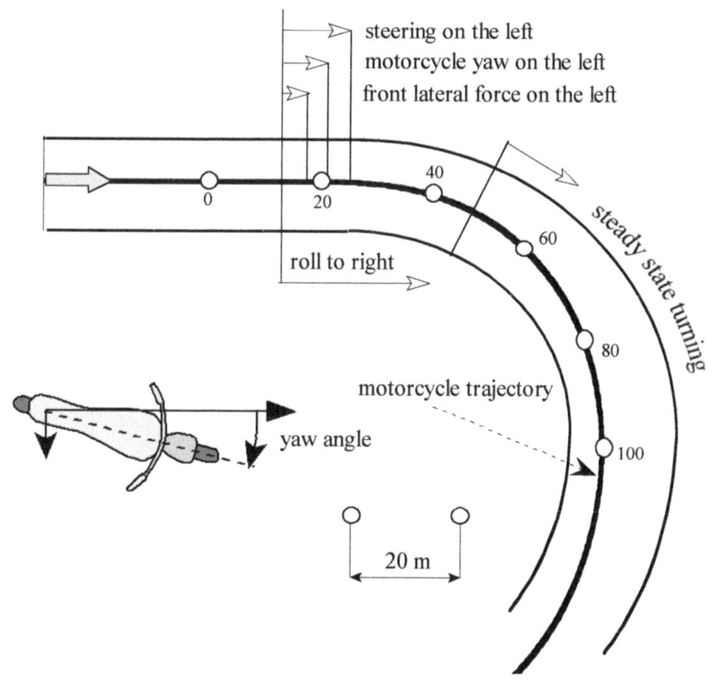

Fig. 8-18 Trajectory when entering fast in a turn.

Figure 8-18 shows the motorcycle traveling at 22 m/s and executing a relatively quick entering in a turn maneuver. The average turning radius of the road is 50 m and the road width is 16 m.

The rider (at approx. 14 m) quickly steers towards the left and the motorcycle immediately yaws to the same side following a path towards the left. The minimum turning radius in this phase is about 100 m. At the same time the motorcycle starts to roll rightwards. The leftward lateral force reaches a maximum value of 50 N after about 2-3 m. Its tilting moment with respect to the mass center (height = 0.6 m) is equal to about 30 Nm.

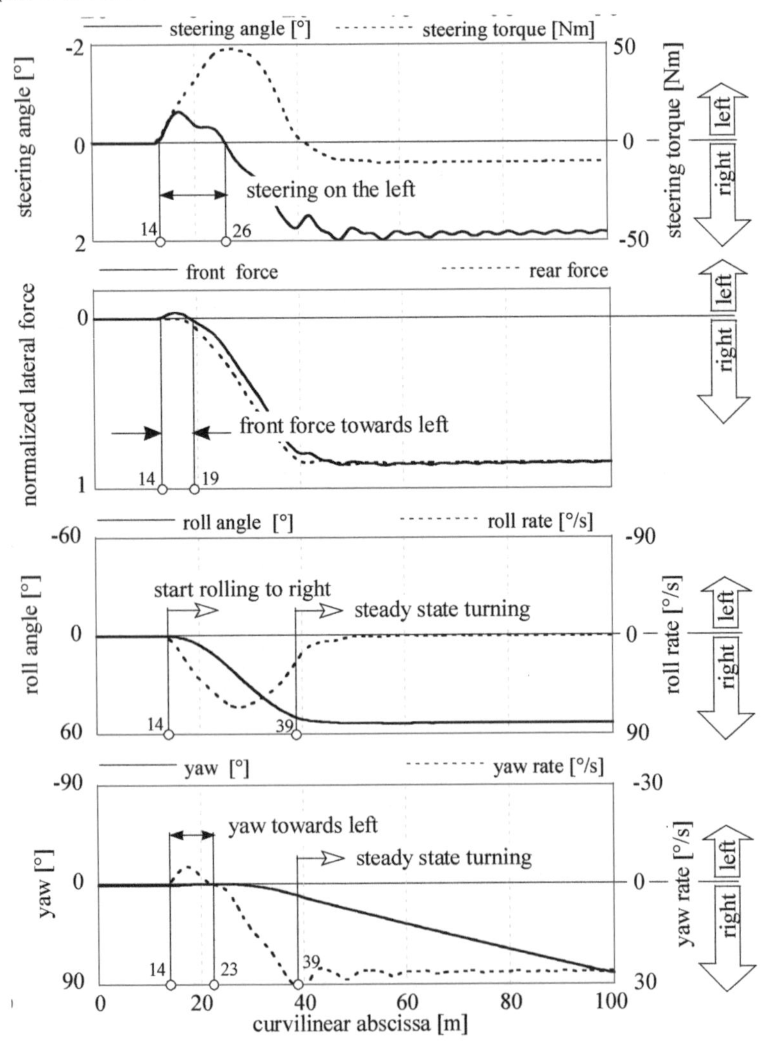

Fig. 8-19 Steering angle, steering torque, lateral force, roll and yaw when entering fast in a turn (computed with FastBike code).

Let us compare this moment with the gyroscopic moment due to front wheel spin-steering motion. The steering rate reaches its maximum value after about 1 m before the lateral force reaches its maximum. If the front wheel spin inertia is equal to 0.6 kgm^2 the gyroscopic roll moment is about 3.5 Nm.

This moment contributes to the generation of the roll motion due to the fact that it is present since the initiation of the maneuver while, depending on the relaxation length of the tire, the lateral force requires more time to reach its maximum value.

Obviously, the gyroscopic contribution becomes more important as the front wheel speed and the steering rate increase.

This example highlights the "out-tracking" techniques that consist of entering a right turn steering for a short time towards the left. The entering phase can be improved by the lateral displacements of the rider's body into the turn neglected in the previous example. Rider lateral displacement causes the motorcycle to lean and can be used to reduce the initial counter-steer.

8.7 The optimal maneuver method for evaluating maneuverability and handling

Motorcycle dynamics is a notoriously thorny topic to deal with, mainly due to the following problems:

- first of all the precise description of vehicle kinematics is complex because of the presence of the steering head;
- secondly, depending on the riding conditions, such as speed for example, motorcycles are naturally unstable vehicles (as we have seen in previous chapters) and the driver's actions are always needed to provide control;
- moreover the personal driving style, the driver's skill and their experience also affect the vehicle's performance.

Thus, it is clear that the whole driver-motorcycle system must be studied if we wish to understand what design parameters influence the vehicle's performance and handling.

A clear way of dealing with the motorcycle dynamics problem is to analyze it using a systematic approach. This means splitting the rider-motorcycle system into subsystems, as shown in Fig. 8-20. It is possible to consider three main subsystems.

The driver, who controls the motorcycle via the steering torque, the brake lever, the throttle, and the movement of his body. These are inputs to the motorcycle system. It is obvious that there exist some limits on these actions. For example the driver cannot exert an infinite value for the steering torque, nor may he exert it without some delay.

The motorcycle itself can be considered as being made up of many other subsystems, depending on the case. For vehicle lateral directional control (usually referred to as "lateral dynamics") two subsystems may be identified:

- the steering subsystem, which is a mechanism that transforms steering torque into a lateral force that acts on the front wheel;
- the vehicle subsystem, which is treated as a rigid body and includes the gyroscopic effect of the wheels, on which wheel-to-ground contact forces are exerted.

Let us suppose that we would like to accomplish a given maneuver such as a U-shaped curve or S-shaped chicane, for example, in the least time possible (minimum time is the objective). The initial and final positions on the track are known, but there are a lot of possible trajectories that can lead the vehicle from the starting point to the final one. We are looking for the fastest one, which best exploits the intrinsic motorcycle characteristics.

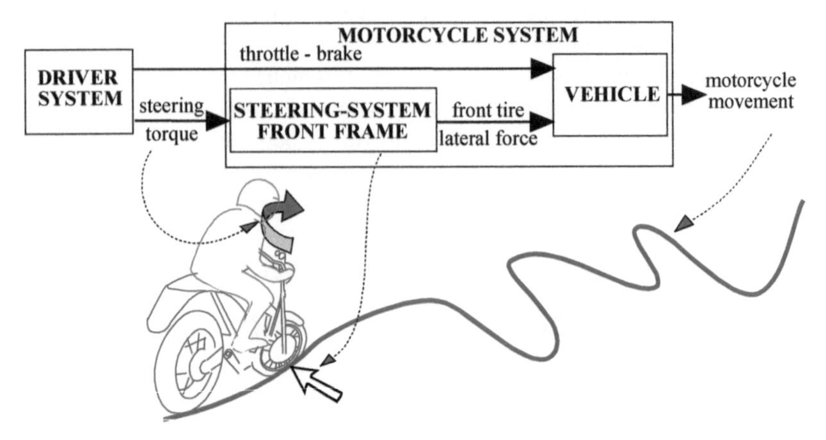

Fig. 8-20 Functional diagram of a motorcycle.

Furthermore, let us suppose that we observe only the motorcycle system without considering the driver's physical and psychological limits such as the maximum torque he is able to apply, or the maximum steering rate he is able to achieve. This situation corresponds to having a motorcycle driven by an ideal perfect driver. The best performance that we get from the motorcycle quantifies its **maneuverability**. In this sense maneuverability is related to the ability of the motorcycle to do complex maneuvers in the shortest time possible. If we also consider the driver's performance limits (we have a real driver riding the motorcycle), the best performance we get from the motorcycle quantifies its **handling**. Thus, handling means the ability of the motorcycle to do complex maneuvers taking into consideration the driver's limits. In other words, a motorcycle, which has better handling than another, is faster and at the same time the driver can ride it with less physical and physiological effort.

It is possible to look at the same problem from the safety point of view. In fact, a motorcycle, which is more maneuverable than another one, is able to accomplish the same maneuver without reaching its limits, for example the tire adherence limits. This means that some margins remain, for example additional lateral force is available, that can be used should a dangerous situation arise.

However, the extra effort that the driver has to exert in order to use these left-over margins (for example the additional tire forces) tells us how easy or how difficult the motorcycle is to drive.

The discussion so far would seem to indicate that with appropriate constraints placed on the system, maneuverability and handling are intrinsic vehicle characteristics, that qualify not only its best performance but also how easily the driver realizes this performance.

Another question that may arise when talking about maneuverability and handling is that the specific trajectory followed, and therefore the forces needed to produce that trajectory, depend on choices made by the rider. From this standpoint, it would not be entirely correct to talk about maneuverability as an intrinsic property of the motorcycle. Instead, we should talk about the overall performance of the motorcycle/rider system. The performance of the system could only be assessed if we assume a specific model for the rider.

However, on the other hand, it seems equally clear that some motorcycles are intrinsically better than others, independent of the rider's driving skills. So how can the intrinsic maneuverability of motorcycles be defined? The answer is to assume a perfect, or ideal rider, capable of choosing the best possible trajectory for a given vehicle. The best trajectory and the necessary driver action together are called the optimal maneuver.

8.7.1 Optimal maneuver

Let us see how it is possible to define the concept of the optimal maneuver with an example. Let us suppose that the desired maneuver is entering a turn where:

- the starting state is steady, rectilinear motion;
- the desired end state is steady, circular motion characterized by a given turning radius.

To move from the starting state to the end state, the lateral force applied to the front wheel (which is assumed to be controlled by the rider through the steering) must vary in specific ways over time. Of course, there are many possible ways of getting the motorcycle to the desired end state, but each solution is characterized by a different series of inputs over time, which affect the system in different ways in transit. Of all possible solutions, the only "optimal" solution is the most "efficient" one, i.e., the one that minimizes a given performance objective.

If the total time required to complete the maneuver is used as the index of efficiency, the "optimal" maneuver will be the one that minimizes that objective. In the example, the solution to the problem (enter a state of steady circular motion as quickly as possible) is given by the time taken to execute that particular maneuver.

Obviously, the "optimal" solution will be different for different motorcycles, as will the minimum time it takes the best possible rider to enter a turn. In short, the performance index associated with the objective to be optimized during the maneuver (time, in this example) can be used to measure the performance of the motorcycle, which quantifies the motorcycle's maneuverability. If we also add some constraints on the driver's physical and psychological effort to the objective, the performance index quantifies motorcycle handling.

8.7.2 An example of an optimal maneuver for an "S" trajectory (chicane)

For our first example of applying the optimal maneuver principle, we will look at a motorcycle following an "S" trajectory (chicane), as shown in Fig. 8-21.

The following constraints are placed on the problem:

- the vehicle must cover the section of track from the centerline at the starting point to the centerline at the end point;
- the starting speed is known, but not the ending speed, since it is determined by

the solution for the maneuver which optimizes the specified performance indices.

The following criteria must all be optimized at the same time, in the order of importance given:

- minimize the time to complete the maneuver;
- prevent the tire reactions from surpassing the edges of the friction ellipse;
- prevent the motorcycle from going over the edges of the track.

Figure 8.21 shows the "optimal" trajectory solution to the problem as a solid line plotted against a dotted line indicating the centerline of the section of track. Note how the solution trajectory "shaves" the turns without overcoming them.

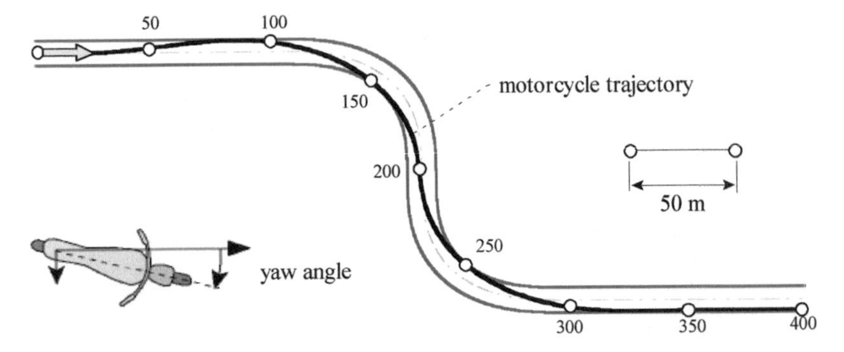

Fig. 8-21 "S" trajectory (chicane).

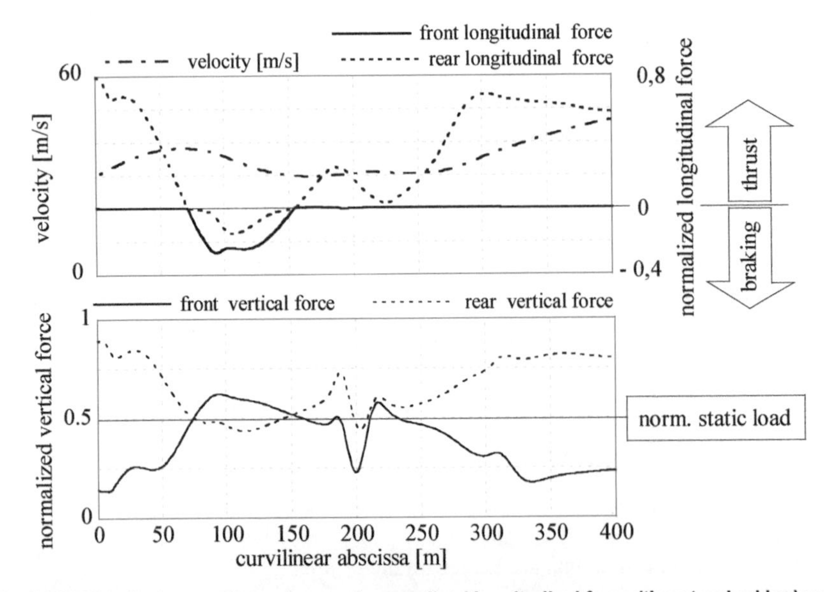

Fig. 8-22 "S" trajectory: motorcycle speed, normalized longitudinal force (thrust or braking) and vertical forces with respect to the weight force.

In Fig. 8.22, note how the speed of the motorcycle decreases before the chicane, and then increases again at the exit. In other words, the "optimal" solution calls for braking when entering the first turn and acceleration when exiting the second.

The vertical reactions exerted on the wheels (Fig. 8-23) illustrate the phenomenon of load transfer, moving first from the rear end to the front end (*braking phase*), and then in the opposite direction (*acceleration phase*). It is necessary to highlight that in the middle of the chicane, quickly tilting from one side to the other, the wheel loads diminish for a short time because of centrifugal acceleration due to roll motion.

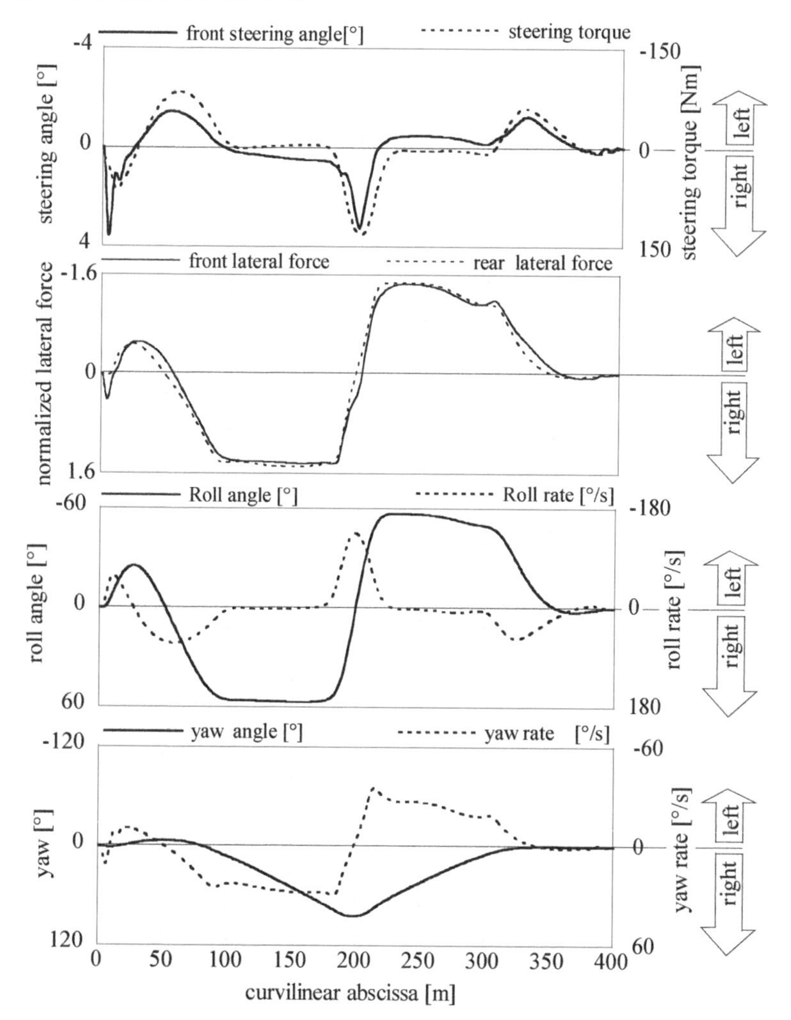

Fig. 8-23 "S" trajectory: steering angle, steering torque, lateral forces, roll angle and roll velocity, yaw angle and yaw velocity.

The evolutions of the steering torque and of the steering angle illustrate the maneuver carried out by the rider (Fig. 8-23). Initially the rider steers quickly toward the right in order to move from the center of the road to the left side of the road. Having reached the left side in about 40 m, the rider steers leftwards to insert the motorcycle into the first turn of the chicane. The steering angle needed for equilibrium during the first turn is very small due to the high value of the roll angle. In the middle of the chicane the quasi-impulsive movement of the handlebars induces the motorcycle to roll quickly from one side to the other and yaw in the opposite direction.

The exit from the second turn is performed more smoothly compared to the entering phase. Increasing the driving force causes a gradual increment in the forward velocity and consequently the increased centrifugal force tilts the motorcycle to the vertical position. In this phase the rider controls the path to be followed by means of the steering angle.

The figure shows what happens to the lateral forces acting on the tires, which have been normalized with respect to current vertical load; in other words, it shows the inclination of the resulting reaction forces on both wheels. This is one of the parameters the optimal maneuver is designed to control. Note that the ratio of lateral to vertical force reaches maximum values of about 1.5 for both front and rear wheels, and is slightly less pronounced for the front wheel. In this specific example, therefore, the motorcycle is well balanced because the front wheel and the rear wheel reach critical grip force at the same time.

In terms of roll angle and roll velocity, Fig. 8-23 clearly shows the motorcycle first leaning over to the left ($\varphi < 0$) and then to the right ($\varphi > 0$), to travel through the first turn, and then back to the left to complete the second turn. Note that the roll maneuver begins right away when the motorcycle is still far from the first turn. Anticipating the maneuver reduces the lateral forces on the tires to some extent and slightly decreases the turning radius of the first turn.

The change in yaw angle and yaw velocity over time shows, especially early on, how complex a maneuver is required to be to start from an initial state of perfect, steady, rectilinear motion and get ready for the next maneuver. This transition occurs in an extremely short space of time, in order to leave as much time as possible for the rest of the maneuver (which is the most important segment).

8.7.3 An example of an optimal maneuver for a "U" trajectory

As a second example of applying the optimal maneuver principle, we will look at a "U" trajectory with the same constraints and optimization criteria as above (Fig. 8-24).

The "optimal" trajectory solution gives relatively little weight to the distance to be kept from the edges of the track. As a result, the motorcycle tends to travel on a trajectory which minimizes the curvature (and/or keeps the curvature constant for as long as possible) using the entire road width. In fact, the motorcycle moves away from the centerline to make a smoother transition where the track turns sharply at the beginning and end of the arc.

The speed and longitudinal thrust graph (Fig. 8-25) shows the motorcycle braking hard before the turn and reaccelerating out of it. As in the first example, the vertical force curves illustrate the load transfer phenomenon caused by initial braking and subsequent acceleration.

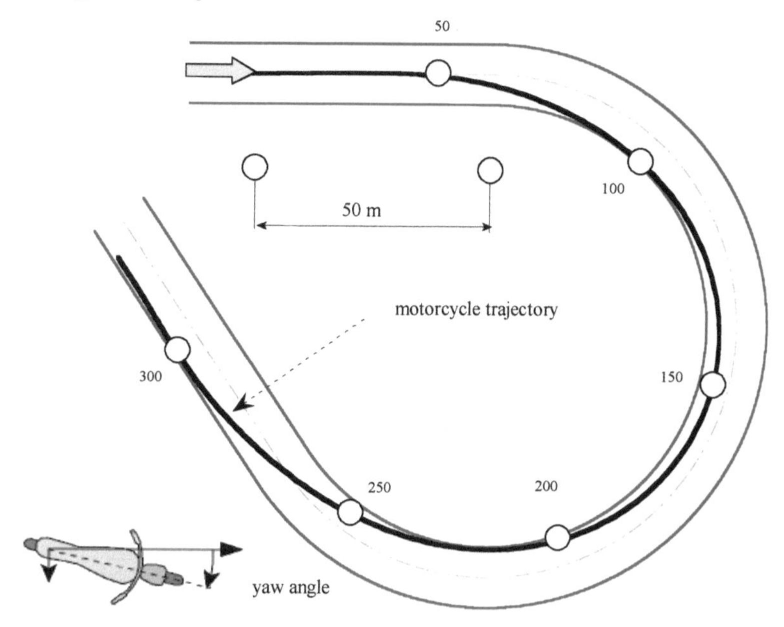

Fig. 8-24 "U" trajectory.

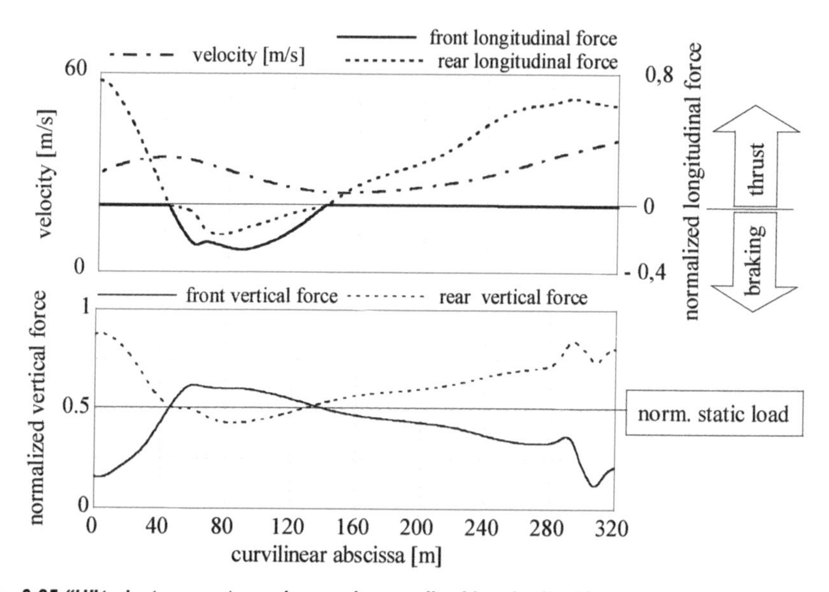

Fig. 8-25 *"U"* trajectory: motorcycle speed, normalized longitudinal force (thrust or braking) and vertical forces with respect to the weight force.

The inclination of the resulting reaction forces (Fig. 8-26) again shows the motorcycle traveling through the U-bend in the track with a nearly steady, circular motion. Here as well, the first part of the graph shows an initial maneuver towards the outside of the turn.

The roll angle graph in Fig. 8-26 shows that the "optimal" solution calls for the motorcycle to lean away from direction of the turn in the first part of the maneuver (*"entry phase"*). The motorcycle then leans over in the direction of the turn and holds steady for almost the entire "U".

The graph of yaw angle and yaw velocity shows how the motorcycle travels through the turn gradually, with a linear increase in yaw angle. It also shows an initial maneuver towards the outside of the turn as part of the "optimal" solution.

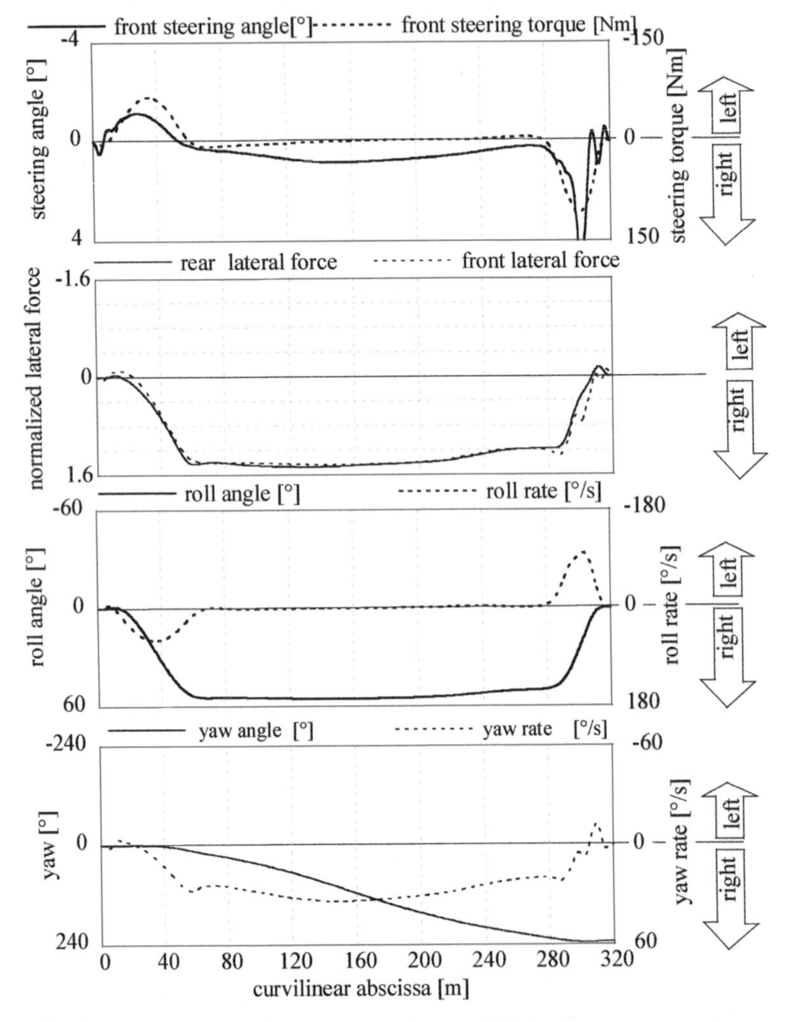

Fig. 8-26 "U" trajectory: steering angle, steering torque, lateral forces, roll angle and roll velocity, yaw angle and yaw velocity.

8.7.4 Influence of the adherence on the trajectory

In this section the effect of different tire adherence is studied considering a U-turn maneuver. The other maneuvers are not reported because they show similar conclusions.

Figure 8-27 is very interesting as it shows the different paths followed by a vehicle as the limit of adherence decreases. As is shown, with reduced adherence the velocity at which the curve is taken decreases. In this case, it makes more sense to drive straight during braking and acceleration phases yielding a path with a sharp curve.

If the adherence becomes smaller than 0.5, the trajectory can no longer touch the inner border. In fact, when adherence is 0.31 the motorcycle goes straight on and touches the external lane border at very low speed.

Figure 8-28 shows that during the approach with high adherence the motorcycle accelerates while with low adherence the motorcycle begins to brake immediately. Figure 8-29 shows that decreasing the adherence decreases the time during which the motorcycle is tilted.

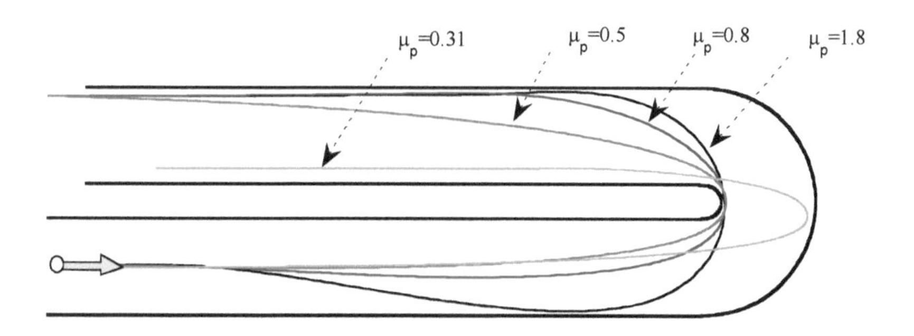

Fig. 8-27 Trajectory comparison carried out with different tire adherence.

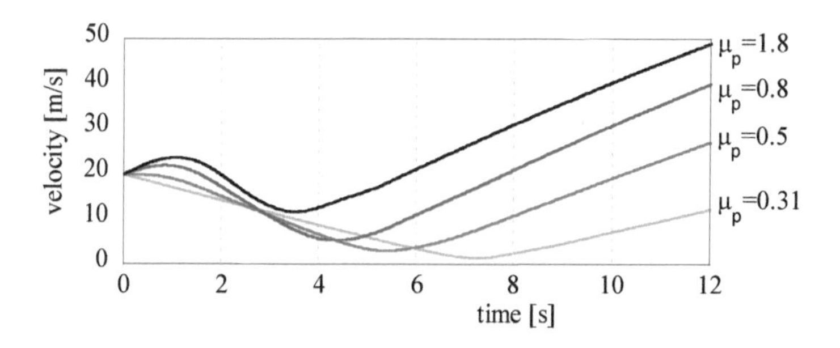

Fig. 8-28 Velocity comparison carried out with different tire adherence.

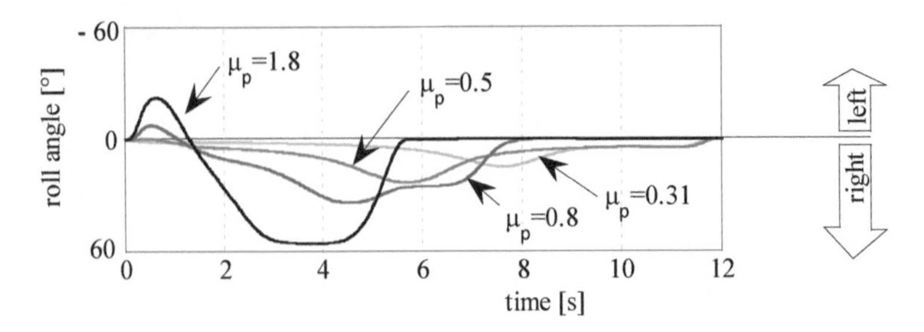

Fig. 8-29 Roll angle comparison carried out with different tire adherence.

8.8 Handling tests

In practice defining the motorcycle's handling quality is not an easy task because it constitutes an overall characteristic determined by different components of the vehicle (engine, brakes, aerodynamics, frame, tires). Moreover, there is a strong subjective involvement in the use and rating of the motorcycle on behalf of the driver, according to the driving style and sensitivity.

Handling is usually associated with the vehicle's response to the control action. First, a prompt response to the control action is required in terms of lateral acceleration and yaw rate. This property however must not decrease the stability and the capability of damping the oscillations that might arise during certain maneuvers. A low sideslip of the motorcycle is also required. A low sensitivity to external disturbances is needed, as well as a uniform response to the control action at different speeds, different tires and road surfaces. Finally, constant feedback between the vehicle and the driver is required, so that the driver is continually aware of the vehicle's dynamic state.

Experimental and simulation tests supply useful data for the comprehension of the actual dynamic behavior of the vehicle. The motorcycle is assumed to be a system with some control inputs (steering angle or torque, forward velocity) and some kinematic and dynamic outputs. The behavior of the motorcycle is thus described by the function that links inputs to outputs in performing typical maneuvers, such as slalom, transient and steady turning, lane change, obstacle avoidance and so on.

The need for different kinds of tests on the motorcycle is a consequence of the fact that it is not possible to divide a generic trajectory into a simple sequence of curves and straights, because the commands given by the driver to perform a maneuver also depend on the previous ones. For example, the commands given (through the steering and the throttle) during a curve of a slalom test are completely different from those given during a steady turning maneuver performed with the same speed and turning radius. In addition, the way the motorcycle is driven along a known path is different from the way it is driven performing the same path for the first time, additionally a premeditated maneuver is different from an emergency one even though the result is the same (for example, the avoidance of a known vs. an unexpected obstacle). As a consequence, different tests have to be planned to study motorcycle behavior in transient and steady state conditions.

8.8.1 Steady turning test

The steady state turning test has proven to be an efficient and quantitative way to assess low frequency and non-transient handling properties of motorcycles and other two-wheelers. Input and output quantities are indeed constant and the steady state vehicle response ratios and gains can be measured in a repeatable and useful way.

The quantities describing the driver's control action are the steering torque and the driver lean angle, since both trajectory (i.e. radius of curvature) and forward velocity are given. In most cases driver control mainly consists of the steering torque, whereas lean angle and body lateral displacement can be considered as secondary control inputs (Fig. 8-30).

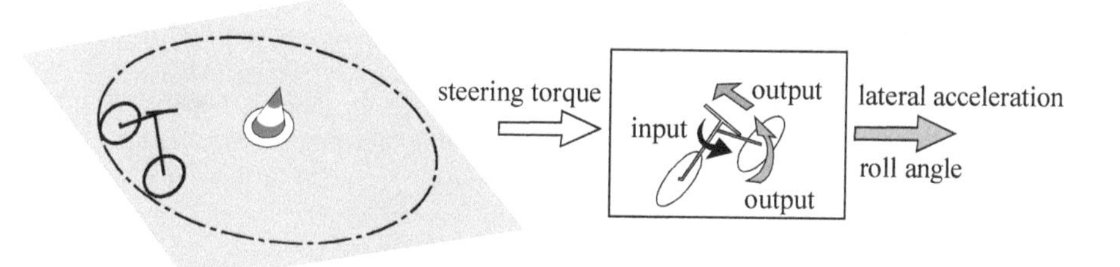

Fig. 8-30 Steady turning test.

Vehicle maneuverability and steering behavior can be quantitatively investigated in steady turning tests.

In general, driver handling feeling is related to the steering effort needed to perform a certain maneuver. In steady turning such feeling is related to the steering torque necessary to follow the required path with the given forward velocity. For a good feeling, little torque should be applied to the handlebar and preferably it should be negative (i.e. away from the curve).

The relation between driver action and vehicle response can be quantified by the ratio between the steering torque and roll angle:

$$\text{roll index} = \tau / \varphi$$

or between the steering torque and lateral acceleration:

$$\text{acceleration index} = \frac{\tau}{V^2 / R_c} \cong \frac{\tau}{g \cdot \tan\varphi}$$

Figure 8-31 shows experimental results in terms of the acceleration index as a function of forward speed for a sport motorcycle. The acceleration index is mainly negative (i.e. negative steering torque, away from the curve). Characteristically for a given radius it transitions from negative to positive (i.e. positive steering torque, towards the curve) as speed increases.

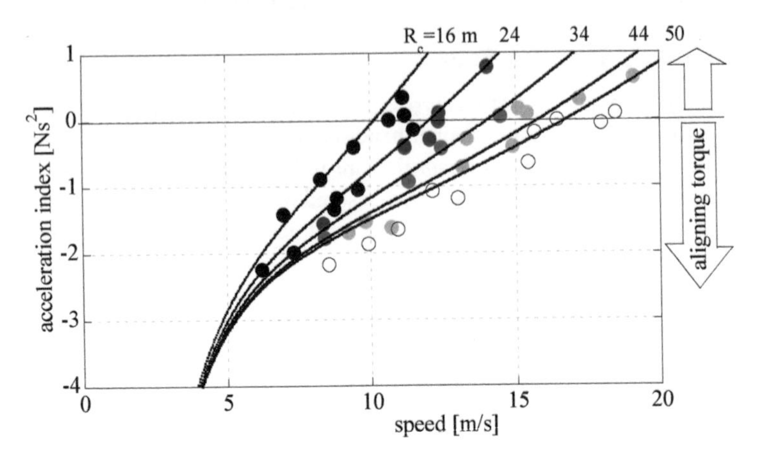

Fig. 8-31 Acceleration index versus velocity for several turning radii.

Negative applied steering torque is preferable because in this situation the motorcycle's turning behavior tends to be stable (Fig. 8-32). In fact, at a sufficient velocity, if the control of the driver is suddenly removed, the motorcycle after some lateral oscillations tends to follow a straight path without capsizing (i.e. the capsize mode is stable). On the contrary, with positive torque, if the control suddenly stopped applying torque the steering angle would decrease and the motorcycle capsizes (i.e. the capsize mode is unstable).

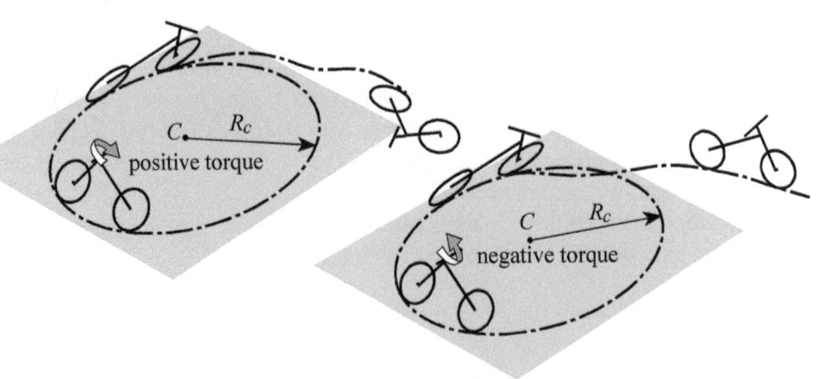

Fig. 8-32 Positive and negative steering torque.

With regard to motorcycle steering behavior, the ratio between the actual turning radius R_c and the ideal turning radius R_{c0} (i.e. associated with the ideal tire behavior) is considered:

$$\xi = \frac{R_{c0}}{R_c} \simeq \frac{\Delta + \lambda_r - \lambda_f}{\Delta} \simeq \left(\frac{p\cos\varphi}{\delta\cos\varepsilon}\right)\left(\frac{\Omega}{V}\right)$$

In particular:

$\xi < 1$ **under-steering** : the actual cornering radius is greater than the ideal one and the motorcycle tends to run on a larger trajectory (i.e. the front sideslip is greater than the rear);

$\xi = 1$ **neutral steering** : the actual cornering radius is equal to the ideal one and the motorcycle follows the kinematic trajectory (i.e. the rear sideslip is almost the same as the front);

$\xi > 1$ **over-steering** : the actual cornering radius is smaller than the ideal one and the motorcycle tends to run on a smaller trajectory (i.e. the rear sideslip is greater than front);

$\xi = \infty$ **critical condition** : the vehicle turns even if the steer angle is null (i.e. $\Delta = 0$): this corresponds to critical speed;

$\xi < 0$ **counter-steering** : in this condition the handlebar must be steered away from the curve (i.e. the rear sideslip is much greater than front and negative steering angle must be adopted to compensate).

Correlations with expert test riders' subjective opinions have shown that the best ratings occur for vehicles with neutral or modest over-steering properties (this trend is in sharp contrast to typical results for automobiles, where small amounts of under-steer are universally preferred).

Let us consider an under-steering motorcycle: since the vehicle tends to expand the curve, the rider, to correct the trajectory, is obliged to increase the roll angle which in turn increases the steering angle (in order to increase the lateral reaction force of the front wheel). When the rotation of the handlebar becomes considerable, the reaction force needed can exceed the friction limit between the front tire and the road surface, with the result that the wheel slides and the rider falls. A motorcycle that is under-steering is therefore dangerous, since vehicle control is very difficult after the front wheel has lost adherence. On the other hand, with an over-steering motorcycle, in cases where the needed reaction force overcomes the maximum friction force between the rear tire and the road plane, the rear wheel slips, but an expert rider, through a counter-steering action, has a better chance of controlling the vehicle's equilibrium and avoiding a fall.

Figure 8-33 shows the steering ratio as a function of velocity for different turning radii from 10 to 50 m. It can be observed that the sport motorcycle has a well-defined over-steering behavior which becomes more and more marked as speed increases. Anyway critical speed is not reached and no counter-steering is evident. Steering ratio fitting has been performed by the simplified expression

$$\xi = \frac{1}{1 - \gamma V^2}$$

where the constant γ depends on the cornering and camber stiffness of the tires. Critical speed is not reached experimentally but is extrapolated by linear fitting ($\simeq 20$ m/s).

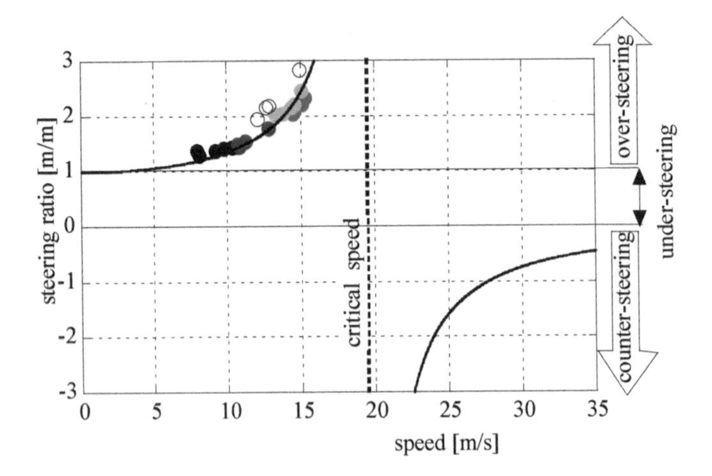

Fig. 8-33 Steering ratio versus velocity for several turning radii.

Figure 8.34 shows experimental tests of the sport motorcycle considered above in a speed – lateral acceleration diagram. Zero steering torque ($\tau = 0$), zero steering torque gradient ($\partial\tau/\partial A = 0$) and zero steering angle ($\xi = \infty$) lines are plotted as forward speed and lateral acceleration vary and different over-steering and counter-steering zones can be identified.

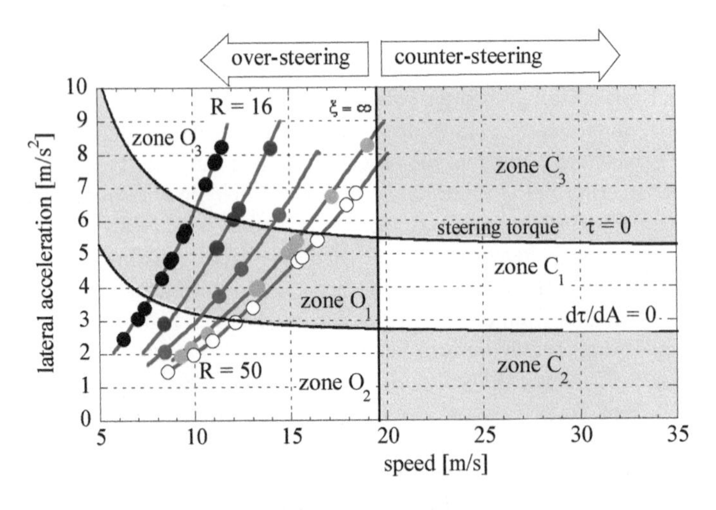

Fig. 8-34 Driving zones.

With regard to over-steering, the O_1 zone is characterized by negative steering torque, positive steering torque gradient and positive steering ratio. Based on the considerations of the previous sections, these conditions are correlated to good handling: the capsize mode is in fact stable, the steering torque decreases approaching zero as lateral acceleration increases (thus requiring lighter rider effort as roll angle increases), and favorable over-steering behavior is achieved. It follows that these combinations of speed and lateral acceleration can be considered a "preferable driving zone".

The O_2 zone is similar to the previous with the exception that the steering torque gradient is negative. This means that steering torque becomes more negative (approaching its relative minima) as lateral acceleration increases, thus requiring greater rider effort as roll angle increases.

The O_3 zone is also similar to the O_1 but the steering torque is positive. This means the capsize mode is unstable, thus requiring even more roll stabilization to be achieved by the rider. Even if this constitutes an inappreciable fraction of the global control being exerted, it effectively makes driving even more difficult and unsafe. Furthermore in this zone the positive steering torque becomes unfavorably larger as the lateral acceleration increases, thus requiring greater rider effort to perform vehicle control at high roll angles.

These three over-steering zones are fully accomplished experimentally.

The C_1 , C_2 , C_3 zones are similar respectively to the O_1, O_2 and O_3 over-steering zones, with the exception that steering ratio is negative. That is counter-steering angle behavior is achieved (i.e. steering angle not in accord with turning direction), which may require a certain amount of experience and skill to be practiced safely and may or may not be perçeptible to the rider.

These three counter-steering zones are not accomplished experimentally.

8.8.2 "U" turn test

The characteristics of a motorcycle's handling are not defined solely on the basis of the value of the torque to be applied under conditions of movement on a stationary turn, but also on the basis of other parameters, such as the torque needed to lean the vehicle to a set angle from the vertical, and the time used to reach the desired angle.

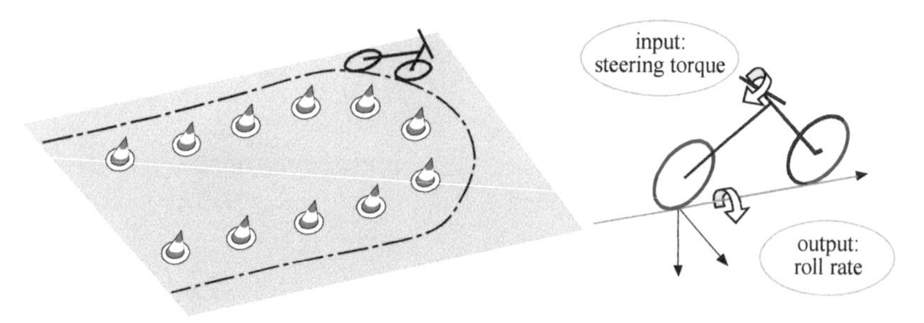

Fig. 8-35 "U" turn test.

J. Koch (1978), after experimental tests in "U" turns, proposed the following index to evaluate the vehicle's capacity to enter a turn:

$$\text{Koch index} = \frac{\tau_{peak}}{V \cdot \dot{\varphi}_{peak}} \quad \left[\frac{N}{rad/s^2}\right]$$

Where τ_{peak} is the peak value of steering torque, $\dot{\varphi}_{peak}$ is the peak value of roll velocity, and V is the forward velocity.

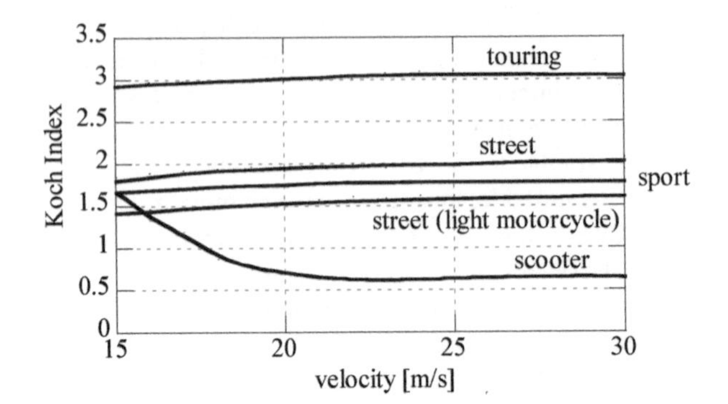

Fig. 8-36 Koch index for different motorcycles versus velocity.

As the forward velocity increases the Koch index tends towards a limit value which depends on the kind of motorcycle and on the turn radius. It is worth noting that all the peak values are reached in transient phases and not in steady-state conditions.

A low value of Koch index highlights that with a high forward velocity there is a high roll speed with a low peak in steering torque; these are the characteristics of motorcycles with good handling.

The transient behavior when entering a turn is mainly influenced by center of mass height, front wheel inertia, front frame inertia with respect to steering axis, frame inertia with respect to rolling axis and yaw axis. Figure 8-36 shows Koch index when entering a "U" turn having a radius equal to 100 m for different types of motorcycles. The figure shows that while increasing the velocity the Koch tends to a limit. The index highlights and quantifies the fact that it is easier to maneuver with a light scooter instead of a heavy touring motorcycle.

8.8.3 Slalom test

In steady-slaloming conditions the rider controls the motorcycle through a periodic action on the steering system and the vehicle reacts with periodic roll, yaw and lateral motion. The slalom frequency is:

$$\nu = \frac{V}{2P}$$

where P is the spacing of the cones.

Therefore the slalom test constitutes a study of the forced response of the system, where the forcing excitation is represented by the steering torque and/or the steering angle. The motorcycle's response changes both in amplitude and phase as a function of speed and slaloming frequency.

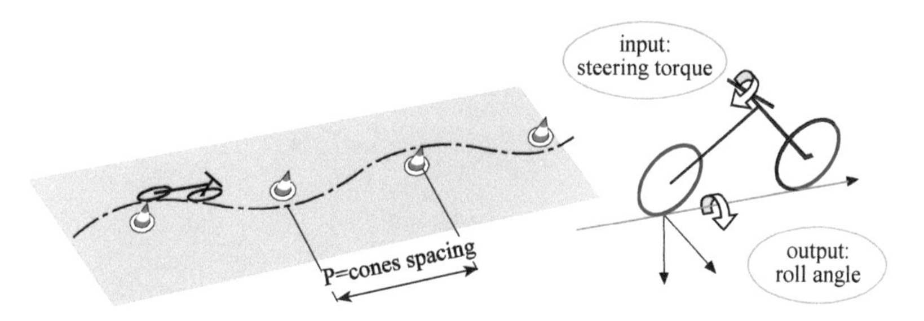

Fig. 8-37 Slalom test.

Figure 8-38 shows a sample of the acquired signals' time histories for a 14 m spacing slalom test performed at low speed. The steering torque is tendentially opposite to the path curvature.

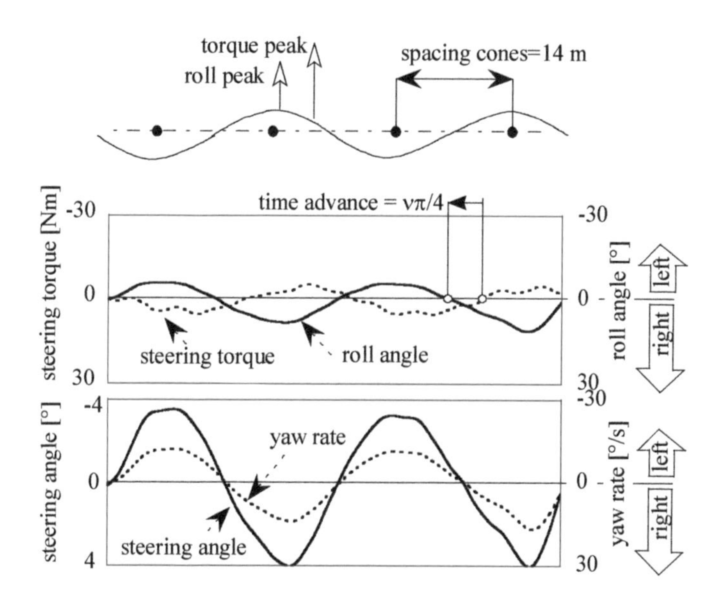

Fig. 8-38 Slalom test at low velocity: V =4.8 m/s.

Figure 8-39 shows the same signals' time histories for a slalom test performed at medium speed. At this particular frequency the steering torque is 180° out of phase with respect to the roll angle. The steering torque amplitude is almost unchanged if compared to the previous plot.

Figure 8-40 shows the slaloming behavior at high speed. It can be observed that the phase between steering torque and roll angle shifts to 90°. The steering torque amplitude is much increased.

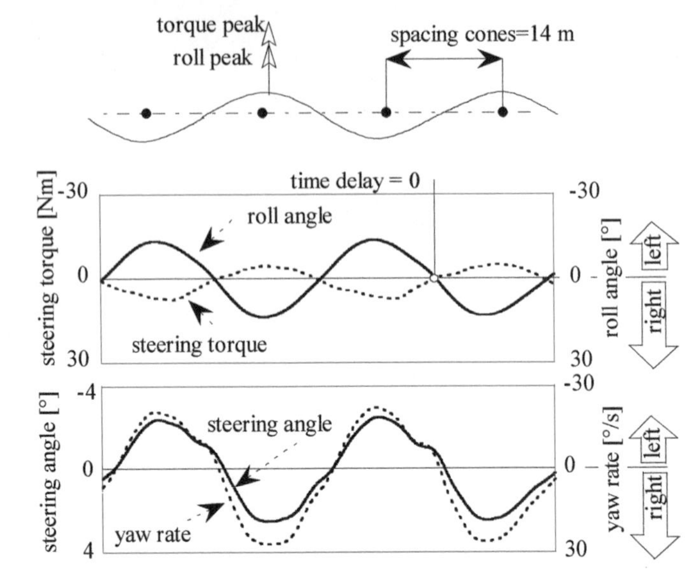

Fig. 8-39 Slalom test at middle velocity: V =7.2 m/s.

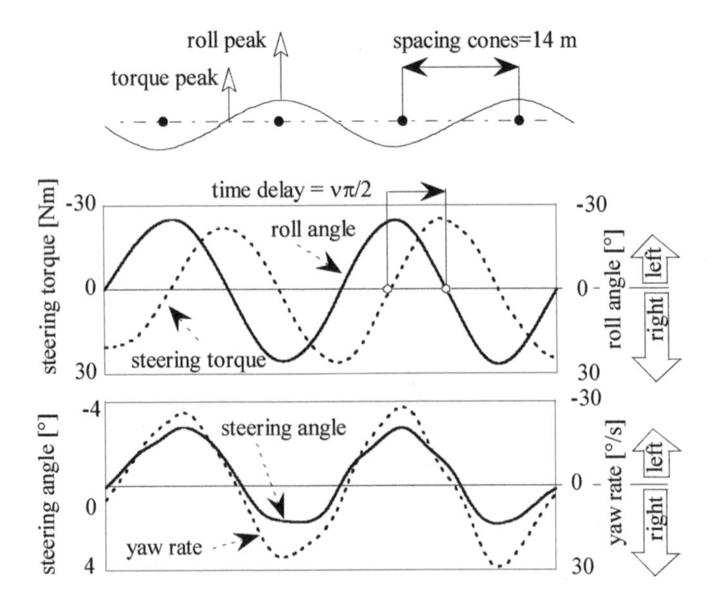

Fig 8-40 Slalom test at high velocity: V = 15.2 m/s.

The most suitable mathematical tool to interpret the results of a slalom test is the transfer function, which makes it possible to describe the system response (roll angle) in relation to the input characteristic (steering torque) as a function of the frequency:

$$\text{roll transfer function} = \frac{\varphi}{\tau}(v)$$

Both the amplitude and the phase of roll transfer function are very interesting. High ratio between roll angle $|\varphi|$ and steering torque $|\tau|$ means that a large motorcycle roll motion is obtained with little steering effort, while a large phase means that the roll angle follows the steering torque with a time lag.

Increasing the frequency, that is increasing the speed if the cone spacing is fixed, the motorcycles need an increased steering effort to follow the slalom path. However, the driver feeling is determined by the phase lag between roll angle and steering torque rather than the maximum steering torque: better handling is associated with motorcycles having a quick response to the steering input. Fig. 8-41 shows an experimental transfer function. The magnitude exhibits a maximum value at about 8 m/s then increases with frequency. The maximum in the magnitude of the roll transfer function corresponds to the minimum effort on the handlebar; at this velocity the weave mode switches from the instability to the stability zone.

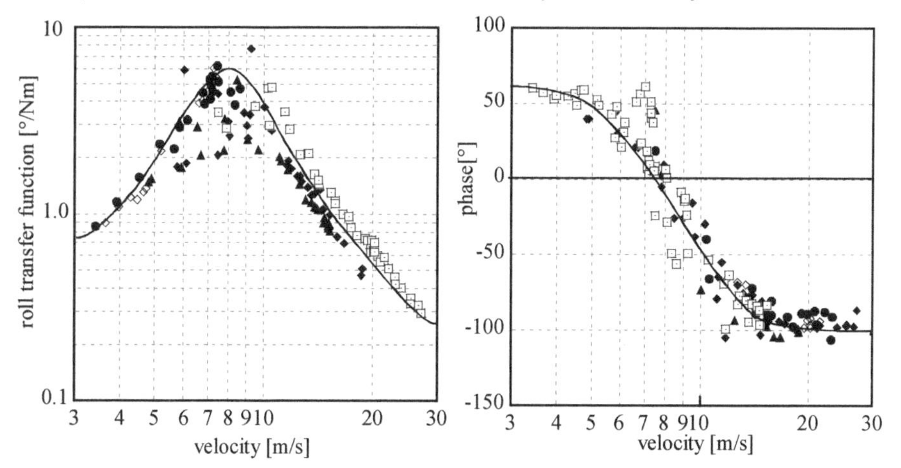

Fig. 8-41 Frequency response function.

8.8.4 Lane change test

The lane change maneuver represents a typical transient maneuver, and strongly depends on driver skill and riding style. Various driving strategies among riders differ from each other depending on the initial counter-steer carried out and on the movement of the rider's body with respect the motorcycle.

Expert riders carry out this maneuver with a high initial *out-tracking* and use their body inclination to remain vertical or even to generate an additional input with respect to steering torque.

In very fast maneuvers, in reality riders tend to apply not only steering torques (i.e. torques parallel to the steering axis), but also rolling torques (i.e. perpendicular to steering axis) to the steering system.

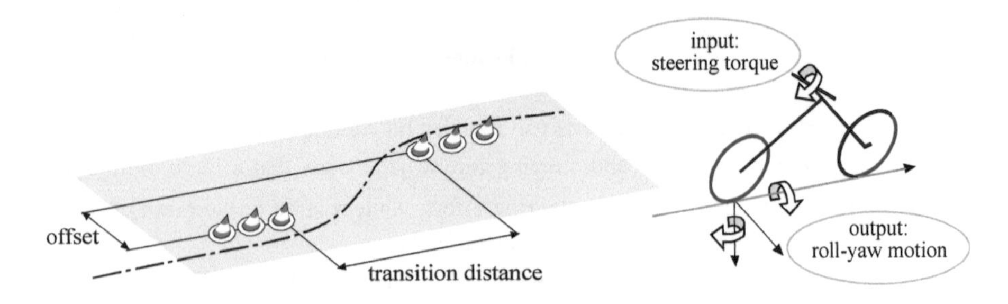

Fig. 8-42 Lane change test.

The phase between yaw velocity and steering torque seems to be the quantity more highly perceived by the rider when carrying out such a maneuver.

Basically the rider imparts some control action, torque, causing the vehicle to roll and yaw. The ratio of the peak-to-peak magnitude of steering torque to the peak-to-peak roll rate is a good indicator of a motorcycle's maneuverability. Normalizing this quantity by velocity we obtain the Lane Change Roll Index:

$$\textit{Lane Change Roll Index} = \frac{\tau_{p-p}}{\dot{\varphi}_{p-p} \cdot V_{avg}}$$

where the subscript p-p indicates peak-to-peak values.

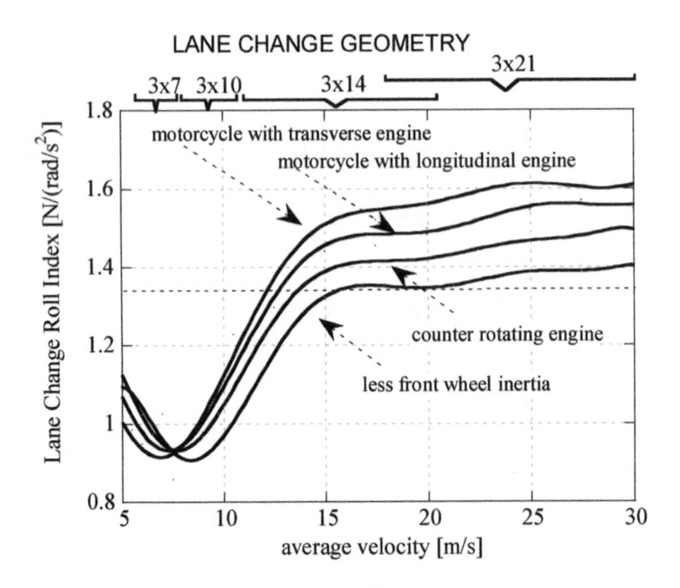

Fig. 8-43 Parameter effects on Lane Change Roll Index for different geometries and speed.

This index represents the effort required of the rider in the form of steering torque to obtain a desired vehicle response in roll rate. Figure 8-43 shows the results of numerical simulation varying critical design parameters.

We can see that by orienting the engine in a counter-rotating direction or by reducing the front wheel spin inertia the value of LC Roll index is decreased. The lower values mean that less effort is needed to perform the lane change, in this case due to the lesser gyroscopic effects.

Also observe the asymptotic nature of the LC Roll index with speed. At low speeds all four vehicle configurations are similar but as speed increases we approach a region where the gyroscopic effects become dominant. In this region the differences between the configurations are obvious. In a similar manner the LC Roll index can be used to contrast the behavior of different classes of motorcycles: touring, sport, cruiser, etc. Typically a scooter will exist at or below 1 while a touring motorcycle can reach or exceed 2.5 N/(rad/s^2).

8.8.5 Obstacle avoidance test

A typical maneuver which causes high roll and yaw speeds is the obstacle avoidance test. In such a maneuver the gyroscopic effect of the front wheel combined with the roll motion has a fundamental role in determining the steering torque that has to be applied by the driver.

The gyroscopic effect causes a torque around steering axis whose value is:

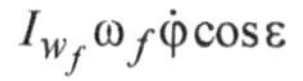

$$I_{w_f} \omega_f \dot{\varphi} \cos \varepsilon$$

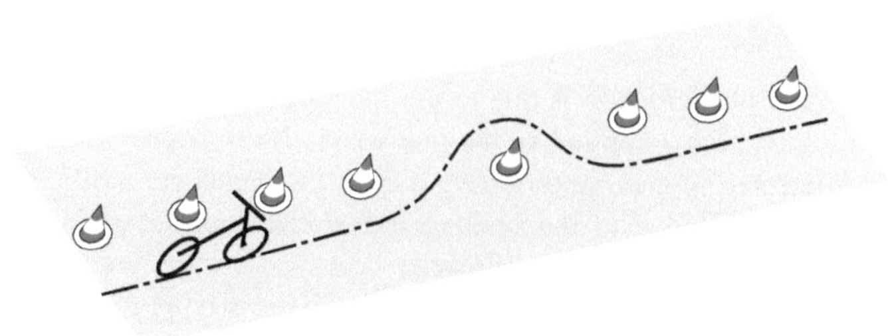

Fig. 8-44 Obstacle avoidance test.

In the previous sections we have highlighted that the total steering torque is determined by a lot of factors acting on the front frame, particularly by vertical and lateral forces. Nevertheless in rapid maneuvers almost the whole effort applied by the driver opposes the gyroscopic moment caused by front wheel and roll motion.

This can be easily seen in Fig. 8-45, where the steering torque applied by the driver is compared with the gyroscopic effect calculated by means of the previous formula.

First of all we observe the out-tracking technique; the driver initially applies a high steering effort to the right to make the vehicle roll rapidly towards the left. After this, there is a steering torque peak to the left, corresponding to the necessary rotation of the steering system, and finally a positive peak corresponding to the final line up of the steering system itself.

The motorcycle handling sensation in such a maneuver is represented by the time lag between the steering torque and the yaw velocity; the shorter this time lag is, the better vehicle handling is.

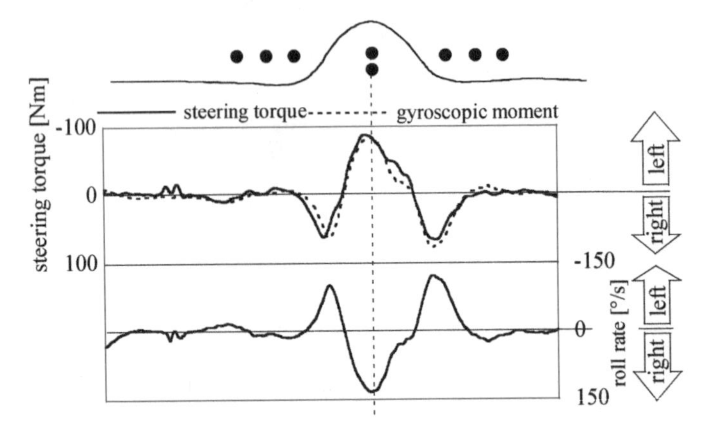

Fig. 8-45 Steering torque applied in a obstacle avoidance test.

8.9 Dangerous dynamic phenomena

8.9.1 High side

This dangerous phenomenon is due to the interaction between the sideslip force with the longitudinal force applied on the rear wheel. It can happen during a braking maneuver while entering a curve or during a thrusting maneuver while exiting from a curve. The high side due to the braking has already been explained in the 2nd chapter with reference to the friction ellipse.

Figure 8-46 shows how the "high-side" fall due to the driving force comes about.

To exit from the curve the rider starts to thrust the rear wheel, therefore the longitudinal driving force increases as does the total friction force (phase 1). The total friction force reaches the limit value, the rear wheel loses grip and therefore the rear of the motorcycle moves outwards (phase 2), The rider stops accelerating, reducing the thrusting force suddenly, and the rear wheel takes grip again (phase 4), The large sideslip, which is still present, generates a lateral force impulse that is not balanced. The result is that the motorcycle is violently twisted and thrown upwards (phase 5).

Tire behavior during a "high-side" may be better understood by looking at Fig. 8-47, which shows the available lateral tire force when a longitudinal driving tire force is present. In both diagrams, the envelope of the families of curves is the friction ellipse. The initial condition is represented by point A in which a lateral force is present and the sideslip angle is about 1.5°. When the driver starts to accelerate the motorcycle, the point moves in the horizontal direction and the sideslip angle increases in order to keep the lateral force constant in the presence of an increasing longitudinal force.

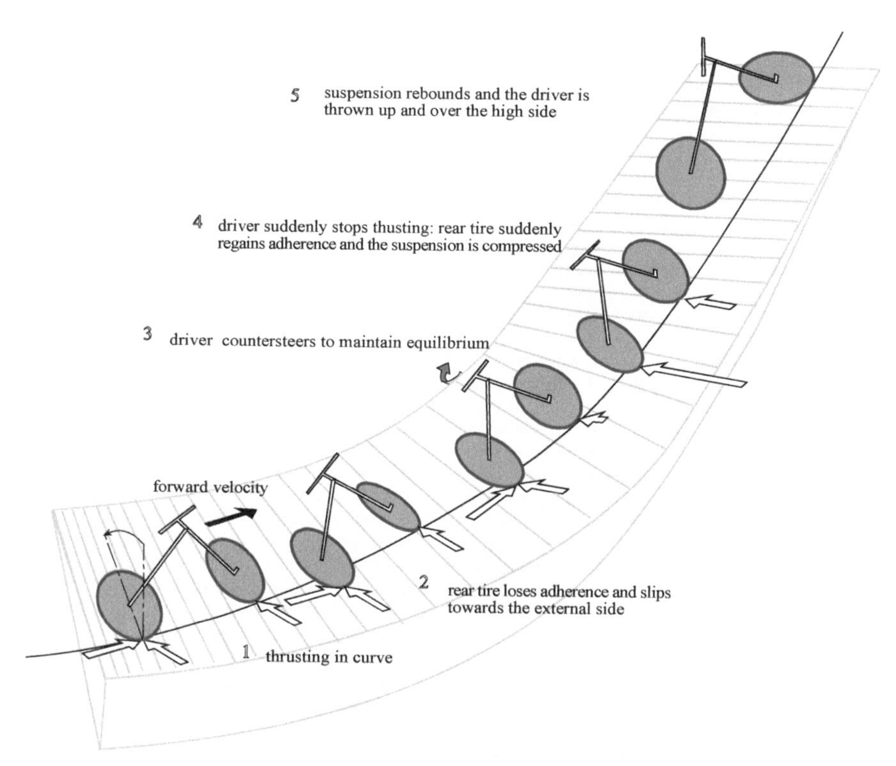

Fig. 8-46 Example of phenomenon known as "high side" due to thrusting maneuver.

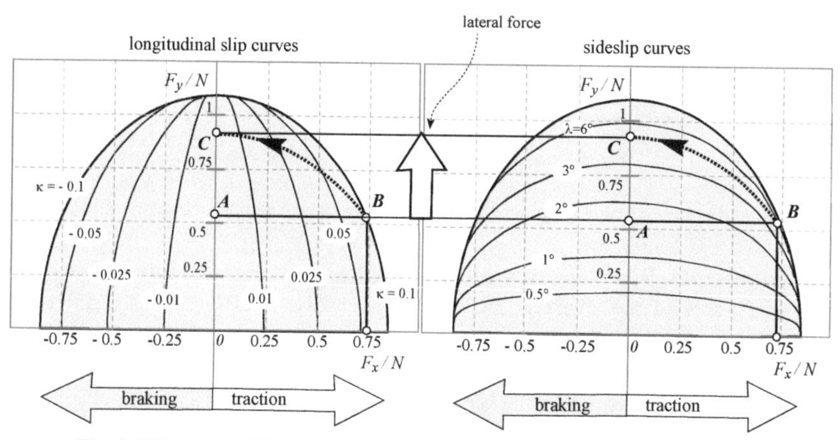

Fig. 8-47 Lateral and longitudinal forces for various values of longitudinal slip κ and sideslip λ.

The loss of grip takes place at point B when the boundary of the friction ellipse is reached; a large sideslip angle (of about 5°) is present. When the rider releases the accelerator the rear wheel takes grip again.

The new condition is represented by point C, where there is still a large sideslip angle but where the longitudinal slip is negligible.

Therefore, the lateral force impulse takes place because the lateral force increases suddenly from the value of point B to the value of point C. The impulse torque, produced by the lateral force, is not balanced, consequently the motorcycle falls.

Fig. 8-48 Example of simulation of the phenomenon known as "high side".

8.9.2 Kick back

The so called "kick back" effect is a phenomenon that concerns motorcycle stability.

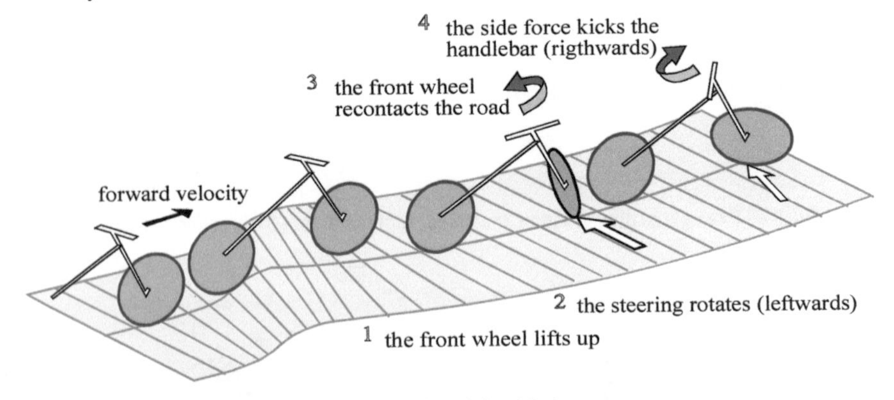

Fig. 8-49 Example of "kick back" phenomena.

Road undulations or transverse joints on the road surface during high speed riding (150-200 km/h) can unload the front wheel causing it to lift from the road surface (phase 1). When the front wheel is unloaded the front assembly is not in equilibrium around the steering axis. The rider moves the handlebar instinctively and the front wheel plane moves out of line with respect to the forward direction of the motorcycle (phase 2).

When the front wheel makes contact with the road surface again (phase 3) the front frame is not in equilibrium with respect to the steering axis. Due to the steering angle a large lateral force is generated. This side force kicks back the handlebar in the opposite direction with respect to the steering angle (phase 4). Consequently the rider can lose control of the motorcycle with dramatic effects.

The kick back effect decreases when using front and rear tires with lower cornering stiffness. Frames with high structural stiffness make the motorcycle's behavior worse.

8.9.3 Chattering

The chatter of motorcycles appears during braking and consists of a vibration of the rear and front unsprung masses at a frequency in the range 17-22 Hz depending on the motorcycles. It appears nearly exclusively in the racing motorcycles and only in some tracks and in some kind of maneuvers. This vibration can be very strong and the unsprung masses acceleration can reach 5 g. It is often observed that the wheel rotation frequency is close to the chatter frequency, which suggests a tire non-uniformity; force variation or run-out, or maybe imbalance. This would normally occur in the mid-corner.

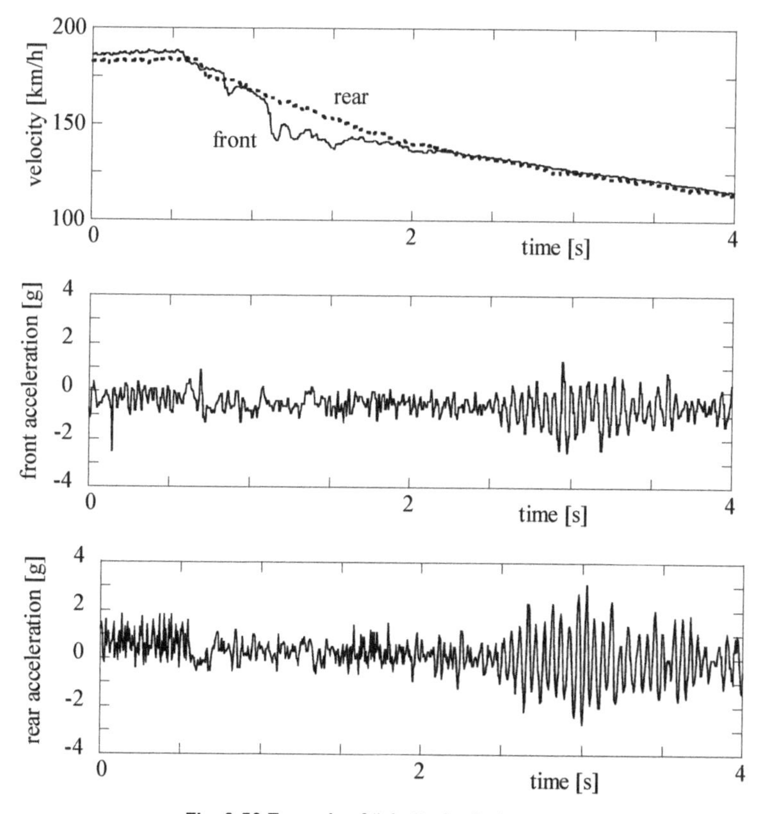

Fig. 8-50 Example of "chattering" phenomenon.

The chatter is an auto-excited vibration and this fact explains why it appears suddenly when the mechanism of auto-excitation is generated.

The suspensions shock absorbers are not able to damp these vibrations. In the presence of chattering motorcycle guidance in limit conditions becomes very difficult.

The mechanism of self-excitation of these vibrations is due to the coupling of the rear wheel unsprung mass resonance oscillations with the fluctuation of the longitudinal frictional force in the contact patch of the tire. The chatter vibrations begin on the rear and appear almost instantaneously on the front due to energy transfer from the rear to the front which occurs when the rear and front hop resonance frequencies are close each other.

Figure 8-50 shows a braking maneuver of a racing motorcycle. The figure highlights that the speed of the front wheel is less of the rear wheel speed, during the braking. During the braking the rear and front unsprung masses begin to oscillate at a frequency of about 20 Hz. The oscillations decrease decreasing the braking rate and disappear completely during the following acceleration phase.

8.9.4 Bounce and weave coupling in cornering

The coupling between the out-of-plane mode "weave" and the in-plane mode "bounce" has been previously analyzed. As we have seen the two modes have similar values for their frequencies and similar modal shapes.

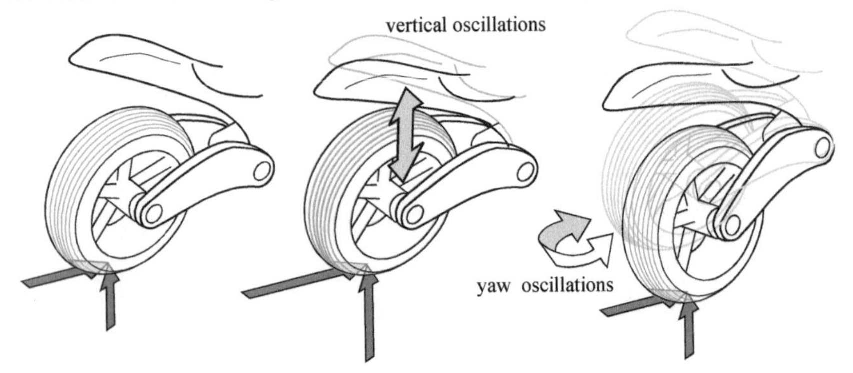

Fig. 8-51 Example of "bounce-weave coupling" phenomena.

This phenomenon is visible in some racing bikes when exiting turns during the acceleration phase. Wide oscillations of the rear frame and movements of the swinging arm are visible. The rider is inclined to slow down to damp out this phenomenon. Fluctuations of the longitudinal slip, depending on the angular position of the swinging arm, can increase the bounce and weave coupling.

8.10 Structural stiffness

Structural stiffness of the motorcycle as a whole and of every single component (in essence front forks, chassis and swingarm) is a key factor in defining the performance with regard to handling and maneuverability of the motorcycle.

Modern motorcycles have frames, swinging arms and forks stiffer than older vehicles. Beyond certain values of lateral and torsional stiffness of the frame, the motorcycle stability properties no longer depend in a significant way on the structural characteristics. High value of stiffness guarantees precision in the trajectory and quick response to the input of the rider but also presents some disadvantages.

For example, vehicles with great frame stiffness are sometimes felt to be nervous by the rider, especially when passing on a transverse bumps and also on wet roads.

Simulation results show that the lateral flexibility of the front fork (or the torsional flexibility of the upper part of the frame near the steering head) stabilizes the wobble mode at high speed and has an opposite effect at low speed, whereas the torsional flexibility of the fork does not appear to have a remarkable influence.

The lateral flexibility of the swinging-arm or of the rear frame slightly stabilizes the weave mode at very high speed whereas the torsional flexibility of the swinging-arm or of the rear frame has a contrary effect.

The motorcycle in steady conditions, both in linear motion and when cornering, is subjected to forces acting in its plane of symmetry whereas in transient conditions it is subjected to lateral forces applied on the wheel contact point with the road and to inertial forces due to lateral accelerations. As an example, when the vehicle is in linear motion and encounters a bump inclined with respect to the surface road, impulsive lateral forces acting on the contact point occur. Consequentially it seems to be appropriate, when measuring the structural stiffness, to apply forces on the contact point between the tire and the surface road.

Lateral forces cause deformations of the vehicle that generate lateral displacements of the wheels and also rotations of the wheel plane. The lateral displacements and the component of the rotation of the wheel around the vertical axis, cause an increase of the tire sideslip and this increases the damping of the structural vibration excited.

If, instead, the deformation of the wheel is a torsion, which could be thought as a torsional rotation about the axis connecting the two contact points with the road, the structural vibration will be less damped since the rotation of the wheel plane does not increase the damping of the vibration excited.

As a result, it is desirable to have moderate lateral flexibility and a high degree of torsional stiffness.

Measurement of the stiffness properties may be carried out using many different approaches.

8.10.1 Structural stiffness of the whole motorcycle

The structure of the motorcycle is excited mainly by the forces generated in the contact patch of the tires. These forces include: the longitudinal force, the vertical load, and the lateral force. In order to highlight the actual behaviour the motorcycle, stiffness should be measured by applying the force in this region, as shown in Fig. 8.52.

Lateral and torsional stiffness, respectively, are expressed by the ratios:

$$K_{rear} = \frac{F}{\Delta l} \qquad Kt_{rear} = \frac{F}{\alpha}$$

The rotation axis is the intersection of the symmetry plane of the rear assembly with the rear wheel plane in the deformed position. If the rotation axis is vertical (angle $\beta = 0$) the deformation is predominantly the flexural type while if the rotation axis is rather horizontal (angle $\beta \approx 90$) the deformation is mainly torsion.

The rotational axis close to the rear contact patch (small value of the arm b) means that there is more plane rotation with respect to lateral deformation. On the contrary the rotational axis far from the rear wheel (large value of the arm b) means more lateral deformation with respect to rotation.

The values for modern motorcycles, without the compliance of the tire, vary in the range of:

- lateral stiffness: $K_{rear} = 0.1\text{-}0.2$ kN/mm
- torsional stiffness: $Kt_{rear} = 1.5\text{-}3.0$ kN/°

With the engine locked instead of the steering head the concepts remain the same. In this case the values of the stiffness are larger with respect to the previous.

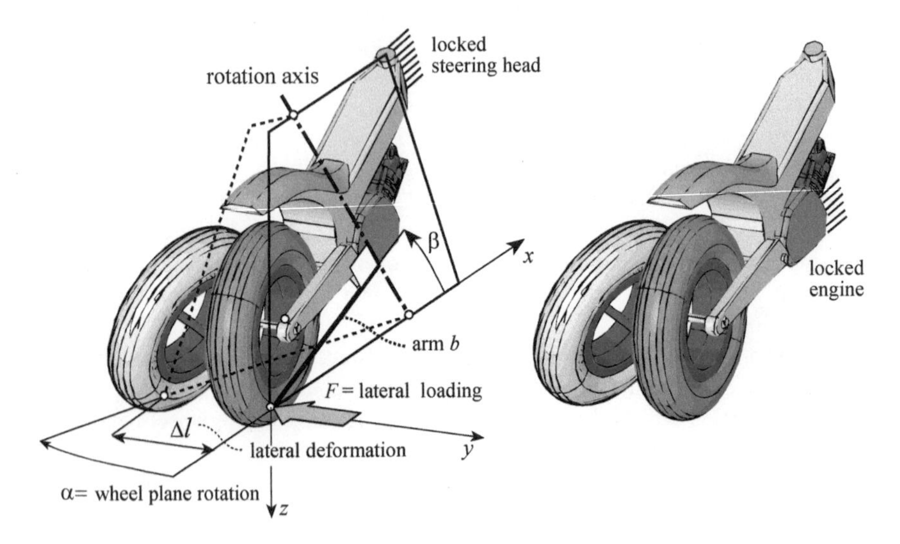

Fig. 8-52 Loading condition for evaluating torsional and lateral stiffness of the rear assembly of the motorcycle.

The stiffness of the front frame can be measured as shown in Fig 8-51. Also in this case there are two possibilities:

- lock the steering head,
- lock the engine.

In the latter case the upper part of the frame contributes to the deformation of the front assembly.

The inclination of the rotational axis is typically tilted by about $\beta \approx 2\varepsilon$ towards front of the motorcycle with respect to a plane perpendicular to the steering axis. Also in this case lateral and torsional stiffness are expressed by the ratios:

$$K_{front} = \frac{F}{\Delta l} \qquad Kt_{front} = \frac{F}{\alpha}$$

The values of modern motorcycle vary in the range:

- lateral stiffness: $K_{front} = 0.08\text{-}0.16$ kN/mm;
- torsional stiffness: $Kt_{front} = 0.7\text{-}1.4$ kN/°.

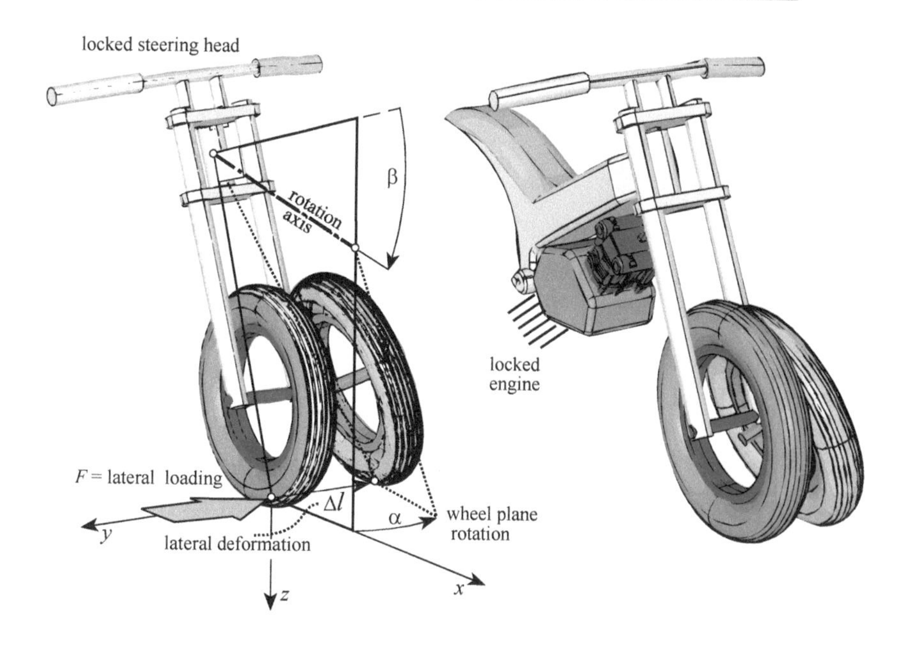

Fig. 8-53 Loading condition for evaluating torsional and lateral stiffness of the front assembly of the motorcycle.

8.10.2 Structural stiffness of the frame

The torsional stiffness of the rear frame member is generally measured with the engine fitted. It is calculated about an axis at a right angle to the steering head and passing through the swinging arm pivot axis and applying a couple (torque) around this axis.

The lateral stiffness can also be represented by the ratio between the force applied along the swinging arm pivot axis and the lateral deformation measured in that direction. The force can be applied with an offset in order to avoid torsional deformation.

Lateral stiffness typically varies depending on the type of frame and on the method of engine attachment.

In some cases the moment is applied on the steering head and the pivot axis of the swinging arm is locked. There are some small differences in the two different measurement procedures due to the asymmetry of the frame.

The values of modern motorcycle (sport 1000 cc.) vary in the range:

- lateral frame stiffness: $K_f = 1 - 3$ kN/mm;
- torsional frame stiffness: $Kt_f = 3 - 7$ kNm/°.
- vertical frame stiffness: $Kz_f = 5 - 10$ kNmm.

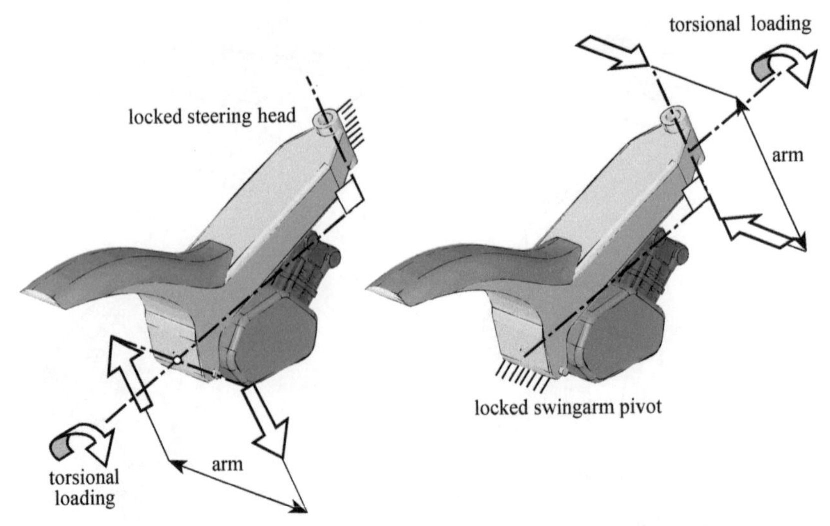

Fig. 8-54 Loading condition for evaluating torsional stiffness of the frame.

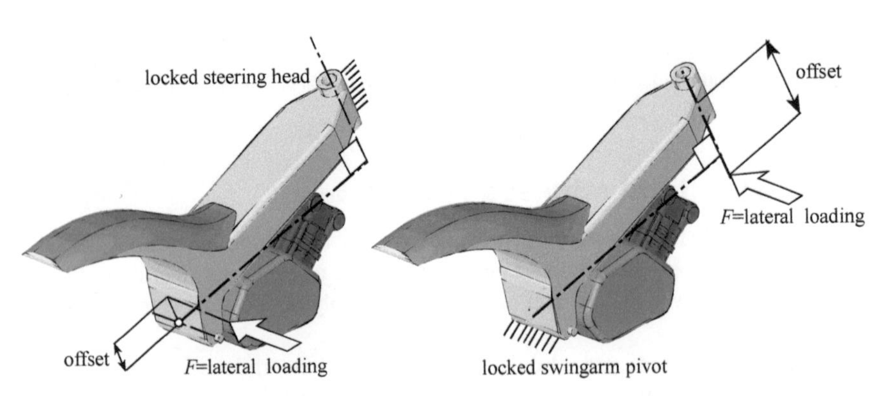

Fig. 8-55 Loading condition for evaluating lateral stiffness of the frame.

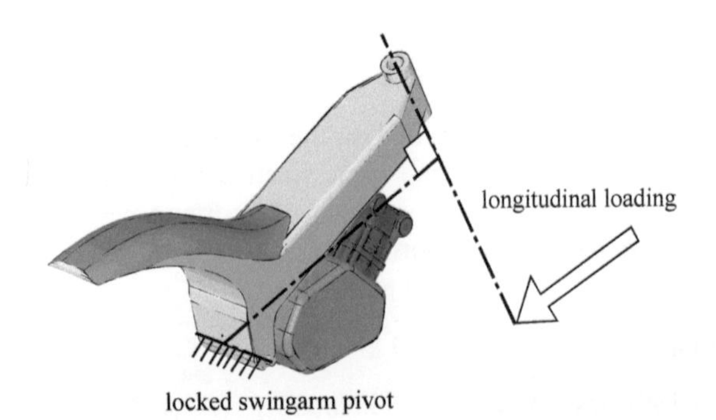

Fig. 8-56 Loading condition for evaluating longitudinal stiffness of the frame.

8.10.3 Structural stiffness of the swingarm

The values of swingarm stiffness are in the range:

- swinging arm lateral stiffness K_s = 0.8-1.6 kN/mm.
- swinging arm torsional stiffness Kt_s = 1-2 kNm/°;

The mono-shock swinging arm is characterized by a greater lateral and a smaller torsional stiffness compared with the classic swinging arm.

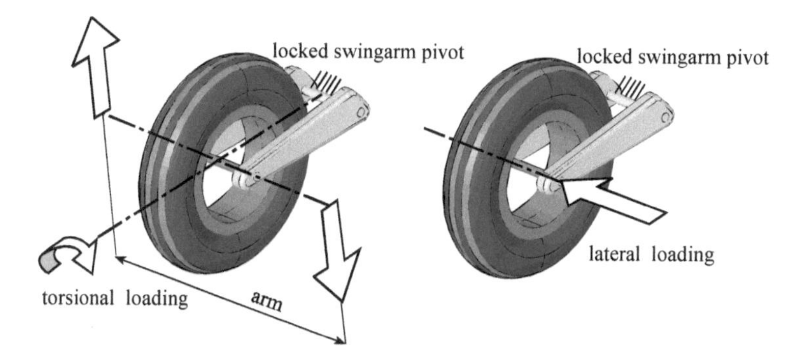

Fig. 8-57 Loading condition for evaluating torsional and lateral stiffness of the swingarm.

8.10.4 Structural stiffness of the front fork

The front fork is the most flexible part of the structural motorcycle.

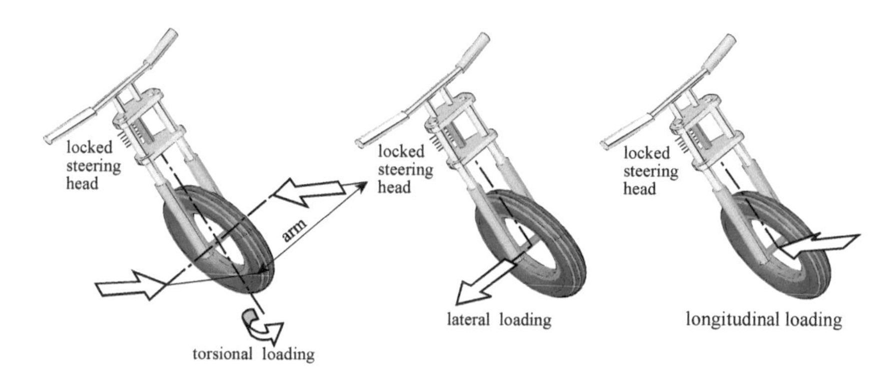

Fig. 8-58 Loading condition for evaluating torsional, lateral and longitudinal stiffness of the fork.

The stiffnesses are in the ranges:

- fork lateral stiffness K_{ff} = 0.07-0.18 kN/mm.
- fork torsional stiffness Kt_{ff} = 0.1-0.3 kNm/°;

Example 6

Calculate the lateral and torsional stiffness of a motorcycle (without the tire compliance) which components have the following values:

- lateral frame stiffness: $K_f = 2.2$ kN/mm;
- torsional frame stiffness: $Kt_f = 6.0$ kNm/°;
- swingarm lateral stiffness: $K_s = 1.4$ kN/mm.;
- swingarm torsional stiffness: $Kt_s = 1.0$ kNm/°;
- wheel lateral stiffness: $K_s = 0.8$ kN/mm.

The geometry of rear assembly is the following:

- frame length: $L_f = 0.85$ m, $\Delta L_f = L_f/3$;
- swingarm length: $L_s = 0.6$ m, $\Delta L_s = L_s/3$;
- wheel radius: $L_w = 0.3$ m, $\Delta L_w = L_w/3$;
- swingarm inclination: $\alpha_s = 8°$;
- frame inclination: $\alpha_f = 30°$;
- torsional axis inclination: $\alpha = -12°$.

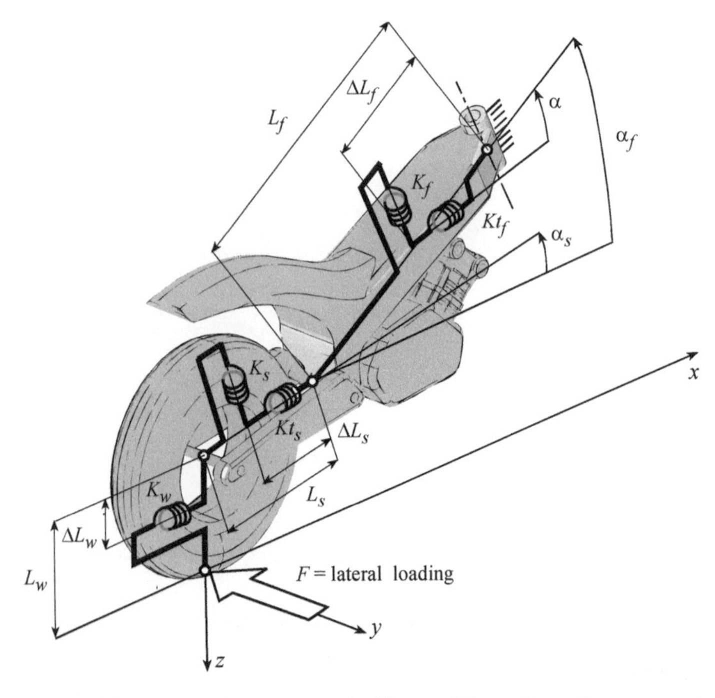

Fig. 8-59 Composition of the structural stiffness of the motorcycle components.

The equivalent lateral stiffness at the contact point is equal to $K_{rear} = 0.16$ kN/mm The rotating axis is inclined, respect to the ground, of an angle equal to 45° and the distance of the rotating axis with respect to the contact patch is equal to 0.76 m. The torsional stiffness split along the x axis and z axis are: $Kt_{rear_x} = 3.6$ kN/° and $Kt_{rear_z} = 4.8$ kN/° respectively. In the experimental test the value should be less because the constraints compliances are not infinitive.

8.11 Experimental modal analysis

Experimental modal analysis was developed in the aeronautical field during the seventies and nowadays it has found an application in two-wheeled vehicles.

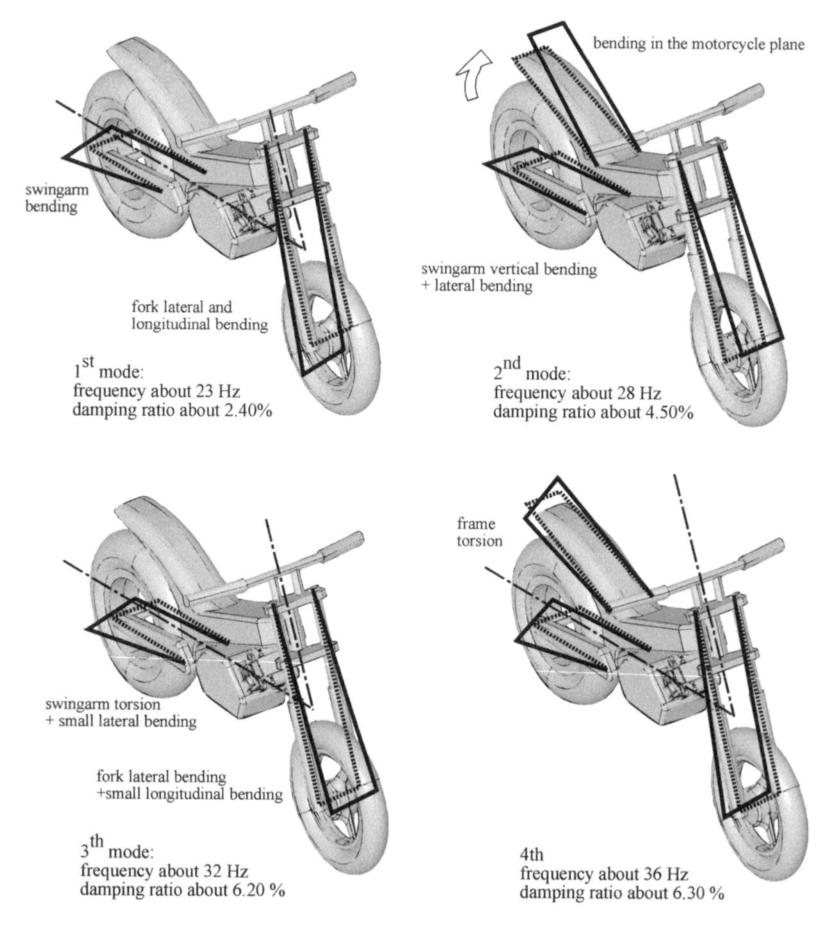

Fig. 8-60 Structural modes of a modern sporting motorcycle (mass equal to 190 kg).

The modal properties of a structure are independent from the excitation and the acquisition point, so a series of experimental FRFs (Frequency Response Functions) can be acquired in laboratory, using an excitation system like an impact hammer with a load cell or a shaker able to generate a frequency sweep, and an acquisition system that usually includes three-axis accelerometers.

The acquisition points are displaced along the whole vehicle, creating a mesh that can be animated after that the modal identification (obtained using specific algorithms able to analyze all the acquired FRFs at the same time) has been performed.

Experimental modal analysis is an important technique that makes it possible to obtain the modal model of complex structures formed by different parts, like a motorcycle (front and rear frame, wheels). This model is given as resonance frequencies, damping and eigenvectors, starting from experimental acquisitions and not from a virtual model like FEM codes.

Figure 8-60 shows the first four structural modes for a modern sporting motorcycle. The first mode shows lateral bending of both front fork and swingarm in the same direction. The second mode shows a bending deformation of the rear part of the chassis and of the fork in the same direction; there is also a remarkable in plane and lateral deformation of the swingarm. The third and the fourth modes are more complicated and involve bending and torsion of the swingarm, of the fork and of the rear part of the chassis while the main part of the chassis has smaller deformations.

8.12 Rigid body properties and Mozzi axis

If we consider the motorcycle as a single rigid body moving in space, subject to translation in the ground plane and rotations about its roll and yaw axes, a number of interesting observations can be made.

Motorcycle motion depends on the kind of maneuver and on the riding style. Every maneuver starts with a variation of the front lateral force generated by the steering motion. If a rightward impulsive force is applied to the front tire contact point, the motorcycle moves with the following leaning and yaw velocity:

$$\dot{\varphi} = -\frac{I_{z_G} h - I_{xz_G}(p-b)}{I_{x_G} I_{z_G} - I_{xz_G}^2} \qquad \dot{\psi} = \frac{I_{x_G}(p-b) - I_{xz_G} h}{I_{x_G} I_{z_G} - I_{xz_G}^2}$$

The front wheel contact point lateral velocity is:

$$\dot{y} = -\frac{(I_{x_G} I_{z_G} - I_{xz_G}^2) - I_{x_G} mb(p-b) + mh(I_{z_G} h + I_{xz_G} b) - I_{xz_G} mh(p-b)}{(I_{x_G} I_{z_G} - I_{xz_G}^2)m}$$

The roll velocity is negative (leftward motion) while the yawing velocity is positive (rightward motion). The roll velocity increases if the roll moment of inertia I_{x_G} is decreased, while the yawing velocity increases, if the yaw moment of inertia I_{z_G} is decreased. Both velocities increase if the corresponding products of inertia (in the inertia tensor) become more negative.

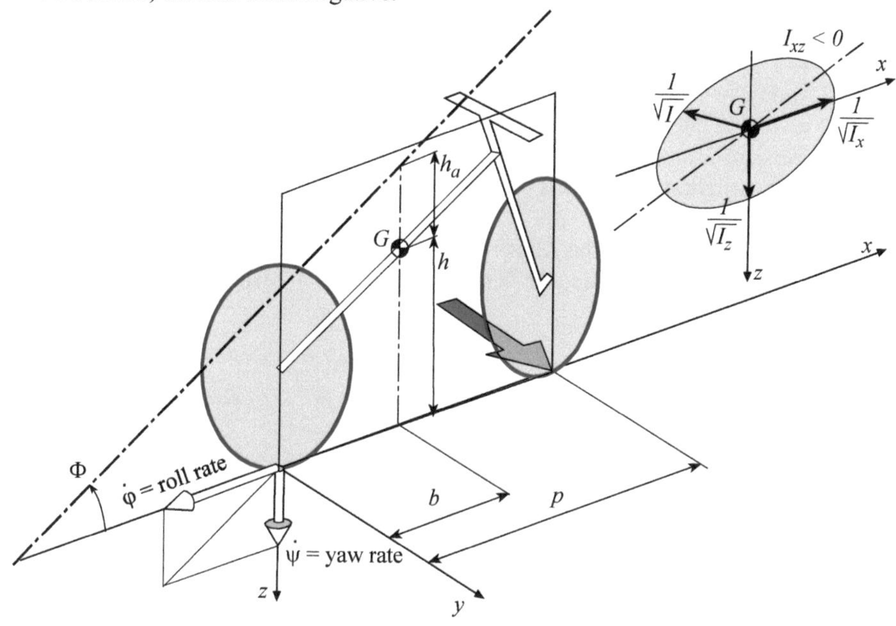

Fig. 8-61 Instantaneous rotation axis.

Knowing the forward velocity, the yaw velocity, and the roll velocity of motorcycle one can define a vector quantity know as the Mozzi axis, or the instantaneous axis of rotation for the rigid body. The Mozzi axis can be used to describe different transient maneuvers made by the motorcycle.

The Mozzi axis (Fig. 8-61) is described in the SAE coordinate system by:

$$y = \frac{\dot{\psi} V}{\dot{\psi}^2 + \dot{\varphi}^2} \qquad z = \frac{\dot{y} + \dot{\psi} x}{\dot{\varphi}}$$

and x is coincident with the body's center of mass location. The angle that the axis makes with the ground plane is:

$$\Phi = \arctan\left(\frac{\dot{\Psi}}{\dot{\varphi}}\right)$$

In terms of the rigid body properties the slope of the instantaneous axis of rotation with respect to the road plane xy is:

$$\Phi = \arctan \frac{I_{x_G}(p-b) - I_{xz_G} h}{I_{z_G} h - I_{xz_G}(p-b)}$$

The angle increases, increasing the roll inertia and decreasing the yaw one (i.e. less roll motion and more yaw).

The vertical distance of the axis from the motorcycle center of mass increases, as the roll inertia increases and product of inertia terms decrease.

People involved in motorcycle design and racing often ask themselves the following questions:

- Where is the instantaneous axis of the vehicle in motion?
- How do yaw and roll rate combine?

Answers to these questions may be found in a mathematical way by making use of the concept of the Mozzi axis. The plots derived from the Mozzi axis theory are useful in highlighting the effects of variations in path, vehicle properties, and riding style.

The geometric loci of the Mozzi trace (the point where the axis intersects the ground plane) and that of the turn center during a slalom maneuver are presented in Fig. 8-62.

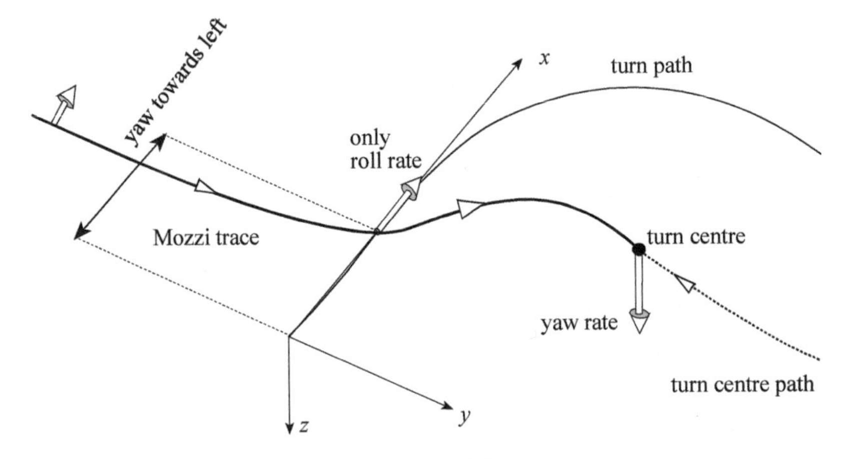

Fig. 8-62 Mozzi trace in a slalom maneuver.

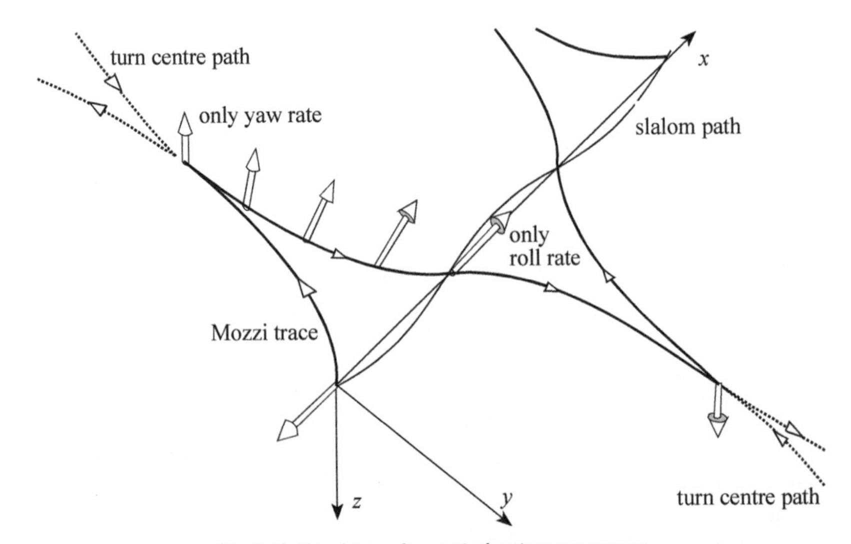

Fig. 8-63 Mozzi trace in a entering turn maneuver.

The locus of the turn center is close to a piecewise linear. It is worth highlighting that the turn center coordinates tend to infinity if the yaw velocity tends to zero. In the sinusoidal slalom this condition happens when the trajectory crosses the x axis.

The locus of the Mozzi trace is a curve with periodic cusps, which always lie on inner side of the path. The loci of the Mozzi trace and the turn center show periodic intersection points that take place when the radius of curvature is close to the minimum value.

Figure 8.63 deals with a motorcycle that is entering a right curve with a counter-steer technique. In this maneuver the Mozzi axis moves from the outside of the path to the center of the curve.

The yaw rate is negative (counter-yaw) at the beginning of the maneuver and positive during the rest of the maneuver. The roll rate reaches a maximum after the beginning of the rotation towards the right and then tends to zero. The Mozzi trace comes from $-\infty$, because the path is almost straight and the yaw rate is negative, then it crosses the path when the yaw rate is zero. Finally, in the steady turning portion of the path, the Mozzi trace tends to coincide with the turn center, because the roll rate is close to zero.

8.13 Dynamic analysis with multi-body codes

Nowadays "multi-body" codes make possible the precise and complete dynamic analysis of a vehicle's operation on the road through the use of computer simulation. MSA *(Multi-body System Analysis)* represents the computer study of movements in mechanical systems as a result of the application of external forces or stresses that act on the system. The spatial systems forming the subject of the study are simulated with rigid bodies and/or flexible elements connected to each other with various types of kinematic and dynamic connections. The external forces and resulting reactions lead to movements of the system's components that satisfy constraint conditions. The MSA codes play a role whose importance is only destined to increase through the use of modern integrated computer-aided design. They make it possible to evaluate and optimize the characteristics and performance of a product even before the prototype phase, thereby assuring the reduction of development costs and the systematic evaluation of alternative designs, and especially reducing the *time to market* for a new product.

Figure 8-64 illustrates by example a model of a scooter that runs in rectilinear motion along a roadway and encounters a step. The chassis of the scooter is further affected by the unbalancing force of the motor.

The computer simulation enables us to represent the characteristics of the springs and the shock absorbers even with non-linear laws. In the same way, the tires can be modeled employing various levels of sophistication.

Figure. 8-65 shows a racing vehicle performing a wheelie during acceleration, generated by a high thrust force. The in-plane modeling can be very accurate and can take account of, for example, the characteristics of the engine torque, the elastic contribution of the gas present in the shock absorbers, the slip characteristics of the tires in terms of the load, etc.

Three-dimensional multibody codes have to be used to simulate the operations of the vehicle out-of-plane. Since the vehicle is unstable, the model requires a control system both for the equilibrium of the vehicle and for the execution of the desired maneuver (Fig. 8-66).

Recent advances in both multi-body software and in control strategies and implementation have yielded models capable of in-depth studies of design changes and parameter variations, which also provide significant insight into driver technique and skill-level. In the near future the validation of a particular vehicle designs can be substantiated before any metal has been cut or wheels have been laced.

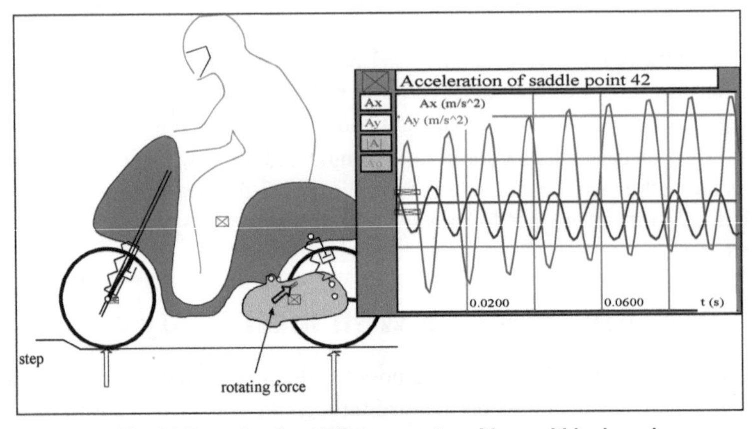

Fig. 64 Example of modeling a scooter with a multi-body code.

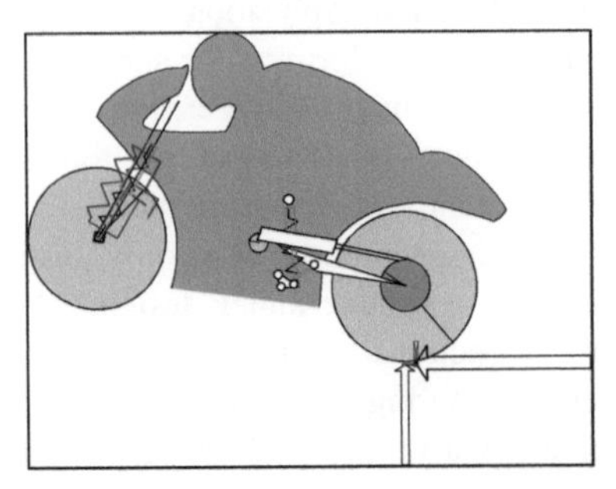

Fig. 8-65 Example of modeling of a racing vehicle with a multibody code.

Fig. 8-66 Example of 3D simulation of the dynamic behavior of a motorcycle.

List of symbols

Coordinate systems

(P_r, x, y, z)	mobile triad with the origin in the rear contact point P_r, according to SAE J670
x	forward and parallel to the longitudinal plane of symmetry
y	lateral on the right side of the vehicle
z	downward with respect to the horizontal plane
(C, X, Y, Z)	mobile triad with the origin in the turn center point C
X	parallel to x axis
Y	parallel to y axis
Z	parallel to z axis
(A_r, X_r, Y_r, Z_r)	triad attached to rear frame
(A_f, X_f, Y_f, Z_f)	triad attached to front frame
r	suffix for parameters of rear frame
f	suffix for parameters of front frame

Kinematics and dynamics parameters

C	turning center point
C	path curvature
R_c, R_{c_r}	path radius of rear wheel
R_{c_f}	path radius of front wheel
V	forward velocity
ξ	steering ratio
δ	steering angle
Δ	kinematic steering angle
Δ^*	effective steering angle
μ	pitch angle of the main frame
ψ	yaw angle of the main frame
σ	thrust chain angle respect to the ground plane
τ	transfer load angle respect to the ground plane
$\tau_{i,j}$	velocity ratio
Ω	yaw angular velocity about the z-axis
ν	frequency
ν_p	natural frequency of pitch vibration
ν_b	natural frequency of bounce vibration

Motorcycle parameters

A frontal area of the motorcycle
a mechanical trail (trail)
a_n normal trail
b longitudinal distance from rear axis to the motorcycle mass center
c viscous damping coefficient of suspension
c viscous damping coefficient of steering damper
C_D aerodynamic drag coefficient
C_L aerodynamic lift coefficient
d fork offset: distance from the center of the front wheel to the steering axis
h height of motorcycle mass center
I_{x_G} moment of inertia of motorcycle about the x-axis through its center of mass (roll inertia)
I_{xz_G} product of inertia of motorcycle about the x-z-axes through its center of mass
I_{y_G} moment of inertia of motorcycle about the y-axis through its center of mass (pitch inertia)
I_{z_G} moment of inertia of motorcycle about the z-axis through its center of mass (yaw inertia)
k suspension stiffness
L swinging arm length
m motorcycle mass
p wheelbase
P tire contact point with the ground
G motorcycle center of mass
r_p radius of drive sprocket
r_c radius of rear sprocket
Δh lowering of the steering head
ε caster angle
ϕ swinging arm angle respect to the ground plane
φ roll angle of the rear frame (camber angle of the rear wheel)
β roll angle of the front frame (camber angle of the rear wheel)
η chain angle respect to the ground plane

Front frame parameters

b_f distance of the mass center of front frame from steering axis
h_f distance of the mass center of front frame from the line passing through the rear wheel center and perpendicular to the steering axis
G_f mass center of front frame
k_f effective (reduced) front suspension stiffness

m_f	front unsprung mass
M_f	front frame mass
I_{w_f}	moment of inertia of the front wheel
I_{x_f}	moment of inertia of front frame about the x_f-axis through its center of mass
I_{xz_f}	product of inertia of front frame about the x_f - z_f-axes through its center of mass
I_{y_f}	moment of inertia of front frame about the y_f-axis through its center of mass
I_{z_f}	moment of inertia of front frame about the z_f-axis through its center of mass

Rear frame parameters

b_r	longitudinal distance of the mass center of rear frame from rear wheel axis
h_r	height of the mass center of rear frame
G_r	mass center of rear frame
k_r	effective (reduced) rear suspension stiffness
m_r	rear unsprung mass
M_r	rear frame mass
I_{w_r}	moment of inertia of the rear wheel
I_{x_r}	moment of inertia of rear frame about the x_r-axis through its center of mass
I_{xz_r}	product of inertia of rear frame about the x_r - z_r-axes through its center of mass
I_{y_r}	moment of inertia of rear frame about the y_r-axis through its center of mass
I_{z_r}	moment of inertia of rear frame about the z_r-axis through its center of mass

Forces and Moments

N_s	static load on the wheel
N	dynamic load on the wheel
N_a	normalized dynamic load on the wheel: ratio of the dynamic load to the motorcycle weight
N_{tr}	dynamic load transfer
F_D	aerodynamic drag force

F_w rolling resistance force
F_P resistant force due to the slope road
F braking force
F_s lateral force on the tire
M elastic moment applied to the swinging arm
M_X overturning moment
M_Y rolling resistance moment
M_Z yawing moment
M_t twisting moment
P power
S driving force
S_a normalized driving force: ratio of the driving force to the motorcycle weight
T chain thrust force
τ steering torque applied by the rider
$\Re$ squat ratio:
μ braking/driving force coefficient (normalized longitudinal force): ratio of the braking/driving force to the vertical load
μ_p, μ_{x_p} braking/driving traction coefficient: the maximum value of the braking/driving force coefficient
μ_{y_p} lateral traction coefficient: the maximum value of the lateral force coefficient

Tires and wheels

a_t tire trail
d rolling friction parameter
f_w rolling resistance coefficient
R outside radius of the tire
ρ radius of torus revolution of the tire
t radius of cross section of the tire
L relaxation length of the tire
k_p radial stiffness of the tire
k_s lateral stiffness of the tire
k_κ longitudinal slip stiffness coefficient: ratio of longitudinal slip stiffness to the vertical load
k_λ cornering (sideslip) stiffness coefficient: ratio of cornering stiffness to the vertical load
k_φ camber stiffness coefficient of the tire: ratio of camber stiffness to the vertical load
K_κ longitudinal slip stiffness of the tire

K_λ	cornering (sideslip) stiffness of the tire
K_φ	camber (roll) stiffness of the tire
R_o	effective rolling radius
λ	sideslip angle of the tire
ω	spin velocity of the wheel about its axis
κ	longitudinal slip of the tire

References

F.J.W. Whipple, *The Stability of the motion of a Bicycle,* Quart. Journal of Pure and Applied Mathematics, 30, 1899.

D. E. H. Jones, *The Stability of the Bicycle*, Physics Today, pp. 34-40, 1970.

R. S. Sharp, *The Stability and Control of Motorcycles,* Journal of Mechanical Engineering Science, 13, 1971.

G. E. Roe, T. E. Thorpe, *Experimental Investigation of the Parameters affecting the Castor Stability of Road Wheels*, Journal Mechanical Engineering Science, Vol. 15, n° 5, 1973.

T. R. Kane, *Fundamental Kinematical Relationships for Single-Track Vehicles*, Int. J. Mech. Sci., 17, 1975.

R. S. Sharp, *A Review of Motorcycle Steering Behaviour on Straight Line Stability Characteristics*, SAE Paper n° 780303, 1978.

D.H. Weir, J.W. Zellner, *Development of Handling Test Procedures for Motorcycles*, SAE paper 780313, 1978.

H. Sakai, O. Kanaya, H. Lijima, *Effect of Main Factors on Dynamic Properties of Motorcycle Tyres*, SAE paper n° 790259, 1979.

M. K. Verma, R. A. Scott, L. Segel, *Effect Of Frame Compliance on the Lateral Dynamics of Motorcycles*, Vehicle System Dynamics, V. 9, 1980.

K. Riedl, P. Lugner, *Naehere untersuchungen zur stationaren kurvenfahrt von einspurfahrzeugen*, Vehicle System Dynamics, V. 11, pp. 175-193, 1982.

R.S. Sharp, *The Lateral Dynamics of Motorcycles and Bicycles*, Vehicle System Dynamics, V. 14, 1985.

G. E. Roe, T. E. Thorpe, *The Influence of Frame Structure on the Dynamics of Motorcycle Stability*, SAE paper n. 891772, 1989.

A.Wiedele, M.Schmieder, *Research on the Power Transfer of Motorcycle Tyres on Real Road Surfaces*, Proc. Society of Automotive Engineers Eighteenth FISITA Congress, Turin, Italy, 1990.

H. B. Pacejka, R S. Sharp, *Shear Force Development by Pneumatic Tyres in Steady State Conditions: a Review of Modelling Aspects*, Vehicle System Dynamics, V. 20, 1991.

H. B. Pacejka, *Tyre Models for Vehicle Dynamics Analysis*, Swets & Zeitlenger, Amsterdam, 1991.

R. Romeva, F. Piera, B. Creixell, *Effect of Sudden Slippage of the Driving Wheel over a Swinging Arm in High Powered Motorcycles. Mechanical Solutions to avoid these Effects*, Inter. Cong. Technische Universitat Graz, 1993.

H. B. Pacejka, J. M. Besselink, *Magic Formula Tyre Model with Transient Properties*, Vehicle System Dynamics, V. 27, 1997.

Katayama, T. Nishimi, T. Okayama, Takumi, A *Simulation Model For Motorcycle Rider's Control Behavior*, SAE 1997.

E. J. H de Vries, H. B. Pacejka, *Motorcycle Tyre Measurements and Models,* Vehicle System Dynamics, V. 28, Suppl., 1998.

V. Cossalter, A. Doria, R. Lot, *Steady Turning Of Two Wheel Vehicles*, Vehicle System Dynamics, V. 31, n° 3, pp. 157-181, 1999.

V. Cossalter, A. Doria, R. Lot, *Steady Turning of Two Wheel Vehicles*, Vehicle System Dynamics, V. 31, n° 3, pp. 157-181, 1999.

R. Berritta, V. Cossalter, A. Doria, R. Lot, *Implementation of a Motorcycle Tyre Model in a Multi-Body Code*, Tire Technology International, UK&International Press, United Kingdom, 1999.

V. Cossalter, A. Doria, R. Lot, *Optimum Suspension Design for Motorcycle Braking*, Vehicle System Dynamics, Vol. 34, 2000.

D. Bortoluzzi, A. Doria, R. Lot, L. Fabbri, *Experimental Investigation and Simulation of Motorcycle Turning Performance*, 3° International Motorcycle Conference, Munchen, Germany, pp. 344-365, 2000.

Fajans J., *Steering in bicycles and motorcycles,* American Journal of Physics, Volume 68, Number 7 , pp. 654-659, 2000.

R. S. Sharp, *Stability Control and Steering Responses of Motorcycles*, Vehicle System Dynamics, V. 35, 2001.

B.R. Davis, A.G Thompson, *Power spectral density of road profiles*, Vehicle System Dynamics, V. 35, pp. 409-415, 2001.

D. Bortoluzzi, R. Lot, N. Ruffo: *Motorcycle Steady Turning: the Significance of Geometry and Inertia*, 7th International Conference and Exhibition, ATA, Florence, 2001.

R. Berritta, V. Cossalter, A. Doria, N. Ruffo, *Identification of Motorcycle Tire Properties by means of a Testing Machine*, 2002 SEM Annual Conference & Exposition on Experimental and Applied Mechanics, Milwaukee, 2002.

V. Cossalter, A. Doria, L. Mitolo, *Inertial and Modal Properties of Racing Motorcycles*, Motorsports Engineering Conference & Exhibition, ,Indianapolis, Indiana, USA,SAE Paper n° 2002-01-3347, 2002.

V. Cossalter, R. Lot: *A Motorcycle Multi-Body Model for Real Time Simulations Based on the Natural Coordinates Approach*, Vehicle System Dynamics, V. 37, n°6, pp. 423-448, 2002.

V. Cossalter, R. Lot, F. Maggio, *A Multibody Code for Motorcycle Handling and Stability Analysis with Validation and Examples of Application*, Meeting: Small Engine Technology Conference & Exhibition, Madison, WI, USA, SAE Paper 2003-32-0035, 2003.

V. Cossalter, A. Doria, R. Lot, N. Ruffo, M. Salvador, *Dynamic Properties of Motorcycle and Sçooter Tires: Measurement and Comparison*, Vehicle System Dynamics, V. 39, n° 5, pp. 329-352, 2003.

V. Cossalter., R. Lot, F. Maggio, *The Modal Analysis of a Motorcycle in Straight Running and on a Curve,* Meccanica Kluwer Academic Publishers V. 39, pp. 1-16, 2004.

V. Cossalter R. Lot, F. Maggio, *On the Stability of Motorcycle During Braking.* Small Engine Technology Conference & Exhibition, Graz, Austria, SAE Paper n°

2004-32-0018, 2004.

R. Lot, *A Motorcycle Tire Model for Dynamic Simulations: Theoretical and Experimental Aspects*, Meccanica, V. 39, pp. 207-220, 2004.

V. Cossalter, A.Doria, *Analysis of Motorcycle Manoeuvres based on the Mozzi Axis*, Vehicle System Dynamics, V. 42, n° 3, pp. 175-194, 2004.

R.S. Sharp, S. Evangelou, D.J.N. Limebeer, *Advances in the modelling of motorcycle dynamics*, Multibody System Dynamics, V 12, n° 3, pp. 251-283, 2004.

R.S. Sharp, D.J.N. Limebeer, *On steering wobble oscillations of motorcycles*, Journal of Mechanical Engineering Science Part C, V. 218, n 12, p 1449-1456, 2004.

R. Lot, V. Cossalter, M. Massaro, *The Significance of Frame Compliance and Rider Mobility on the Motorcycle Stability*, International Conference on Advances in Computational Multibody Dynamics-ECCOMAS, Universidad Politecnica de Madrid, 2005.

V. Cossalter, A. Doria, *The Relation between Contact Patch Geometry and the Mechanical Properties of Motorcycle Tyres*, Vehicle System Dynamics, V. 43, Suppl., pp. 156-167, 2005.

V. Cossalter, A. Doria, *The Instantaneous Screw Axis of Two-Wheeled Vehicles in Typical Manoeuvres*, XIX IAVSD Symposium International Association for Vehicle System Dynamics, Milano, 2005

V. Cossalter, A. Doria, S. Garbin, R. Lot, *Frequency-domain Method for Evaluating the Ride Comfort of a Motorcycle*, Vehicle System Dynamics, V. 44, n° 4, pp. 339–355, 2006.

V. Cossalter, J. Sadauckas, *Elaboration and quantitative assessment of manoeuvrability for motorcycle lane change*, Vehicle System Dynamics, Vol. xx, No. x, pp. xxx–xxx, 2006.

V. Cossalter, R. Lot M. Massaro, *The influence of Frame Compliance and Rider Mobility on the Scooter Stability*, Vehicle System Dynamics:, Vol. xx, No. x, pp. xxx–xxx, 2007

V. Cossalter, R. Lot, M. Peretto, *Motorcycles Steady Turning*, Journal of Automobile Engineering, Vol. xx, No. X, pp. xxx–xxx, 2007

Books and Reports

G. Pollone, Il veicolo, Leprotto&Bella Editore, Torino, Italy, 1970.

ISO, Draft Standard ISO/TC 108/WG9, *Proposals for Generalized Road Inputs to Vehicles*, 1972.

J. Koch, *Experimentelle und Analytische Untersuchungen des Motorrad-Fahrer Systems*, Dissertation, Berlin, 1978.

C. Koenen, *The dynamic behaviour of a motorcycle when running straight ahead and when cornering*, Doctoral Dissertation, Delft University, 1983.

B. Bayer, *Das Pendeln und Flattern von Kraftradern*, Institut fur Zweiradsicherheit,

Bochum, 1986, (in German).

T. D. Gillespie, *Fundamentals of Vehicle Dynamics*, SAE, Warrendale, 1992.

H. W. Bonsch, *Einfuhrung in die Motorradtechik*, Motorbuch Verlag, Stuttgart, Germany, 1993, (in German).

J. Bradley, *The Racing Motorcycle*, Broadland Leisure Publications, York, England, 1996.

G. Genta, *Motor Vehicle Dynamics*, World Scientific Publishing, Singapore, 1997.

ISO 2631, Mechanical vibration and shock, *Evaluation of human exposure to whole-body vibration*, International Organization for Standardization, 1997.

R. Lot, *Studio della stabilità e della maneggevolezza di veicoli a due ruote*, Doctoral Dissertation, University of Padova, Italy, 1998, (in italian).

F. Maggio, *Modi di vibrare della motocicletta: Accoppiamenti tra modi laterali e verticali*, Degree Report, University of Padova, Italy, 2001, (in italian).

ISO 5349, Mechanical vibration, *Measurement and evaluation of human exposure to hand-transmitted vibration*, International Organization for Standardization, 2001.

H. B. Pacejka, *Tire and Vehicle Dynamics*, Butterworth Heinemann, Oxford, 2005.

Index

Notes